Quantum Computing with Superconducting Qubits and Cavities

Quantum Computing with Superconducting Qubits and Cavities

Concepts and Technology

Hasan Padamsee
Newman Laboratory
Cornell University
Ithaca, New York, USA

Tanay Roy
Superconducting Quantum Materials and Systems Center
Fermi National Accelerator Laboratory
Batavia, Illinois, USA

Alexander Romanenko
National Quantum Information Science Research Center
Fermi National Accelerator Laboratory
Batavia, Illinois, USA

Anna Grassellino
National Quantum Information Science Research Center
Fermi National Accelerator Laboratory
Batavia, Illinois, USA

WILEY

Published by John Wiley & Sons, Inc., Hoboken, New Jersey.

Library of Congress Cataloging-in-Publication Data applied for:

Hardback ISBN: 9781394294589

Cover Design: Wiley
Cover Images: Alexandr Dubynin/Getty Images, Tanay Roy, with permission from Fermi Forward Discovery Group, LLC as manager and operator of Fermi National Accelerator Laboratory under Contract No. 89243024CSC000002 with the United States Department of Energy

Set in 9.5/12.5pt STIXTwoText by Straive, Chennai, India

Contents

Preface

Why a Book on Superconducting Qubits and Cavities—and Why Now?

Quantum computing is advancing at a breathtaking pace, and while many excellent books, reviews, and tutorials exist, they remain challenging for newcomers who need an accessible introduction to this emerging technology. Some resources often launch directly into dense mathematics or specialized engineering details, leaving entrants without the conceptual grounding required to appreciate the field's significance. What is missing is a resource that explains *why* quantum computing matters, *how* its principles work, and *what* makes superconducting qubits and microwave cavities such important platforms for the next generation of quantum technologies.

This book aims to fill such gaps. It begins by demystifying the foundations of quantum physics—wave–particle duality, superposition, entanglement, and interference—concepts that are essential yet difficult to internalize. To anchor these ideas, we trace some of the development of quantum theory and the evolution of quantum information science, showing how today's breakthroughs grew from decades of conceptual and technological progress. This context prepares readers to later appreciate the potential of quantum computing, from factoring large numbers and searching massive databases to simulating molecules, materials, and many-body systems far beyond the reach of classical computation.

With a foundation in place, we introduce the qubit and the central challenge of quantum information processing: maintaining coherence in the presence of noise. We explain how quantum algorithms manipulate fragile states through sequences of gates. We bring in the generalization of the qubit, the qudit, to discuss some its merits and challenges. We then motivate the need for quantum error correction, showing how decoherence and operational errors accumulate and why logical qubits—built from hundreds or thousands of physical qubits—are essential for fault-tolerant computation. This naturally leads to a discussion of the hardware platforms capable of supporting demanding requirements.

Among these platforms, superconducting qubits have emerged as one of the most promising. Their rapid progress is driven by rapidly improving coherence, mature microfabrication lithography techniques, compatibility with commercial microwave electronics, and the ability to operate at millikelvin temperatures where thermal noise is suppressed. We review the physics of transmon qubits based on Josephson junctions, the remarkable improvements over the past two decades, and the materials and fabrication challenges that continue to shape the field. This sets the stage for a deeper exploration of the complementary role of superconducting microwave cavities.

Here, the story becomes especially compelling. Three-dimensional superconducting cavities open new pathways for quantum computation by harnessing long-lived bosonic modes whose large Hilbert spaces support qubits encoded in Fock states, higher-dimensional qudits, and a hierarchy of increasingly powerful error-resilient encodings. When coupled to transmon ancillas, these cavities form hybrid architectures that combine exceptional coherence with fast control. This enables compact representations of quantum information, reduces gate counts, and supports error detection schemes such as dual-rail encodings, followed by more advanced bosonic codes including binomial and cat codes, and ultimately Gottesman–Kitaev–Preskill (GKP) codes that approach the threshold of fault-tolerant operation. In such systems, the cavity mode serves as a robust quantum memory capable of hosting coherent superpositions and interference between probability amplitudes, while the transmon provides conditional operations, syndrome extraction, and measurement. By supporting multimode operation within a single device and reducing hardware overhead relative to large 2D transmon arrays, 3D cavity–transmon architectures offer a promising and scalable platform for both quantum storage and computation. It is not enough to demonstrate that a handful of qubits can perform a clever algorithm. Those are proofs of principle—necessary, inspiring, but preliminary. The true challenge is to take these fragile, finely tuned systems and multiply them—by tens, hundreds, and eventually thousands—without losing the very quantum properties that make them powerful.

The narrative broadens to the materials science underpinning these achievements. Loss mechanisms in niobium oxide, aluminum oxide, and dielectric substrates—two-level systems, quasiparticles, interstitial impurities—are examined alongside emerging alternatives such as tantalum, and improved silicon or sapphire substrates. We show how decades of research in superconducting RF cavities and quantum circuits have converged, enabling hybrid architectures.

All of this progress is occurring against a backdrop of impressive global investment on the scale of many billions of dollars. National quantum initiatives in the United States, United Kingdom, European Union, China, Canada, and many other countries are accelerating research and development. In the United States, five major QIS centers—including SQMS, where the authors of this book play leading roles—are driving advances in materials, devices, algorithms, system integration, and scaling. The convergence of scientific opportunity, technological capability, and international momentum makes this an ideal moment for a comprehensive, accessible book that unifies the physics, engineering, and emerging applications of superconducting qubits and cavities.

Ultimately, this book is not a "quantum computing for dummies" guide, nor is it a narrow technical monograph. It is designed for IT specialists, engineers, researchers, industry practitioners, policymakers, and students who want to understand the principles, challenges, and promise of quantum technologies without being overwhelmed by formalism but also without omitting the key mathematical principles that any serious understanding of quantum technologies requires. By combining conceptual clarity with explanations where needed, it offers a coherent picture of a field on the cusp of transforming computation.

We are witnessing the birth of a technological revolution with quantum at its core. This book aims to equip readers with the understanding needed to participate in that revolution—inviting scientists, engineers, students, newcomers, and curious observers alike to appreciate the extraordinary possibilities that superconducting qubits and cavities are beginning to unlock for the unfolding world of quantum technologies.

1

Foundations

1.1 Overview

Quantum computing (QC) represents a radical shift in the way we process information. Rooted in the principles of quantum mechanics, it challenges the classical framework of computation by harnessing the superpowers of the quantum world: *superposition, entanglement*, and *interference.* With these powers, quantum computers may someday surpass the most powerful supercomputers for a certain class of problems. Unlike classical bits, which can only be 0 or 1, *qubits* can exist in superpositions of both states. This property is vividly represented on the *Bloch sphere* and provides the foundation for algorithms. Although a single qubit does not store more extractable information than a classical bit, systems of many qubits can represent and process information in ways that scale exponentially with the number of qubits. When many qubits are entangled, the resulting system can represent information that grows exponentially with the number of qubits.

This chapter introduces the conceptual and mathematical underpinnings of QC, explains how qubits differ from classical bits, and highlights the central challenges of *coherence* and *error correction.* It also situates QC within its historical trajectory—from Feynman's early vision in the 1980s to breakthroughs such as Shor's algorithm, Grover's search, and the emergence of various QC platforms such as superconducting and ion-trap systems. By contrasting classical and quantum logic, the chapter underscores why certain problems remain intractable for classical machines but could become solvable within the quantum paradigm.

QC is a transformational computing paradigm that utilizes the principles of quantum mechanics to process information [1]. In 1981, physicist Richard Feynman proposed that classical computers might struggle to simulate quantum systems efficiently, suggesting that a "quantum computer" could potentially handle this task better [2]. "Nature isn't classical, dammit, and if you want to make a simulation of nature, you'd better make it quantum mechanical," Feynman quipped.

Quantum systems, like atoms and molecules, can exist in a *superposition* of multiple states simultaneously. In classical computing, bits are binary, taking on either a 0 or a 1 at any given time. The bit is just a switch with two possible configurations, "on" or "off." In a quantum computer, the fundamental unit of information is a *qubit*, which can be in a state of $|0\rangle$, $|1\rangle$,or both at the same time, thanks to superposition. The notation "$|\rangle$," known as "*ket*," is used to distinguish quantum states from classical states. When a qubit is in superposition, it holds a blend of states, described mathematically as a linear combination of $|0\rangle$ and $|1\rangle$ as $\alpha|0\rangle + \beta\,|\,1\rangle$. The complex numbers α and β—called probability amplitudes—define the probability of being in the basis states $|0\rangle$ and $|1\rangle$, i. e., $P(|0\rangle) = |\alpha|^2$ and $P(|1\rangle) = |\beta|^2 = 1 - |\alpha|^2$. By harnessing multiqubit superposition states, a quantum computer can evolve in a state space whose size grows exponentially with the number of qubits. This allows quantum algorithms to process many possible outcomes simultaneously in ways that classical computers cannot, even though the final measurement still yields only a single classical result.

The *Bloch sphere* (Figures 1.1 and 1.2) provides a convenient way to visualize the state of a single qubit. Pure states correspond to points on the surface of the sphere, while mixed states lie inside it. (We cover *mixed states* in Section 4.5.) The north and south poles represent the computational basis states $|0\rangle$ and $|1\rangle$, respectively, and every other point on the surface represents a pure superposition of these basis states. Because the Bloch sphere is continuous, a qubit can occupy infinitely many distinct pure states, unlike a classical bit, which has only two possible values. Nevertheless, a measurement of a single qubit yields only one classical bit of information.

A generic single qubit can thus be conveniently expressed as $|\psi\rangle = \cos(\theta/2)|0\rangle + e^{i\phi}\sin(\theta/2)$, which can be represented as a Bloch vector (see the green arrow in Figure 1.2) characterized by two angles: a polar angle θ indicating the inclination from the north pole and the azimuthal angle ϕ showing rotation from the horizontal x-axis [3].

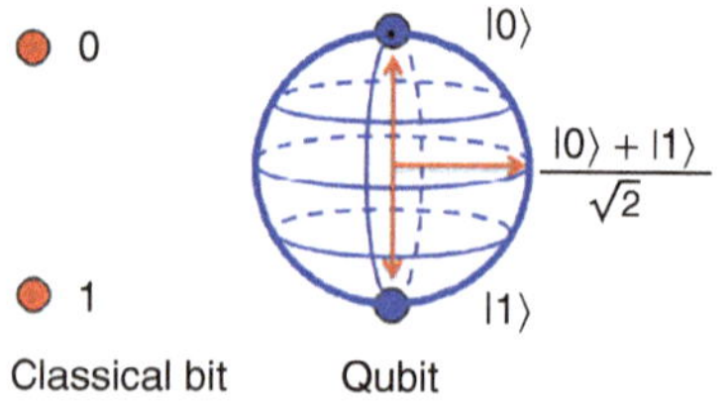

Figure 1.1 Comparison between a classical bit and a qubit. Classical bits are restricted to discrete values 0 and 1, represented as distinct red dots. In contrast, a qubit is represented on the Bloch sphere, where its state can exist as any superposition of $|0\rangle$ and $|1\rangle$. The equator represents equal superpositions such as $(|0\rangle + |1\rangle)/\sqrt{2}$, while the poles correspond to the pure basis states $|0\rangle$ and $|1\rangle$.

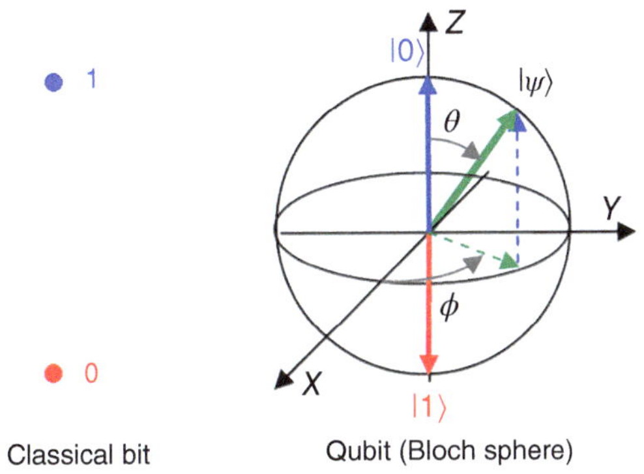

Figure 1.2 A qubit, represented on the Bloch sphere, can exist in an arbitrary superposition of $|0\rangle$ and $|1\rangle$. The general qubit state $|\psi\rangle$ is characterized by two angles: θ (polar angle) and ϕ (azimuthal angle), defining its position on the sphere's surface.

For successful QC, it is necessary to keep qubits in their superposition states for a long time in order to process the information. The longer the time they stay in the superposed state, the higher the *coherence. Decoherence* turns superpositions into mixed states. A mixed state is *not* a superposition. It represents classical uncertainty about which pure state the system is in.

The coherence time of the qubit is a key performance measure. Quantum computation has to be completed before decoherence kicks in and scrambles the qubits. The quantum state of superposition is rather fragile. The slightest disturbance, such as vibration, change in temperature, and interaction with the environment, can cause qubits to tumble out of superposition before their job has been done. A central challenge in developing quantum computers is to increase the coherence time of qubits in order to increase the number of computations possible. *Quantum error correction* (QEC) [4, 5] is a way around decoherence but requires more physical qubits and introduces more complexity. By contrast, the zeros and ones in a classical computer so rarely switch by accident that little error correction is needed.

Besides superposition states, where qubits still have their own identities, multiple qubits can interact and become correlated in a nonclassical way known as *entanglement*. Entanglement is a unique quantum phenomenon where two or more particles or qubits become "linked" so the state of one directly affects the state of the other, even when they are separated by large distances, such as at the opposite ends of a galaxy. Measuring the state of one qubit will instantly determine the state of the other without having to test it. Entanglement enables qubits to work together in a way that is not possible with classical computers. Like superposition, quantum entanglement is also fragile, subject to decoherence due to the coupling of the quantum system to the environment.

Superposition and entanglement both expand the computational space available to a quantum computer, but they do so at very different scales. Superposition allows a single quantum system to explore many classical possibilities at once. In a register of qubits, superposition creates a vast space of possible computational paths that coexist simultaneously. Many inputs are represented in parallel, and operations act on all of them at once. When qubits become entangled, their states can no longer be described independently; information is encoded in the relationships between qubits, i.e., in the correlations between them. Once qubits become entangled, the system can occupy the full 2^n-dimensional space of joint states.

A single-qubit pure state is specified by two real parameters, usually written as the Bloch-sphere angles $\theta$and ϕ. If you have N independent (i.e., unentangled) qubits, each one requires two parameters, so the full product state needs only $2N$ real numbers. In contrast, a general pure state of four qubits assigns an independent complex amplitude to each of the 16 computational basis states. This gives 32 real numbers, and removing one degree of freedom for normalization $((|\alpha|^2 + |\beta|^2 = 1)$ and one for global phase ($|\phi\rangle$ and $e^{i\phi}|\Psi\rangle$ leaves $32 - 2 = 30$ real parameters. In general, an arbitrary pure state of N qubits requires $2^{N+1} - 2$ real parameters. The additional parameters—beyond the $2N$ needed for separable states—encode correlations among the qubits that cannot be attributed to any qubit individually; these are precisely the degrees of freedom associated with entanglement (see Section 2.18).

Superposition enables quantum parallelism and interference (see below), but on its own, it does not yield a general computational advantage over classical computation. A superposed state allows a quantum operation to act on all basis states simultaneously, yet a measurement still returns only a single outcome. Without correlations among qubits, this parallelism cannot be harnessed to extract useful information. Superposition supplies the raw material—many amplitudes evolving in parallel—but entanglement provides the structure that allows those amplitudes to interact through interference in a coordinated way. Genuine quantum speedup arises when superposition is combined with entanglement, creating global correlations and interference patterns that cannot be efficiently simulated by classical systems.

Interference in QC allows quantum algorithms to manipulate probabilities in favor of correct or desired solutions. When some contributions to a qubit amplitude are positive and others are negative, the contributions interfere destructively and cancel so that the net amplitude is zero and the corresponding outcome is never observed. Similarly, the contributions can interfere constructively and increase the likelihood of an outcome. The strategy for programming a quantum computer is to arrange a pattern of constructive and destructive interference so that the contributions to qubit amplitude for the wrong answer cancel out, and the contributions reinforce each other for the right answer. When the qubit is finally measured, the act of measurement forces the qubit to "collapse" down to one of the basis states $|0\rangle$ or $|1\rangle$ corresponding to a classical bit with a definite 0 or 1 value.

1.2 How QC Differs from Classical Computing

Classical bits operate independently of each other. Operations are generally performed sequentially, although parallel processing is possible depending on the architecture of the hardware. The classical computer accomplishes two main tasks using bits—storage and processing. Long strings of ones and zeros are stored in memory registers for any number, letter, or symbol by using the binary code with ones and zeros. Computers process the stored numbers with operations like "add" and "subtract." They do more complex calculations by stringing together simple operations into a series of operations called an *algorithm*. Multiplying is done as a series of additions, for example.

Classical computing uses logic gates (AND, OR, NOT, etc.) to perform operations on bits. Calculations use a set of single-bit and two-bit gates. One example of single-bit gate is the NOT gate. Here the output of the gate operation is 1 when the input is 0, and the output is 0 when the input is 1. One example of a two-bit gate is the AND gate. It has output 1 only when both inputs are 1; otherwise, the output is 0. Logic gates compare patterns of bits, stored in registers, and then turn them into new patterns.

For many tasks, classical computers are efficient and sufficient. Classical computing is well established with mature technology and widely used in everyday applications, from smartphones to supercomputers. Error correction is relatively straightforward because a bit is either 0 or 1, and errors can be easily detected and corrected. However, they struggle with certain complex problems, like factoring large numbers or simulating quantum systems, where the time required can grow exponentially with the size of the input.

QC, similarly, uses *quantum gates* that operate on single or multiple qubits, creating and destroying superposition and entanglement. The strategy a quantum algorithm follows is to arrange the sequence of gates (Figure 1.3) in a pattern of constructive and destructive interference so that the contributions to qubit amplitude for the wrong answer cancel out, and the contributions reinforce each other for the right answer. When the qubits are finally measured, the system is guaranteed or highly likely to collapse to the state corresponding to the correct answer. The entire calculation must take place before the natural decoherence of the superposed state.

Quantum logic is performed by a set of single-qubit and two-qubit gates (Sections 2.6 and 2.12). Single-qubit operations translate an arbitrary quantum state from onc point on the Bloch sphere to another point by rotating the Bloch vector a certain angle about a particular axis. For example, the Hadamard (H) gate will transform a qubit from the $|0>$ state to an equal superposition state $(|0\rangle + |1\rangle)/\sqrt{2}$, which is equivalent to rotating the vector by 180° about an axis in the yz plane inclined by 45° from the z-axis (Figure 1.4). Another family of

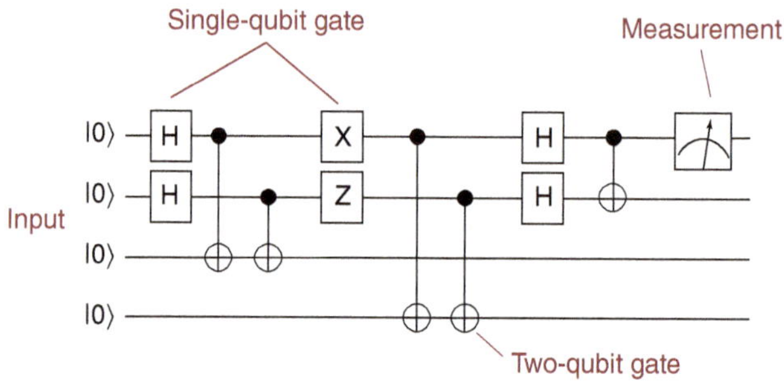

Figure 1.3 Schematic of a quantum computing operation sequence. It demonstrates the fundamental steps of quantum computation: state preparation, entanglement, manipulation, and measurement. The computation begins with all qubits initialized in the $|0\rangle$ state. Single-qubit gates (Hadamard H, Pauli-X, and Pauli-Z) create superpositions and modify individual qubit states. Two-qubit gates (controlled operations, represented by dots and circles) generate entanglement between qubits. The circuit continues with further single-qubit and two-qubit gates, evolving the quantum state through *unitary transformations* (Section 2.13). Finally, a measurement is performed on selected qubits, collapsing the superposition and extracting classical output values.

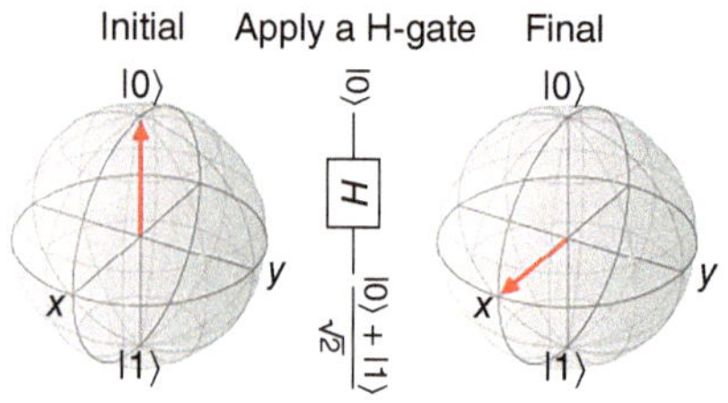

Figure 1.4 Effect of applying a Hadamard (H) gate to a qubit. Initially, the qubit is in the $|0\rangle$ state, aligned along the positive z-axis of the Bloch sphere. After the application of the Hadamard gate, the qubit is rotated to an equal superposition state $(|0\rangle + |1\rangle)/\sqrt{2}$, represented by a vector lying along the positive x-axis. The Hadamard gate transforms basis states into superpositions, playing a key role in quantum computation.

single-qubit gates are called the phase gates, which perform a rotation of the Bloch vector about the z-axis. For example, a rotation by 90° is known as the T-gate.

Two-qubit quantum-logic gates are generally *conditional* gates that take two qubits as inputs. Typically, the first qubit is the *control* qubit, and the second is the *target* qubit. For example, two qubits can be entangled using the *controlled-NOT* (CNOT) gate. The gate flips the second qubit only if the first qubit is 1. If the control qubit is 0, nothing happens to the target.

Now imagine the control qubit is not just 0 or 1, but a superposition of the two:

$$\frac{(|0\rangle + |1\rangle)}{\sqrt{2}},$$

and the target qubit starts in state $|0\rangle$. When we apply the CNOT gate, the target qubit flips only in the part of the superposition where the control qubit is $|1\rangle$. The result is

$$\frac{(|0\rangle|0\rangle + |1\rangle|1\rangle)}{\sqrt{2}},$$

which is an *entangled state*—the two qubits are now linked, and measuring one determines the state of the other. This is how a *Bell pair* (Section 2.15) is created [6]. The state is a two-qubit quantum state. Here the two-qubit state $|0\rangle\,|\,0\rangle$ means: the first qubit is in state $|0\rangle$ and the second qubit is in state $|0\rangle$. We often write this more compactly as $|00\rangle$. Similarly, $|1\rangle\,|\,1\rangle = |\,11\rangle$. So, the Bell state can also be written as

$$\frac{|00\rangle + |11\rangle}{\sqrt{2}}$$

The two qubits are perfectly correlated: they are both 0 or both 1. They are also called maximally entangled because each qubit by itself is completely random, yet the two together are perfectly correlated. Entanglement becomes partial whenever the qubits are still quantum-correlated, but noise has weakened the perfect Bell-state correlation due to photon loss, dephasing, or imperfect gates. We will discuss gates and entanglement more completely in Chapter 2.

A set of qubit operations {H, T, CNOT} form a *universal* set, meaning that any type of multiqubit operation can be constructed using these elementary gate operations. Also note that all quantum gates are *reversible*, a property known as *unitarity* (Section 2.13).

Quantum computers can solve specific problems such as factoring large numbers (Section 6.3) or searching unsorted databases (Section 2.9) much faster than classical computers [7, 8], by leveraging quantum parallelism and entanglement. Problems that are unfeasible for classical computers could be solved in a reasonable time by quantum computers. However, one major challenge is that quantum computers are quite error prone (Section 2.19) due to interaction with the environment and require QEC in order to perform meaningful, reliable, large-scale computations. A wide range of QEC schemes (Section 2.23) are being developed, all of which work by encoding a small amount of logical information into a larger collection of physical qubits. This redundancy allows the system to detect and correct errors—such as those caused by decoherence—before they accumulate and corrupt the logical state.

QC is still in developmental stages with many technical challenges to overcome. While there have been significant advances in building small-scale quantum processors, practical and large-scale quantum computers are not yet available for most applications. It is generally accepted that quantum computers will not replace classical computers but will perform different calculations that classical computers simply cannot. Quantum computers will be practical for certain types of problems that can take advantage of *quantumness*. Thus, a utility-scale QC architecture will most likely have a hybrid architecture where classical and quantum processing happen in conjunction to bring the most efficient solution to currently intractable problems. For a comprehensive treatise on quantum information processing (QIP), we refer to the following books, reviews, and tutorials [1, 9–18].

1.3 Brief History

In 1985, David Deutsch developed a theoretical model for a quantum computer, formalizing how qubits could process information using superposition and entanglement. In 1992, David Deutsch and Richard Jozsa introduced the Deutsch–Jozsa algorithm [19], the first example of a quantum algorithm demonstrating speedup over classical algorithms (Section 2.8). In 1994, Peter Shor introduced Shor's algorithm (Section 6.3) for factoring large numbers [7], demonstrating that a quantum computer could theoretically solve certain problems (like breaking RSA encryption) much faster than classical computers. Shor's approach posed a threat to RSA encryption and sparked interest in quantum cryptography. In 1996, Lov Grover developed Grover's algorithm [20], which showed that quantum computers could search unstructured databases more efficiently, underscoring the potential speedup for unsorted data searches (Section 2.9). Between 1995 and 1996, Peter Shor and Andrew Steane developed error correction codes for QC, addressing the challenge of decoherence and helping to stabilize quantum systems against errors. In 1998, scientists at Los Alamos National Laboratory and MIT created a two-qubit quantum computer using nuclear magnetic resonance (NMR) technology to mark one of the first physical realizations of a quantum processor. In 2001, IBM and Stanford built a seven-qubit quantum computer that ran a simplified version of Shor's algorithm, factoring the number 15. Over the next few years, small quantum computers with up to seven qubits were built using various approaches, including trapped ions and superconducting circuits. In 2007, Canadian company D-Wave announced the first "commercial" quantum computer, though it was somewhat controversial due to its approach of quantum annealing for optimization problems. This specific type of quantum computation is not suitable for general-purpose applications that require gate-based operations.

Through 2010, companies like IBM and Google made significant strides in building scalable qubit systems, particularly with superconducting qubits. In 2011, IBM and Google launched cloud-based QC platforms, making QC more accessible for experimentation and research. This allowed researchers and developers worldwide to experiment with real quantum processors. Leading companies and research institutions began building quantum computers using dozens to hundreds of qubits, with plans to achieve fault-tolerant quantum computers. In 2019, Google claimed "quantum supremacy," [8, 21] using a 53-qubit quantum computer called Sycamore to perform a task in 200 seconds that, they estimated, would take a classical supercomputer 10,000 years. This milestone sparked both excitement and debate within the community.

In the 2020s, the race for error-corrected, scalable quantum computers accelerated, with IBM, Google, Honeywell, and other tech giants announcing breakthroughs. Quantum processors expanded to over 100 qubits, and new architectures like superconducting circuits, trapped ions, and photonic systems were explored.

It is not our goal to review the status of the many players, but a snapshot of the state-of-the-art can be served by mentioning one of the leaders IBM's Heron processor based on 156 high-quality qubits, with 99.5% fidelity allowing complex algorithms to run successfully. The previous processor Osprey had 433 qubits but with lower fidelity [22]. *Qubit fidelity* is a measure of how close a real, noisy qubit state is to the ideal state intended. It provides a single number between 0 and 1 to capture how much the qubit has been distorted by decoherence, control errors, or imperfect gates.

Today, QC is still in the *NISQ* (*Noisy Intermediate-Scale Quantum*) era [23], where hardware limitations and error rates pose challenges. However, the field is rapidly advancing, and researchers are hopeful that large-scale, fault-tolerant quantum computers will be realized within the coming decades.

1.4 How to Compare Computing Performance? (Complexity Theory)

Complexity theory [24, 25] is used to evaluate how the performance changes when different algorithms solve problems as they scale in size. Complexity theory uses Complexity Class diagrams (Figure 1.5) and *Big O notation* to compare algorithms in terms of the resources that computers require—mainly time and memory.

Time complexity theory categorizes algorithms by how their run times grow with input size. If an algorithm takes the same time to finish no matter how large the input is, we call it constant time, complexity $O(1)$. If doubling the input size doubles the run time, it is a linear time, with complexity $O(n)$. Sorting a list by simply scanning through it might be an example. An algorithm that takes time proportional to n^2 as input size n grows will have a complexity of $O(n^2)$, showing that as n doubles, the run time quadruples. Some algorithms take exponentially longer $O(2^n)$ as the input grows, like finding solutions to certain puzzles. Exponential time algorithms quickly become impractical for large inputs. Similarly, *space complexity* is the amount of memory or storage an algorithm needs as the

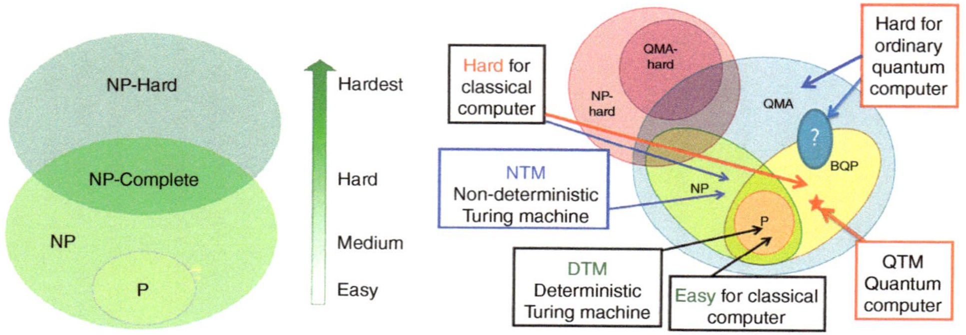

Figure 1.5 (a) Complexity class diagram showing how all P problems are also in NP. NP-complete problems are the intersection of NP and NP-hard; NP-hard includes problems that are not in NP, such as certain undecidable problems. P is the class of problems that can be solved efficiently (in polynomial time) by a deterministic classical computer. Surrounding P is NP, the class of problems for which a proposed solution can be *verified* efficiently by a classical computer, even if finding that solution may be difficult. Inside NP lies the important subset NP-complete. These are the hardest problems in NP: if any one of them could be solved efficiently, then *all* NP problems could be solved efficiently. Finally, NP-hard problems lie outside or above NP. (b) Relationship between classical and quantum computational complexity classes. As in (a), the innermost region, P contains problems efficiently solvable by a DTM and represents tasks that are easy for classical computers. The larger class NP includes problems whose solutions can be efficiently verified by a classical computer, corresponding to computation on a NTM. The region NP-hard contains problems at least as difficult as the hardest problems in NP, many of which are believed to be intractable for classical computation. The quantum complexity class BQP (bounded-error quantum polynomial time) denotes problems efficiently solvable by a quantum Turing machine (QTM), capturing the computational power of fault-tolerant quantum computers. BQP overlaps with NP and P but also contains problems believed to be classically hard, such as integer factoring and discrete logarithms, illustrating the potential for quantum advantage. The class QMA (quantum Merlin–Arthur) represents problems whose solutions can be efficiently verified by a quantum computer given a quantum proof, serving as the quantum analog of NP. Problems labeled QMA-hard are believed to be difficult even for quantum computers, marking fundamental limits of quantum computation (adapted from Ref. [12]).

input size increases. For instance, an algorithm that only needs a fixed amount of memory, regardless of the input size, has constant space complexity $O(1)$.

Quantum computers have the potential to offer significant advantages over classical computers for certain types of problems, particularly in terms of complexity.

P-Class Problems (polynomial time) can be solved by a classical computer in polynomial time. A polynomial time solution is one that can be found by an algorithm whose run time grows at most polynomially with the input size, i.e., the time required to solve the problem increases by a power of n (like n^2 or n^3), rather than exponentially (like 2^n). Hence, the resources required to solve P-class problems grow reasonably with the size of the input. Examples include sorting numbers, finding the shortest path in a graph, and basic arithmetic operations. Quantum computers do not offer a major advantage for P-class problems.

NP-Class Problems (nondeterministic polynomial time) are those for which a given solution can be *verified* in polynomial time, if we are provided with a "guess" or potential solution. However, finding such a solution might not be feasible in polynomial time for large inputs with particular algorithms. The classic example is the *traveling salesman problem* (TSP), where given a route, we can quickly verify its total distance, but finding the shortest route that visits each city once and returns to the starting city may be computationally hard as the number of cities grows. Quantum computers offer advantages for specific cases within NP-class. TSP problems can benefit from quantum algorithms.

NP-hard problems are harder than problems in NP-class, and NP-complete problems are the hardest problems within NP. *Quantum annealing* and *adiabatic QC* (AQC) [26] (Section 1.6) may offer heuristic approaches that find solutions faster than classical techniques. These heuristic methods aim to find "good enough" or approximate solutions for complex optimization problems, rather than guaranteeing the absolute optimal solution. Since they belong to the NP-class, the solutions can be verified in polynomial time. For example, finding the shortest route to visit all cities (not just checking if a route exists below a certain length) is NP-hard. In the knapsack problem, you have a set of items, each with a weight and a value, and a knapsack (or bag) with a weight capacity. The goal is to select a subset of items to maximize the total value without exceeding the knapsack's weight limit. This problem is widely studied in operations research and computer science because it models real-world scenarios like resource allocation, budget optimization, and packing. For the 0/1 knapsack problem, each item can be either included (1) or excluded (0) from the knapsack, making it a binary choice for each item. This problem is NP-complete because finding the optimal solution becomes exponentially hard as the number of items grows.

Figure 1.5(a) shows only the classical complexity classes [12]. Quantum computers provide dramatic advantages for certain structured problems in NP (like factoring and period finding), modest speedups for some problems in P, and no known advantage for NP-complete or NP-hard problems. Figure 1.5(b) [12] emphasizes that QC expands the set of efficiently solvable problems beyond classical capabilities, while still leaving large regions—particularly NP-hard and QMA-hard problems—believed to remain computationally intractable even with quantum resources.

From Figure 1.5(b), QC shines in the region that is the part of BQP that lies outside classical P. QC provides moderate advantage in the overlap of BQP with P. Here, quantum computers offer quadratic speedups, not exponential ones. They do not shine in the NP-hard region. At best, they offer heuristic or quadratic improvements. The QMA-hard region represents problems believed to be difficult even for quantum computers.

The *Turing machines* used in the figures can be understood as follows: a Turing machine is a simple abstract model that defines the meaning of computation. It consists of an infinite tape used as memory, a read–write head that moves along the tape, and a set of rules that determine how the machine updates symbols and changes the internal states step-by-step.

A deterministic Turing machine (DTM) follows exactly one rule at every step. Given the same input, it always performs the same sequence of operations and produces the same output. This model represents ordinary classical computers and defines the class P problems that can be solved efficiently by classical computation. A nondeterministic Turing machine (NTM) allows multiple possible moves from a given state. It explores many computational paths at once and accepts if any path leads to a solution. NTMs are not physical machines; they are theoretical tools used to define NP, the class of problems whose solutions can be efficiently verified, even if finding them may be difficult. DTMs model how classical computers actually compute, while NTMs help us understand why some problems are easy to check but hard to solve.

Every operation performed by a modern digital computer can be reduced to the same logical structure embodied in a Turing machine. Its tape represents computer memory, storing bits and intermediate results. Its read–write head acts like the processor, reading a symbol, writing a new one, and moving to the next memory location. Its finite control corresponds to the computer's instruction set and program counter, determining which rule to apply at each step. The machine's current state and the symbol it reads uniquely determine the next action.

Adding two binary numbers such as

$$101 + 011$$

is done using a sequence of bit operations and carry rules. For example, if the machine reads $1 + 1$, it writes 0 and switches to a "carry" state. If it reads $1 + 0$ while carrying, it writes 0 and keeps the carry, and so on. Eventually, the tape contains the correct sum.

1.5 QC Architectures (Examples)

Any physical system with two distinguishable basis states for 0 and 1 can be used as a qubit. Several QC architectures are under development, each with unique methods for creating, manipulating, and measuring qubits, with unique advantages and trade-offs. The future [27] of QC will likely involve many specialized architectures tailored to different applications.

Superconducting qubits [28–33]—a primary topic of this book—are made from electrical circuits using superconducting metals like niobium, aluminum, or tantalum. These devices use three elementary components: linear inductors, linear capacitors, and nonlinear inductors. The nonlinear inductor is the most important component formed by a thin insulating layer like aluminum oxide sandwiched between two superconductors, known as the *Josephson junction (JJ)* [34–36] (Section 3.12). The parameters of these elements are chosen in such a way that the operating frequency lies in the microwave range (2–10 GHz). The devices are cooled to near absolute zero (10–20 mK) to minimize thermal disturbance and ensure that the qubit dominantly stays in its "ground state," represented as $|0\rangle$. The first excited state $|1\rangle$ along with the ground state form a qubit. Some of the strengths for this platform are fast gate speeds and well-understood fabrication methods based on semiconductor fabrication, with demonstrated progress toward scalability.

A superconducting qubit behaves like an *artificial atom* because it stores energy in quantized states and responds to electromagnetic fields coherently, but it is artificially built from macroscopic [37] circuit elements like JJs and capacitors. While atoms interact with photons inside an optical cavity, referred to as *cavity quantum electrodynamics* (cavity QED), in *circuit QED* [38–45], a superconducting qubit (the artificial atom) interacts with microwave photons inside a superconducting resonator. Figure 1.6 shows the layout exemplifying a multiqubit superconducting transmon architecture (Section 3.15) with resonator coupling (Section 3.26), suitable for circuit QED experiments, qubit control, and dispersive readout (Section 3.30), topics to be covered in Chapter 3.

The superconducting qubit sits at an amazing crossroads. It obeys the same fundamental laws as electrons in atomic hydrogen, yet it is big enough to see under a microscope, human-fabricated, and engineered with intent. It takes the abstract ideas of quantum mechanics—superposition, entanglement, quantized energy—and places them into a tangible device. That is why these qubits are called artificial atoms: not just as an analogy, but because they truly extend the concept of the atom into a domain shaped by human creativity and technology.

Superconducting quantum processors are rapidly growing in qubit count and functionality, demonstrating many orders-of-magnitude improvement in coherence over the past decades [33, 46]. A complex dilution refrigerator

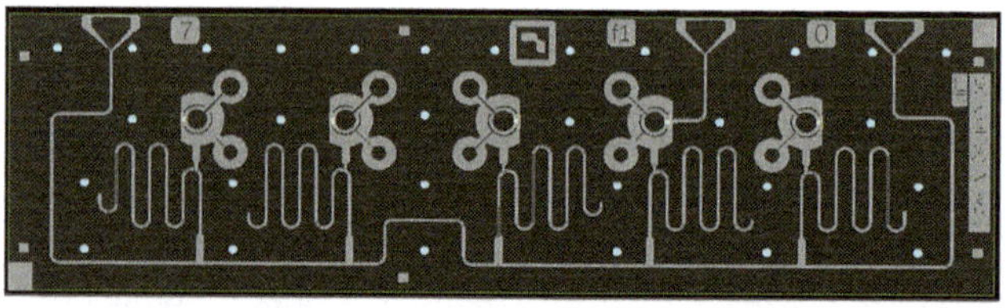

Figure 1.6 Superconducting qubit chip with coplanar waveguide circuitry and transmon qubits. Illustration of a planar superconducting quantum circuit implemented on a chip substrate. The pale metallic structures represent patterned superconducting film forming the control, readout, and coupling circuitry; the green background indicates the underlying insulating substrate. The central repeated units are transmon qubits, each identifiable by a pair of large "paddle"/island electrodes connected through a small Josephson junction region (located in the narrow constriction between the islands). Each qubit is coupled capacitively to a meandered coplanar waveguide resonator, shown as serpentine $\lambda/4$ structures below the qubit islands, which serve as readout and/or coupling resonators. Microwave control and readout are routed via the larger coplanar feed lines running horizontally across the chip. The small square and rectangular alignment markers around the periphery are used for lithographic registration (with permission from SQMS, Fermilab).

(Section 3.46) is required to maintain high coherence at low temperature (typically below 20 mK). Google, IBM, Rigetti, IQM, Origin Quantum, and others are developing superconducting qubits. Bosonic qubits (Chapter 4), which combine trapped microwave photons in cavities and superconducting qubits for their manipulation and readout, are under development by companies such as Alice&Bob, AWS, and Nord Quantique.

In trapped-ion quantum computers [47–51], ions are trapped and manipulated in carefully shaped electromagnetic fields, with quantum states encoded in the energy levels of each ion (Figure 1.7). Laser pulses are used for quantum operations. The strengths here are high-fidelity operations and long coherence times, seconds, or longer—an eternity on quantum timescales. But gate operations are slow compared to superconducting qubits, and complex systems are needed to control multiple ions. Each ion is a nearly identical qubit, with energy levels defined by fundamental physics, giving trapped ions exceptional uniformity and stability. IonQ, Oxford Ionics, Quantinuum, and other groups are worldwide developers.

Photonic quantum computers (Figure 1.8) use photons as qubits [52–54]. Quantum states are based on several possible degrees of freedom, or physical properties, of the photons. It can be their phase, amplitude, frequency, path, photon number (aka Fock states), and spin orbital momentum. The most common is: photons with different polarizations, such as horizontal and vertical. These polarization states can be one way to represent the two states of a qubit. The states are

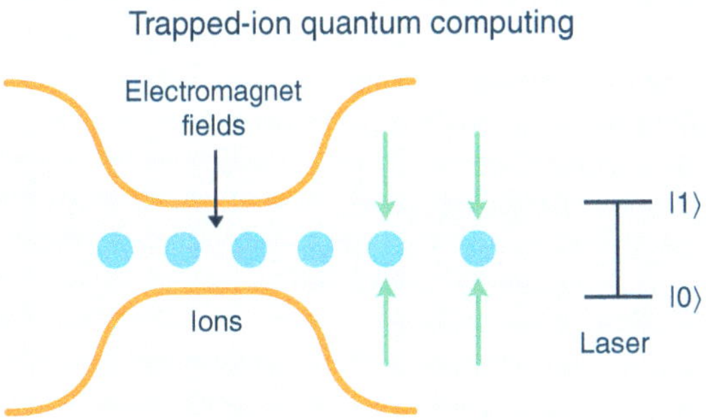

Figure 1.7 Ions are trapped and manipulated in electromagnetic fields, with quantum states encoded in the energy levels of each ion.

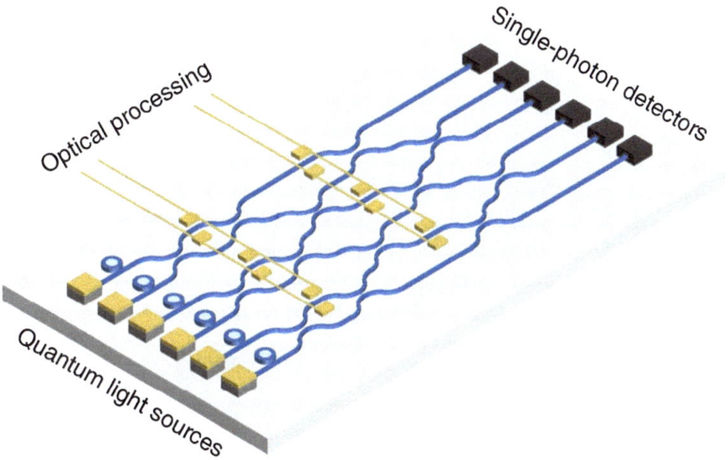

Figure 1.8 Schematic illustration of an on-chip photonic quantum processor, where single photons serve as qubits propagating through lithographically patterned optical waveguides. At the lower left, the yellow blocks represent quantum light sources—such as heralded single-photon sources, spontaneous parametric downconversion (SPDC) sources, quantum dots, or defect-based emitters—each generating individual photons prepared in well-defined quantum states. These photons are injected into the blue integrated waveguides, which route them across the chip. The central region labeled *optical processing* contains a network of interferometers, beam splitters, and tunable phase shifters that implement the logic of photonic quantum computing. Interference between indistinguishable photons produces entanglement and multi-photon correlations, while programmable phase delays and couplings realize quantum gates and reconfigurable circuits. At the upper right, the black blocks denote single-photon detectors—such as superconducting nanowire detectors—which measure the output modes. Detection patterns encode the result of the quantum computation [55] / arXiv / CC BY 4.0.

manipulated with beam splitters, phase shifters, polarizers, and other optical devices. Unlike matter-based qubits that must be well isolated, photons ignore most environmental noise, gliding through fibers and free space with astonishingly low loss, carrying fragile quantum states across kilometers. Photons can be entangled and transmitted over long distances, ideal for networking. The strengths here are room-temperature operation and suitability for quantum communication. Photonic qubits can maintain coherence for microseconds to milliseconds (and even longer in quantum memories) because they hardly dephase, but the dominant limitation is photon loss; photonic gate operations occur on picosecond to nanosecond timescales, with practical system-level rates set by detectors, sources, and feed-forward rather than the optics. The challenges are precise photon generation and detection, and scaling. Photons don't naturally interact with each other, making it harder to implement two-qubit gates. Developers are Xanadu, PsiQuantum, and others.

In other platforms, photonic qubits are typically encoded in optical degrees of freedom—such as path (which waveguide the photon occupies), polarization, time bin, or frequency—with logical $|0\rangle$ and $|1\rangle$ represented by distinct optical modes. Because photons interact weakly with their environment, these qubits exhibit low intrinsic dephasing. Computation arises from quantum interference rather than strong particle–particle interactions.

Quantum dot qubits [56–58] are made from nanoscale semiconductor particles that confine electrons or electron holes, creating discrete energy levels that allow discrete quantum states (Figure 1.9). The spin state of an electron in a quantum dot can also represent a qubit, with spin-up $|\uparrow\rangle$ and spin-down $|\downarrow\rangle$ states corresponding to $|0\rangle$ and $|1\rangle$. Microwaves, magnetic fields, and optical pulses control

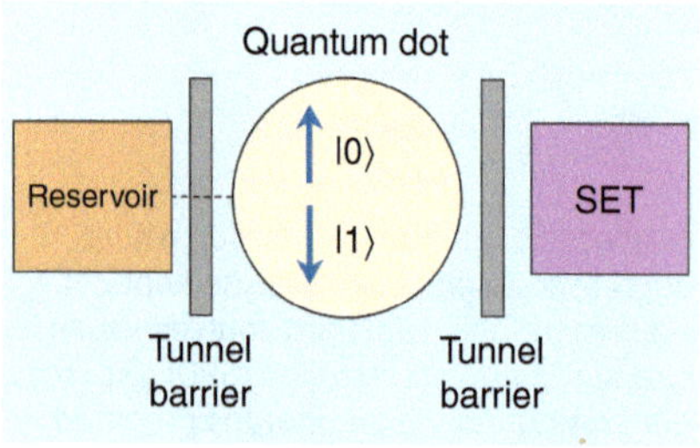

Figure 1.9 In a quantum dot, the reservoir is a large collection of electrons connected to the quantum dot through tunnel barriers. Electrons tunnel between the reservoir and quantum dot when allowed. The SET is a sensitive charge detector. It can sense when an electron enters or leaves the dot by detecting current changes. Its role is as a readout to detect the charge state of the quantum dot without disturbing it. When an electron tunnels into or out of the quantum dot, the SET detects a blip. For a spin qubit, magnetic fields and microwave pulses manipulate the electron spin, and the SET reveals the state $|0\rangle$ or $|1\rangle$.

the states for quantum operations. A region of semiconductor, only tens of nanometers across, confines one or a few electrons so that they occupy discrete, tunable quantum energy levels. Lithography and gate voltages shape, move, and entangle these "designer atoms." In gallium arsenide devices, spin coherence reaches microseconds with echo techniques, while in isotopically purified silicon or Si/SiGe heterostructures, the absence of nuclear spin noise allows T_1 lifetimes of seconds and T_2 (Section 2.3) times from tens of microseconds up to milliseconds with dynamical decoupling. At the same time, their gate operations are fast: single-qubit spin rotations are routinely performed in 1–50 ns, and two-qubit exchange gates typically take tens to a few hundred nanoseconds.

The strengths here are compatibility with semiconductor manufacturing, enabling integration with classical computing systems. They open a pathway to integration of millions of qubits on a chip along with cryogenic control electronics. Quantum dot qubits require millikelvin range temperatures to operate reliably. Other challenges are fabricating quantum dots with precisely controlled sizes and coupling strengths. Developers include Intel and QuTech.

Neutral atom systems [59–64] use individual laser-cooled atoms, typically alkali or alkaline earth metals (rubidium, cesium, strontium, or ytterbium), trapped in optical tweezers that are highly focused laser beams that create potential wells (Figure 1.10). The qubit states are typically encoded in the electronic or hyperfine states of the atoms, with quantum states controlled by laser pulses. These arrays of atoms can be rearranged dynamically for various computations.

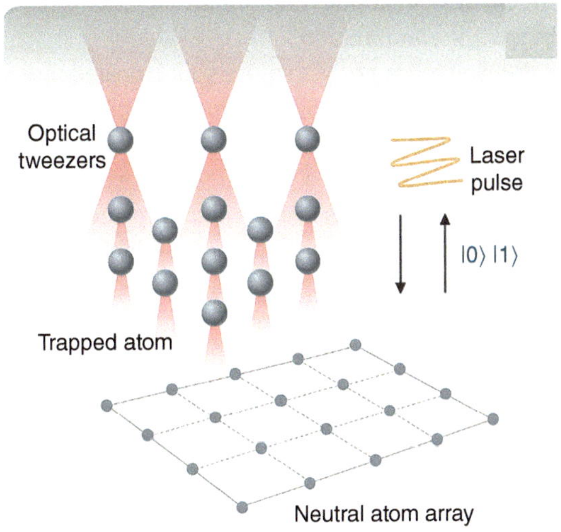

Figure 1.10 A 2D array of neutral atoms individually trapped in optical tweezers formed by tightly focused laser beams. Each atom encodes a qubit, with quantum states $|0\rangle$ and $|1\rangle$ corresponding to two distinct internal electronic or hyperfine states. Laser pulses control transitions between these states, enabling qubit initialization, manipulation, and entanglement. The spatial arrangement of atoms is flexible, allowing dynamic reconfiguration of the qubit grid for different computational tasks.

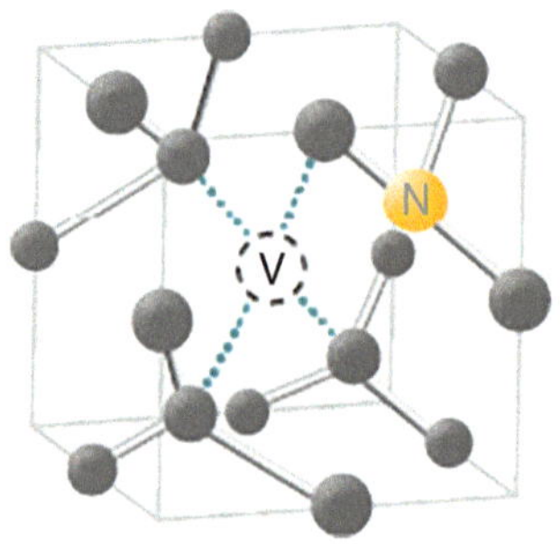

Figure 1.11 Diagram of a nitrogen-vacancy (NV) center in a diamond lattice. Carbon atoms are shown as filled gray circles, with one carbon replaced by a nitrogen atom (yellow) and an adjacent vacant carbon site. The NV pair forms the basis for NV center qubits.

Neutral-atom quantum systems have reached a remarkable moment in the history of physics: the ability to see, hold, and command individual atoms. With exquisite control, lasers move these atoms across space, arrange them into programmable arrays, and excite them into highly interactive Rydberg states. What was once the realm of thought experiments, i.e., manipulating nature one atom at a time, has become an experimental reality. Light itself becomes a pair of hands, placing atoms in position at will and choreographing their quantum states.

Neutral-atom arrays can be scaled up to hundreds or thousands of atoms, making them promising for large-scale quantum systems [65]. The internal states of neutral atoms, especially in their ground states, have long coherence times—seconds to minutes! The spatial arrangement of atoms in a 2D or 3D grid allows for highly customizable and flexible connectivity between qubits. This enables efficient implementation of certain quantum algorithms, especially those that benefit from high qubit connectivity. The trapping arrays can be reconfigured in real time, enabling different arrangements of atoms based on the requirements of specific computations.

Neutral-atom arrays require sophisticated optical systems to trap and manipulate individual atoms. This includes precise alignment of laser beams, stabilization of optical tweezers, and control of laser frequencies. Such infrastructure may be harder to scale up. Cooling atoms to extremely low temperatures (often in the microkelvin range) is necessary to trap and control them effectively. This requires advanced laser cooling techniques, which can be technically demanding. Operational systems have been assembled with 100–300 atoms in simulation mode.

One of the biggest challenges for this platform is that neutral-atom qubits can literally disappear from background gas collisions, heating from optical traps, or imperfect rearrangement. Qubit loss means the algorithm must be restarted or the lattice refilled. Another challenge is the slow two-qubit gates, typically 0.5–1 μs as compared to 50–100 ns for superconducting two-qubit gates. Developers are QEra, Atom Computing, Infleqtion, Pasqal, Planqc, and others.

Nitrogen Vacancies (NVs) in Diamonds [66–68]: A diamond crystal is a lattice of carbon atoms. An NV center is a defect where one carbon atom is replaced by a nitrogen atom (Figure 1.11). An adjacent lattice site is vacant (missing a carbon atom). This defect creates localized electronic states. An NV center has an extra electron (negatively charged NV configuration), and this electron's spin is used as the qubit. Typically, the qubit uses two of three spin states, for example:

$$|0\rangle \equiv |m_s = 0\rangle, |1\rangle \equiv |m_s = -1\rangle.$$

The energy difference between $|m_s = 0\rangle$ and $|m_s = -1\rangle$ corresponds to a frequency of about 2.87 GHz. The spin qubit is manipulated using microwave pulses. By choosing the duration and phase of the pulse, quantum gates can be implemented. NV qubits operate at room temperature, unlike many platforms that require dilution refrigerators, and have long coherence times. But it is difficult to place NV centers precisely in diamond with atomic accuracy. Commercial products support about 10 qubits at the present stage.

This is a remarkable marriage of the everyday transparent gemstone with the extraordinary quantum bit. Hosting a single defect makes it sensitive enough to feel a single nearby nuclear spin. The same spin that stores information is exquisitely sensitive to magnetic and electric fields, strain, and temperature, allowing nanoscale magnetometry. Companies involved are Quantum Brilliance, Turing Inc., XeedQ, and Saxon.

1.6 QC Paradigms

1.6.1 Gate-Based QC, AQC, Measurement-Based QC

Gate-based QC is the paradigm most widely developed and used [69]. It is analogous to classical computing, where operations are performed using logic gates, to manipulate qubits and perform computations. Quantum algorithms, like Shor's algorithm [70] for factorization, and Grover's algorithm [20] for database searches, are designed in this framework. This model is scalable but requires precise control over the large number of quantum gates and qubits. Current gate-based quantum computers are sensitive to decoherence and noise, leading to errors. QEC methods make it more robust to noise. However, QEC (Section 2.23) and fault-tolerant computing require a large overhead of physical qubits.

AQC [26, 71–73] works by slowly evolving the quantum state of a system from an initial "easy-to-prepare" ground state to the ground state of the harder problem, which encodes the solution to a computational problem. It is based on the adiabatic theorem, which states that if the evolution is slow enough, the system will remain in its ground state where coherence is best. AQC slowly transforms the initial problem into the target problem by adjusting parameters in the quantum system over time.

The AQC has been applied to optimize traffic flow and routing in networks for urban transportation planning or airline scheduling. Similarly, AQC can help find efficient solutions in managing resources within warehouses and supply chains. In finance, portfolio optimization involves balancing risk and return across a range of assets, an inherently complex and high-dimensional problem to find better portfolio allocations under specific constraints compared to classical algorithms. AQC has applications in simulating molecular interactions, where understanding the lowest-energy structure of molecules like proteins is critical for drug discovery and exploring biochemical compounds. Companies like D-Wave are continuously improving quantum annealing devices, but maintaining high coherence across a large number of qubits remains a challenge.

1.7 Measurement-Based Quantum Computing

Measurement-based quantum computing (MBQC) [74, 75], also known as the one-way quantum computer, operates by preparing an entangled resource state (often a cluster state) and then performing a sequence of measurements on individual qubits. An entangled resource state is a large web of qubits prepared with collective entanglement. Instead of performing long sequences of gates, the computation simply measures individual qubits in chosen bases (Section 2.7). Measurements steer the flow of quantum information through the entangled network, and classical feed-forward of the measurement outcomes determines how the remaining qubits are measured next. An important example of a resource state is the *cluster state* formed by arranging qubits on a regular lattice—a line, square grid, or 3D array—and entangling each qubit with its neighbors. Each measurement effectively rewires the remaining network, propagating quantum information through the lattice and implementing logical gates on the unmeasured qubits.

This model has gained traction because it is resource-efficient and highly compatible with certain quantum architectures. MBQC is used in QEC codes, like surface codes (Section 2.23), to suppress noise in quantum systems. MBQC is beneficial for distributed QC and quantum networks. Here multiple quantum systems are connected across distances to perform computations, share information, or enable quantum communications. Quantum key distribution (QKD) systems are pioneering implementations of quantum networks. QKD ensures highly secure data transmission since eavesdropping disturbs the quantum states, alerting parties to any intrusion while destroying the original information. Distributed quantum sensors with arrays of entangled sensors can perform tasks like weak magnetic field sensing, sensitive gravimetry, and positioning with unparalleled precision. These could enable new types of scientific measurements that detect

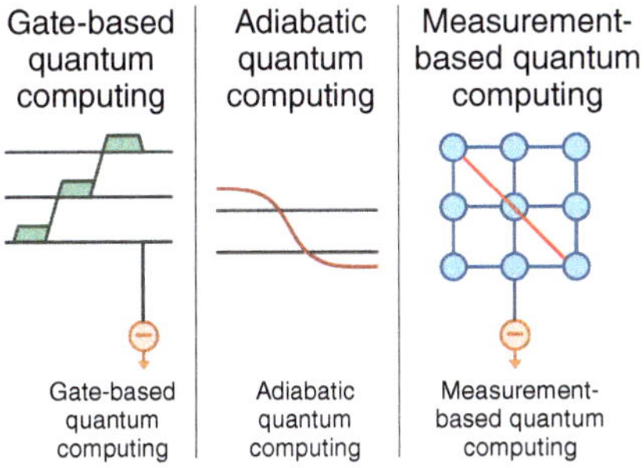

Figure 1.12 Comparison of three major paradigms of quantum computation: gate-based, adiabatic, and measurement-based, illustrating, in schematic form, the conceptual differences among three leading approaches to QC. The simple schematics emphasize how each model conceptualizes the flow of quantum information through the computation. (Left) In *gate-based QC*, quantum information is processed through a discrete sequence of unitary gates applied step-by-step. The green "staircase" segments represent successive gate operations that move the system state through different levels in Hilbert space, emphasizing the digital, pulsed nature of control. The downward orange symbol indicates that the final state is read out only after the prescribed gate sequence is completed. (Center) In *adiabatic quantum computing*, the computation proceeds not by discrete gates but by slow, continuous evolution of the system Hamiltonian. The red, smoothly curving line illustrates how the quantum state is guided adiabatically from an initial energy configuration to a final one that encodes the solution, while the nearly parallel black lines represent the instantaneous energy levels during this process. The lack of "steps" highlights that the control is analog and gradual rather than pulsed. (Right) In *measurement-based quantum computing*, the computation is driven by successive measurements on a highly entangled *cluster state*. The blue circles denote qubits arranged on a lattice, and the blue lines indicate entanglement links between neighboring qubits. The red diagonal line represents a sequence of single-qubit measurements carried out across the entangled lattice; these measurements, together with adaptive classical feed-forward, effectively implement the computation. The orange readout symbol again signifies extraction of the final classical result.

weak gravitational changes or magnetic fields. IBM and other organizations have tested MBQC in small-scale quantum processors. NASA has explored distributed quantum sensing applications for space-based positioning and Earth observation, while DARPA (Defense Advanced Research Projects Agency) has invested in entangled sensor networks for precise navigation systems. Figure 1.12 qualitatively compares the three paradigms.

1.8 Where Quantum Computers Might Shine

Quantum computers naturally simulate quantum systems more efficiently. When we say *naturally*, we are pointing to a profound idea: a quantum computer is already quantum. Classical computers struggle with quantum systems because they must explicitly keep track of enormous amounts of information. A molecule made of just a few dozen electrons lives in a quantum state described by an astronomical number of parameters, growing exponentially with system size. 50 quantum particles → 2^{50} amplitudes (~10^{15} numbers!). 300 particles → more numbers than atoms in the observable universe. Simulating quantum systems (e.g., molecules, materials, or biological processes) is extremely resource-intensive on classical computers, as the required resources grow exponentially with the size of the system. A classical machine has to represent all of those amplitudes as numbers in memory and must update them one by one. Such a bookkeeping task becomes impossible. A quantum computer, by contrast, speaks the same "language" as the system being simulated. The evolution of qubits is governed by the same unitary transformations (gate operations) and interference effects that govern electrons in atoms or spins in a magnet. Unitary gates are transformations dictated by Schrödinger's equation.

When a quantum computer simulates a molecule, the complex amplitudes and entangled structures of the molecule are encoded in the physical state of the qubits themselves, not laboriously tracked as a list of numbers. One quantum system stands in for another, using one piece of quantum matter to model the behavior of another piece of quantum matter. Exponentially exploding classical bookkeeping is replaced with the natural evolution of another quantum system. A quantum computer simulating quantum mechanics is similar to using water to simulate water, instead of trying to simulate the motion of the ocean by individually tracking every water molecule with marbles on a table.

For example, imagine trying to understand how electrons move through a high-temperature superconductor. The electrons do not behave independently; they are entangled in a dense, correlated quantum fluid in which the motion of one electron is inseparable from the motion of many others. To describe that faithfully on a classical computer requires tracking an exponentially large wavefunction. A quantum computer, however, can place its qubits into the same richly entangled patterns that the electrons themselves occupy and then let them evolve under a model Hamiltonian. In doing so, it becomes a laboratory for phases of matter that cannot be easily created in the real world.

Consider complex chemical reactions, such as the catalytic processes inside enzymes or the binding of drugs to proteins. These processes are governed by the quantum structure of many-electron molecules, where subtle interference between electronic configurations determines whether a reaction proceeds or stalls. A quantum computer can directly represent the electronic wavefunction. In the long run, that capability could allow us to predict reaction rates, design new catalysts, and discover materials for energy storage or carbon capture by simulating the chemistry at its native quantum level. It could revolutionize fields like material science, chemistry, and drug discovery.

Once we accept that a quantum computer can naturally inhabit vast superposition spaces and manipulate entanglement as a native resource, those same capabilities can be recruited for problems outside physics. Shor's algorithm harnesses interference in large state spaces to uncover hidden periodicity, turning the structure of quantum mechanics into a method for factoring large integers exponentially faster than known classical methods. Thus, quantum computers can potentially break encryption schemes (Section 6.3). In cryptography, many encryption systems, like RSA, rely on the difficulty of factoring large numbers. *RSA encryption* secures data transmission for online communications and financial transactions. It is named after its inventors, Rivest, Shamir, and Adleman, who developed it in 1977. Classical computers require exponential time to factor large numbers as the size of the numbers increases. For example, factoring a 2048-bit number (which is common in RSA encryption) could take classical computers billions of years. A quantum computer running Shor's algorithm could eventually factor a 2048-bit number in a matter of hours or days, once such a high-quality quantum computer is available (possibly in a decade or two). This possibility poses a significant threat to current cryptographic systems, leading to the need for the development of quantum-resistant encryption algorithms as part of post-quantum cryptography.

Grover's algorithm uses amplitude amplification—a quantum way of reinforcing the right answer—to search unsorted databases quadratically faster than any classical search. Searching through an unsorted database for a specific item classically takes time proportional to the number of items in the database. Grover's algorithm can search an unsorted database in a time proportional to the square root of the number of items, providing a quadratic speedup over classical search algorithms.

In both cases, what first emerged as a tool for understanding nature's quantum systems became a general computational resource. The same features that let a quantum computer imitate the behavior of electrons and atoms also allow it to reshape the landscape of mathematics and computer science. Similarly, other real-world problems, such as supply chain management, financial modeling, and traffic flow optimization, involve finding the best solution from a vast number of possibilities. These are often NP-hard problems, which are computationally intractable for classical computers. Quantum algorithms, like the Quantum Approximate Optimization Algorithm (QAOA) [76, 77], could provide better approximations to optimal solutions faster than classical algorithms (Chapter 6), potentially transforming industries that rely on solving large-scale optimization problems. QAOA uses a combination of quantum operations to find a "good enough" or approximate solution rather than the perfect one. The algorithm can be adjusted to find increasingly good solutions by changing the number of operator repetitions, or by fine-tuning the parameters with the classical optimizer.

1.9 Conclusion

QC stands at the forefront of a technological revolution, one that promises to redefine the boundaries of computation, simulation, and optimization. QC leverages phenomena such as superposition, entanglement, and interference to tackle problems that remain intractable for even the most powerful classical supercomputers. While still in its early stages of development, QC has already shown theoretical and practical promise across a diverse range of domains. Quantum algorithms are being developed to provide a dramatic boost to many disciplines—including fundamental physics, chemistry, finance, logistics, artificial intelligence, and materials science (Chapter 6).

The coming chapters offer a window into a future where classical limitations give way to quantum possibility. Chapter 2 emphasizes the conceptual leap from classical to QC: bits versus qubits, deterministic gates versus unitary quantum operations, and the fragile but powerful potential of superposition, entanglement, and interference. It also introduces the practical challenges—decoherence, error correction, and hardware scalability—that define the field's current "noisy intermediate-scale" stage. Importantly, it situates QC not as a replacement for classical systems but as a complementary tool, uniquely suited to problems like cryptography, quantum simulation, and large-scale optimization.

2

Quantum Computing Basics

2.1 Overview Introduction

Chapter 2 lays the foundation for understanding how quantum computers differ fundamentally from classical computers. It begins with the familiar territory of classical bits—binary units of zeros and ones—before moving to their quantum counterpart, the qubit. Unlike a bit, a qubit can exist in superpositions of states, described mathematically with Dirac's bra–ket notation and represented geometrically on the Bloch sphere.

The chapter introduces the algebra of quantum mechanics—bras, kets, inner and outer products—and shows how they enable the description of quantum states and transformations. Building from this, quantum gates are presented as reversible, unitary operations that manipulate qubits. Core gates like the Pauli operators, Hadamard, S, and T gates are introduced as the building blocks of quantum circuits.

From single-qubit operations, the discussion advances to two-qubit systems, tensor products, and controlled operations such as CNOT and CZ, which enable entanglement—Einstein's "spooky action at a distance." With entanglement, quantum states gain correlations with no classical analog, opening the door to exponential computational possibilities.

The chapter also introduces the principles of measurement, quantum algorithms (Deutsch and Deutsch–Jozsa, Grover), and the role of unitarity in ensuring reversibility and probability conservation. Finally, it broadens the horizon beyond qubits to qudits, where higher-dimensional systems promise richer computation and shorter circuits.

2.2 Bits and Qubits

For classical computing, the most basic piece of information is called a *bit*, which represents a "yes–no" answer to a question. Since we are dealing with a two-state system, we can represent information using a base 2 or *binary* numbers. A single digit of binary number can be represented using 0 ("no" answer) or 1 ("yes" answer) containing *one bit* of information. In order to express more information, one needs to increase the number of *bits*. For example, a *two-bit* number can represent $2^2 = 4$ possible states using the binary numbers "00," "01," "10," and "11." Classical computers utilize these expressions to represent the four decimal numbers 0, 1, 2, 3. Table 2.1 shows the binary representation.

Now, one cannot express decimal 4 or higher using a two-bit binary number as there is not enough bits. To represent the numbers 0–7, we need three bits, because we have $2^3 = 8$, as shown in Table 2.2. Thus, three bits can store eight different values.

The most basic word or block of information in classical computing is called a *byte*, which is a string of eight bits linked together. Since $2^8 = 256$, a byte can

Table 2.1 Binary representation of four numbers 0–3.

Decimal	Binary
0	00
1	01
2	10
3	11

Table 2.2 Binary representation of eight numbers 0–7.

Decimal	Binary
0	000
1	001
2	010
3	011
4	100
5	101
6	110
7	111

store 256 different values. However, it is important to understand that at any given instance, one *byte* contains only *one* of the 256 possible values.

A qubit is fundamentally different from a bit and allows information processing in new and interesting ways. A qubit can exist in the state $|0\rangle \rightarrow$ or the state $|1\rangle$, but it can also exist in a *superposition* state, which is a linear combination of the states. Here we use the *ket* notation for the state of a qubit, as in quantum physics. A superposition state is written as

$$|\psi\rangle = \alpha|0\rangle + \beta|1\rangle,$$

where α and β are complex numbers. When a qubit is measured, it is only found to be in the state $|0\rangle$ or $|1\rangle$. Here $|\alpha|^2$ tells us the probability of finding $|\psi\rangle$ in state $|0\rangle$, and $|\beta|^2$ tells us the probability of finding $\psi\rangle$ in state $|1\rangle$. Here $|\alpha|^2 = (\alpha)(\alpha^*)$, where α^* is the complex conjugate of α, and similarly for β. Recall that the complex conjugate of complex number $z = x + iy$ is $z^* = x - iy$. Since the probabilities must sum to one, we have

$$|\alpha|^2 + |\beta|^2 = 1$$

The *Hilbert space* describing n qubits has 2^n basis states. A general quantum state of n qubits is a superposition of all 2^n basis states. For n qubits, the computational basis states are

$$|00\ldots 0\rangle, |00\ldots 1\rangle, \ldots, |11\ldots 1\rangle$$

A general quantum state is

$$|\psi\rangle = \sum_{x=0}^{2^n-1} \alpha_x|x\rangle$$

with complex amplitudes α_x.

A quantum computer "stores 2^n bits of classical information." Quantum operations act on all 2^n amplitudes at once. Interference can amplify correct answers and cancel wrong ones. Entanglement allows correlations, introduced in Chapter 1 and to be discussed later. Measurement gives you access to all 2^n values simultaneously. But when you measure, you get just one outcome, not 2^n.

We can also express the qubit's state in a column vector format

$$|\psi\rangle = \begin{pmatrix} \alpha \\ \beta \end{pmatrix}$$

Using the properties of vector addition and scalar multiplication, we have

$$|\psi\rangle = \begin{pmatrix} \alpha \\ \beta \end{pmatrix} = \begin{pmatrix} \alpha \\ 0 \end{pmatrix} + \begin{pmatrix} 0 \\ \beta \end{pmatrix} = \alpha\begin{pmatrix} 1 \\ 0 \end{pmatrix} + \beta\begin{pmatrix} 0 \\ 1 \end{pmatrix}$$

which allows us to identify the vectors for $|0\rangle$ and $|1\rangle$ as

$$|0\rangle = \begin{pmatrix} 1 \\ 0 \end{pmatrix}, \quad |1\rangle = \begin{pmatrix} 0 \\ 1 \end{pmatrix}$$

We can express any vector in the space V in terms of a linear expansion on a *basis set* where the *dimension* of a vector space V is equal to the number of elements in the basis set. Since the vectors above span the 2D vector space, the set is a basis set. A quantum state can be written as a linear combination of a basis set with complex coefficients of expansion.

2.3 Coherence and Decoherence

The superposition state carries two kinds of information: *population* information and *phase* information, which together define the quantum information. The population of a state is just the probability of finding the system there: population of $|0\rangle = |\alpha|^2$, population of $|1\rangle = |\beta|^2$. When $|\alpha|^2 = 1, |\beta|^2 = 0$, the qubit is fully in $|0\rangle$. If $|\alpha|^2 = |\beta|^2 = 1/2$, the qubit is in equal superposition.

When $|\psi'\rangle = \frac{1}{\sqrt{2}}(|0\rangle - |1\rangle)$, the population is also the same in both states as above, but there is difference in their relative phase, which gives us the *phase* information. The state is physically different.

When there is *decoherence*, there are two processes: energy relaxation (with decoherence time T_1) and *dephasing* (with time T_2). In the first process, the populations change. T_1 measures how fast the excited state $|1\rangle$ decays to $|0\rangle$. In the second process, the populations stay fixed, but phase is lost.

T_1 is controlled by how easily the qubit can lose its energy into the environment. If the environment can "absorb" the qubit's transition energy, the qubit relaxes. T_2 measures how fast the relative phase between $|0\rangle$ and $|1\rangle$ is lost. Physically, it means how long the Bloch vector keeps pointing in a definite direction. T_2 is controlled by how stable the qubit's transition frequency is over time. These two times are related by

$$\frac{1}{T_2} = \frac{1}{2T_1} + \frac{1}{T_\phi}$$

where T_ϕ is the pure dephasing time. Energy decay automatically causes dephasing, which sets a fundamental limit: $T_2 \leq 2T_1$.

When a qubit is in superposition state $|\psi\rangle = \alpha\,|\,0\rangle + \beta\,|\,1\rangle$, if there is energy relaxation, the excited state $|1\rangle$ decays to $|0\rangle$, at the rate $1/T_1$. Hence,

$$|\beta(t)|^2 = |\beta(0)|^2 e^{-t/T_1}$$

So,

$$\beta(t) = \beta(0)e^{-t/(2T_1)}$$

The population decays exponentially with time constant T_1, but the amplitude β must decay as the square root. But coherence depends on the excited-state *amplitude*, while population decay depends on its *square*. Therefore, the phase coherence time T_2 (also called decoherence time or transverse relaxation time) obeys $T_2 = 2T_1$. In real systems, there is also pure dephasing T_ϕ, which makes T_2 become shorter. Generally $T_2 \leq 2T_1$, according to

$$\frac{1}{T_2} = \frac{1}{2T_1} + \frac{1}{T_\phi}$$

Here we touch upon the various loss factors that influence T_1, to discuss (Chapter 5) in more detail later. Some of these factors are radiation into transmission lines, readout resonators, and other modes. Material losses [78] are another big factor as from *two-level system (TLS)* in surface oxides, or dielectric interfaces, lossy substrates, and defective JJ barriers. In addition, there can be quasiparticles from broken Cooper pairs in the superconducting qubit when it absorbs energy at the qubit frequency. Finally, there are thermal photons from the residual microwave photons in the environment.

The main limiter of T_2 is low-frequency (*1/f*) noise arising from charge noise (transmons are designed to suppress this), flux noise, or magnetic noise. Low-frequency noise is called *1/f* noise because its power spectrum scales inversely with frequency, meaning slow fluctuations dominate. This is a signature of many microscopic fluctuators acting across a wide range of timescales—and a primary cause of dephasing in quantum systems. It means that at 10 Hz, the noise is ~10× larger than at 100 Hz, and at 1 Hz, it's ~100× larger than at 100 Hz. Recent evidence [79] suggests that factors like JJ impurity content may play a defining role for T_2.

2.4 Bra–Ket Basics

In quantum mechanics, the *bra–ket notation* (or Dirac notation) is a convenient way to represent states, inner products, and operators. This notation was developed by physicist Paul Dirac, and it provides a clear and powerful framework for describing quantum states and the linear algebra that underlies quantum mechanics. *Ket vectors* represent quantum states as column vectors, as we have just used above.

The *conjugate transpose* (or Hermitian conjugate) of *a ket* $|\psi\rangle$ is denoted by $\langle\psi|$ (called a *bra*). A bra is a row vector that represents the dual of a ket. If $|\psi\rangle = \begin{pmatrix}\alpha\\ \beta\end{pmatrix}$, then $\langle\psi| = (\alpha^* \quad \beta^*)$, where α^* and β^* are the complex conjugates of α and β, respectively.

The *inner product* $\langle\psi \mid \phi\rangle$ between two states $|\psi\rangle$ and $|\phi\rangle$ is a complex number that represents the overlap between the two states. For $|\psi\rangle = \begin{pmatrix}\alpha\\ \beta\end{pmatrix}$ and $|\phi\rangle = \begin{pmatrix}\gamma\\ \delta\end{pmatrix}$,

$$\langle\psi|\phi\rangle = (\alpha^* \quad \beta^*)\begin{pmatrix}\gamma\\ \delta\end{pmatrix} = \alpha^*\gamma + \beta^*\delta$$

If $\langle\psi \mid \phi\rangle = 0$, the states $|\psi\rangle$ and $|\phi\rangle$ are *orthogonal*.

The *outer product* $|\psi\rangle\langle\phi|$ of two states is an operator formed by multiplying a *ket* with a *bra*.

This product produces a matrix. If $|\psi\rangle = \begin{pmatrix} \alpha \\ \beta \end{pmatrix}$ and $|\phi\rangle = \begin{pmatrix} \gamma \\ \delta \end{pmatrix}$, the outer product

$$|\psi\rangle\langle\phi| = \begin{pmatrix} \alpha \\ \beta \end{pmatrix} (\gamma^* \quad \delta^*) = \begin{pmatrix} \alpha\gamma^* & \alpha\delta^* \\ \beta\gamma^* & \beta\delta^* \end{pmatrix}$$

For examples of inner and outer products, when $\psi\rangle = \frac{1}{\sqrt{2}} \begin{pmatrix} 1 \\ 1 \end{pmatrix}$ and $|\phi\rangle = \frac{1}{\sqrt{2}} \begin{pmatrix} 1 \\ -1 \end{pmatrix}$, the *inner product*

$$\langle\psi|\phi\rangle = \frac{1}{\sqrt{2}}(1^*) \cdot \frac{1}{\sqrt{2}} \cdot 1 + \frac{1}{\sqrt{2}}(1^*) \cdot \frac{1}{\sqrt{2}} \cdot (-1) = \frac{1}{2} - \frac{1}{2} = 0$$

as these states are orthogonal.

The outer product is a *projection operator* that projects onto the state $|\psi\rangle$:

$$|\psi\rangle\langle\phi| = \frac{1}{\sqrt{2}} \begin{pmatrix} 1 \\ 1 \end{pmatrix} \cdot \frac{1}{\sqrt{2}} (1 \quad 1) = \frac{1}{2} \begin{pmatrix} 1 & 1 \\ 1 & 1 \end{pmatrix}$$

Later we use the *measurement operators* based on outer products to evaluate the effects of superposition and entanglement.

Dirac's bra–ket notation grew out of an historical moment in physics, when the new quantum theory was still young and its mathematical language unsettled. In the late 1920s, Schrödinger's wave mechanics and Heisenberg's matrix mechanics were both known to describe the same physics, but they looked completely different mathematically. One treated quantum systems as continuous waves in space; the other as infinite matrices acting on abstract state vectors. Dirac realized that underneath these two formalisms, there was a single, deeper structure. What the subject needed was a clearer, more flexible language to talk about quantum states and the operations acting on them. He wanted a notation that would place states and operations on an equal footing and to make linearity obvious. Instead of representing states as columns or rows of numbers tied to a specific coordinate system, Dirac treated them as abstract vectors in a Hilbert space.

Dirac's terminology had a sense of humor to it. He deliberately split the word *bracket* into "bra" and "ket" so that the inner product of two states literally formed a "bra–ket." What seems playful was actually a profound reorganization of the mathematics: it made linear algebra operations intuitively readable. Operators naturally nestled between bras and kets, as $\langle\phi|\hat{A}|\psi\rangle$, and eigenstates, completeness relations, and orthogonality could be written in a compact form that physicists could manipulate.

2.5 Gates

Classical computers manipulate bits through logic *gates* that transform the value that a bit holds. Some examples of classical logic gates are OR, AND, NOT, NAND, etc. In classical computers, the NOT gate takes one input and reverses its value. For example, it changes the 0 bit to a 1 bit, and vice versa. This is like a light-switch flipping a light from ON to OFF, or from OFF to ON. The other classical gates, like OR, NOT, and NAND, act on two input bits and results in one. Consequently, these gates are not reversible.

Quantum gates modify the state of qubits. However, quantum gates are more versatile, as those can perform a continuum of operations, contrary to the classical ones that are discrete. One important difference is that quantum gates are always reversible, independent of whether the gate acts on a single or multiple qubits. This requires the number of outputs to be always equal to the number of inputs. A sequence of such quantum gates makes a quantum circuit, which is eventually measured to obtain classical bits.

Quantum gates must be reversible because the laws of quantum mechanics are fundamentally reversible. The evolution of an isolated quantum system is governed by Schrödinger's equation, which preserves information: if you know the quantum state now, you can, in principle, run the clock backward and recover exactly what the state was in the past. This property is expressed mathematically by saying that time evolution is *unitary*, and *unitarity* implies reversibility. If a gate were not reversible, it would destroy information, which is not allowed for a closed quantum system.

In classical computing, logic gates like AND or OR throw information away. If you know the output of an AND gate is 0, you do not, in general, know what the inputs were—multiple inputs map to the same output. In Quantum logic, every distinct input state must map to a distinct output state so that the mapping can be undone, as required by unitarity.

Reversibility is also intimately connected to interference. When paths meet again, phases determine whether they will add constructively or cancel destructively. Phase information must remain intact when the paths recombine. Losing information destroys the phase relations between alternatives, turning superpositions into mixtures, and mixtures do not preserve their phases to interfere as waves. The system has one path or the other, but not both at once in a phase-related way that will cause interference.

It is worth adding that there is irreversibility in *measurement*. When a qubit is measured, its state collapses to genuinely lose information.

2.6 Single-Qubit Gates

The simplest quantum operations to consider are for a single qubit. Single-qubit gates are mainly used for individual qubit rotations and phase adjustments. As we discussed in Chapter 1, the state of a qubit can be represented on the Bloch sphere. A single-qubit gate is essentially a rotation of the Bloch vector about an axis passing through the center.

A set of operators of fundamental importance in quantum computation is known as *Pauli operators*. An *operator* is a mathematical rule that can be applied to a function to transform it into another function. The Pauli operators are sometimes denoted by

$$\{\sigma_0, \sigma_1, \sigma_2, \sigma_3\} \text{ or } \{\sigma_0, \sigma_x, \sigma_y, \sigma_z\} \text{ or } \{I, X, Y, Z\}.$$

The first Pauli operator is the identity operator so that it leaves states unchanged including the basis states

$$\sigma_0 \,|\, 0\rangle = |\, 0\rangle, \ \sigma_0 \,|\, 1\rangle = |\, 1\rangle, \ \sigma_0|\psi\rangle = |\,\psi\rangle$$

In other words, the "identity gate" performs no rotation on the state of the qubit. The next Pauli operator denoted by $\sigma_1 = \sigma_x = X$ acts as follows:

$$\sigma_1|0\rangle = |1\rangle, \ \sigma_1|1\rangle = |0\rangle.$$

X is also known as the *NOT operator*. A quantum X gate is similar to its classical counterpart in that a qubit in a definite state $|0\rangle$ will become $|1\rangle$,and vice versa. When the qubit is in a superposition of basis states, then the superposition also flips.

$$\sigma_x\,(\alpha|0\rangle + \beta|1\rangle = \alpha|1\rangle + \beta|0\rangle$$

The results of the operators $\sigma_2 = \sigma_y = Y$ and $\sigma_3 = \sigma_z = Z$ are

$$\sigma_2|0\rangle = -i|1\rangle, \ \sigma_2\,|\,1\rangle = i\,|\,0\rangle \text{ and } \sigma_3\,|\,0\rangle = |\,0\rangle, \ \sigma_3\,|\,1\rangle = -|\,1\rangle$$

The X gate performs a π rotation about the x-axis. Similarly, the Y gate and Z gate perform a π rotation about the y- and z-axes.

Just as a *ket* can be represented as a column vector, operators can be represented by matrices meaning that the action of an operator on a vector is a simple matrix multiplication.

With respect to the computational basis, the matrix representation of the Pauli operators X, Y, and Z are given by the following Pauli matrices:

$$X = \begin{pmatrix} 0 & 1 \\ 1 & 0 \end{pmatrix}, \quad Y = \begin{pmatrix} 0 & -i \\ i & 0 \end{pmatrix}, \quad Z = \begin{pmatrix} 1 & 0 \\ 0 & -1 \end{pmatrix}$$

The Z gate is also known as phase-flip gate and is crucial for error correction and phase shift operations. The Y gate is similar to the X gate but includes a phase flip, so it performs a bit flip and a phase flip simultaneously.

The identity operator is represented by a two-by-two identity matrix as

$$I = \begin{pmatrix} 1 & 0 \\ 0 & 1 \end{pmatrix}$$

Each one of these matrices is its own Hermitian conjugate. Consequently, these gates have the property that applying them twice gives the identity matrix.

$$X^2 = Y^2 = Z^2 = I$$

Wolfgang Pauli introduced these to describe electron spin in quantum mechanics. When quantum information science emerged, decades after Pauli, qubits were recognized as the same kind of two level systems (TLS) Pauli had described. The same matrices naturally reappeared, not as abstract spin operators, but as quantum logic operations.

A set of such operations is *universal* for single qubits if you can build any possible single-qubit unitary transformation (up to a global phase) from them. Any single-qubit unitary can be written (up to phase) as a rotation on the Bloch sphere. This means with just rotations around two different axes, we can generate any single-qubit operation. On the Bloch sphere, this means you can rotate the qubit to any point.

To illustrate more completely, it is possible to rotate the qubit by an arbitrary angle θ about any axis. For example, a θ rotation about the y-axis can be expressed as

$$Y(\theta) = e^{-i\sigma_y\theta/2} = \begin{pmatrix} \cos\frac{\theta}{2} & -\sin\frac{\theta}{2} \\ \sin\frac{\theta}{2} & \cos\frac{\theta}{2} \end{pmatrix}.$$

Similar definitions apply for rotations about the x- or z-axis:

$$X(\theta) = e^{-i\sigma_x\theta/2} = \begin{pmatrix} \cos\frac{\theta}{2} & -i\sin\frac{\theta}{2} \\ -i\sin\frac{\theta}{2} & \cos\frac{\theta}{2} \end{pmatrix}, \quad Z(\theta) = e^{-i\sigma_z\theta/2} = \begin{pmatrix} e^{-\frac{i\theta}{2}} & 0 \\ 0 & e^{\frac{i\theta}{2}} \end{pmatrix}.$$

In general, it is possible to rotate the qubit about an arbitrary axis $\sigma_{arb} = c_x\sigma_x + c_y\sigma_y + c_z\sigma_z$, with $|c_x|^2 + |c_y|^2 + |c_z|^2 = 1$. The matrix representation of such an operation can be calculated by evaluation $e^{-i\sigma_{arb}\theta/2}$.

The most general single-qubit operation takes the form

$$U(\phi,\theta,\omega) = \begin{pmatrix} e^{-i(\phi+\omega)/2}\cos\frac{\theta}{2} & -e^{-i(\phi-\omega)/2}\sin\frac{\theta}{2} \\ e^{-i(\phi-\omega)/2}\sin\frac{\theta}{2} & e^{i(\phi+\omega)/2}\cos\frac{\theta}{2} \end{pmatrix},$$

which can be decomposed using three basic rotations as

$$U(\phi,\theta,\omega) = Z(\omega)Y(\theta)Z(\phi) = Z(\omega+\pi/2)Y(\theta)Z(\phi-\pi/2).$$

By utilizing $U(\phi, \theta, \omega)$, one can traverse the entire single-qubit space as shown in the Bloch sphere picture below.

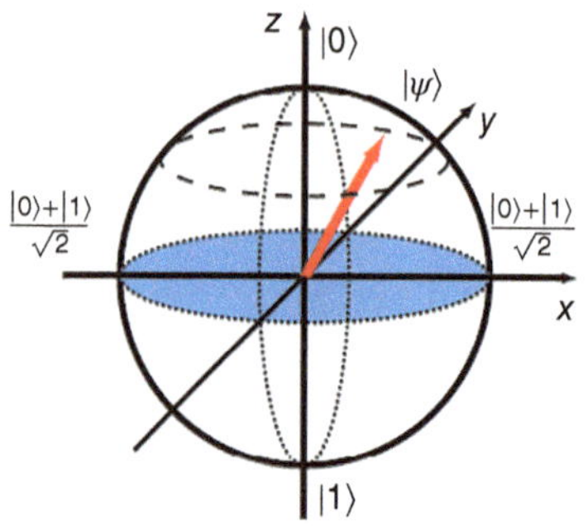

Figure 2.1 Bloch sphere representation of a general single-qubit state $|\psi\rangle$. The north and south poles correspond to the computational basis states $|0\rangle$ and $|1\rangle$, while the equatorial plane represents equal superpositions such as $(|0\rangle \pm |1\rangle)/\sqrt{2}$. The red arrow indicates the qubit state vector $|\psi\rangle$, defined by rotations about the x-, y-, and z-axes. This picture illustrates how unitary operations correspond to rotations of the state vector on the sphere.

In other words, one can perform universal single-qubit operations using the set of $\{Y(\theta), Z(\theta)\}$ or $\{X(\theta), Z(\theta)\}$. One particularly useful decomposition for superconducting qubit platforms is given by [80]

$$U(\phi, \theta, \omega) = Z(\omega)X\left(\frac{-\pi}{2}\right)Z(\theta)X\left(\frac{\pi}{2}\right)Z(\phi)$$

as, in practice, it requires calibration of the $X(\pi/2)$ gate only. The Z gate is performed using software and doesn't require actual calibration. There are also other sets that form a universal set.

The single-qubit gate known as the *Hadamard gate* is very important in QC. It puts the qubit in a superposition state by transforming the base states:

$$|0\rangle \rightarrow \frac{1}{\sqrt{2}}|0\rangle + \frac{1}{\sqrt{2}}|1\rangle, \quad \text{and} \quad |1\rangle \rightarrow \frac{1}{\sqrt{2}}|0\rangle - \frac{1}{\sqrt{2}}|1\rangle$$

The former state is often denoted as $|+\rangle$ and the latter as $|-\rangle$. As a result, if the qubit starts in a definite $|0\rangle$ or $|1\rangle$ state, the Hadamard gate puts each into a superposition of $|0\rangle$and $|1\rangle$ states.

On the Bloch sphere (Figure 2.1), every point on the equatorial plane represents an equal magnitude superposition of $|0\rangle$ and $|1\rangle$; only the relative phase changes as the vector goes around the equator. Any equatorial plane qubit can be written as

$$\frac{|0\rangle + e^{i\phi}|1\rangle}{\sqrt{2}}$$

Hence, if the qubit vector points along the $+Y$ or $-Y$ directions, the superposition states would be

$$\frac{|0\rangle + i|1\rangle}{\sqrt{2}} \quad \text{and} \quad \frac{|0\rangle - i|1\rangle.}{\sqrt{2}}$$

Applying the Hadamard gate two times in a row to a qubit is equal to applying the identity matrix. In other words, at output, the qubit will be in the exact same state it was at the input to the first of the two Hadamard gates.

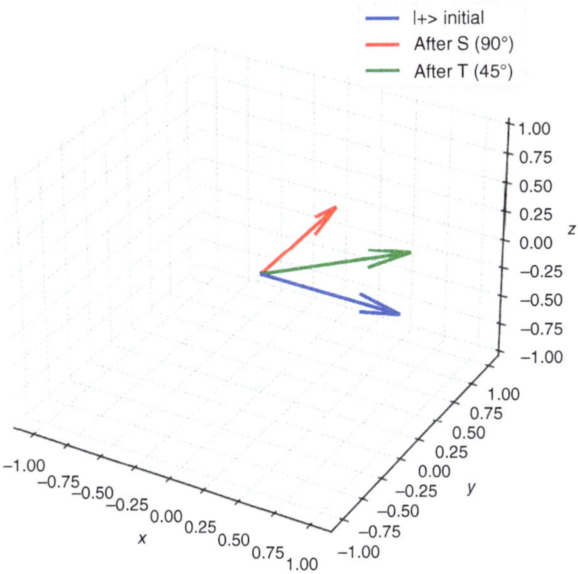

Figure 2.2 Bloch sphere representation of the effect of the S and T gates on the $|+\rangle$ state. The initial state $|+\rangle$ (blue) lies on the equator of the Bloch sphere along the $+x$-axis. Application of the *S gate* applies a 90° phase rotation around the z-axis, moving the state vector to the $+y$-axis (red). Application of the *T gate* applies a 45° phase rotation around the z-axis, moving the state vector to an intermediate position between $+x$ and $+y$ (green).

The matrix associated with a Hadamard gate is

$$H = \frac{1}{\sqrt{2}}\begin{pmatrix} 1 & 1 \\ 1 & -1 \end{pmatrix} = \frac{X+Z}{\sqrt{2}}$$

The Hadamard gate is one of the most important operations in QC because it is the doorway from the classical world into the quantum one. A classical bit can only be 0 or 1, but when a qubit passes through a Hadamard gate, it is rotated from one of these definite states into an equal superposition of both. The Hadamard gate turns the qubit into something profoundly quantum: a state that represents multiple possibilities simultaneously. The ability to create superposition is what allows quantum computers to explore many computational paths in parallel, the foundation of quantum speedups.

The Hadamard lets us both prepare quantum states and later read out interference among the paths they took. Many celebrated algorithms—Deutsch–Jozsa and Shor's—rely on chains of Hadamard gates to spread information across many qubits and then recombine in just the right way so that the correct answer emerges with high probability.

Another important single-qubit gate is the S-phase gate (Figure 2.2) represented by the matrix:

$$S = \begin{pmatrix} 1 & 0 \\ 0 & i \end{pmatrix}$$

$$S = \begin{pmatrix} 1 & 0 \\ 0 & i \end{pmatrix} \equiv Z\left(\frac{\pi}{2}\right)$$

It multiplies $|1\rangle$ by a phase of i (a 90° phase shift), while leaving $|0\rangle$ unchanged. The S gate is a rotation around the z-axis by $\pi/2$ (90°).

In Section 2.12, we discuss how the Pauli operators (I, X, Y, Z) together with H and S are generators of the *Clifford group* gate set. These are the gates that respect the internal geometry of quantum information. A Clifford gate is any quantum gate that maps Pauli operators to Pauli operators under conjugation.

$$SXS^{\dagger} = Y, \quad SZS^{\dagger} = Z$$

Clifford gates cannot create arbitrary quantum phases. The T gate rotation applies a finer phase rotation that Clifford gates cannot. It is represented by the matrix

$$T = \begin{pmatrix} 1 & 0 \\ 0 & e^{i\pi/4} \end{pmatrix} \equiv Z\left(\frac{\pi}{4}\right).$$

The T gate rotation is smaller than that for S gate (π/4 instead of π/2), so it gives *finer control.*

The set {H, T} forms *a universal* set as one can approximate any *single-qubit* unitary to arbitrary precision by applying a sequence of these two gates. By contrast, the S gate combined with H does *not* give universality. By composing H and T repeatedly, you can approximate **any** single-qubit unitary, as for example: $HZH = X$. Full universality for multiqubit computation requires the addition of an entangling gate such as CNOT, which is covered in Section 2.10.

Table 2.3 summarizes the properties and matrix representations of single- and two-qubit gates.

2.7 Measurement Basics in QC

Measurement is usually done in the *computational basis*, which consists of two basis states:

$$|0\rangle = \begin{pmatrix} 1 \\ 0 \end{pmatrix} \text{ and } |1\rangle = \begin{pmatrix} 0 \\ 1 \end{pmatrix}.$$

These represent the two possible outcomes when measuring a single qubit. Measurement operators are called *projectors* and are associated with each possible

Table 2.3 Properties and matrix representations for gates.

Gates and rotations in Bloch sphere	Circuit representation	Matrix representation (up to a global phase)	Truth table
I gate: no rotation is performed.		$I = \begin{pmatrix} 1 & 0 \\ 0 & 1 \end{pmatrix}$	$\|0\rangle \rightarrow \|0\rangle$ $\|1\rangle \rightarrow \|1\rangle$
X gate: rotates the qubit state by π about the x-axis.		$X = \begin{pmatrix} 0 & 1 \\ 1 & 0 \end{pmatrix}$	$\|0\rangle \rightarrow \|1\rangle$ $\|1\rangle \rightarrow \|0\rangle$
Y gate: rotates the qubit state by π about the y-axis.	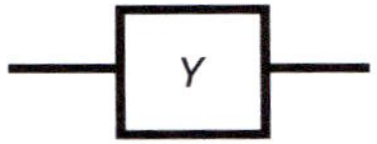	$Y = \begin{pmatrix} 0 & -i \\ i & 0 \end{pmatrix}$	$\|0\rangle \rightarrow i\|1\rangle$ $\|1\rangle \rightarrow -i\|0\rangle$
$Y/2$ gate: rotates the qubit state by $\pi/2$ about the y-axis.	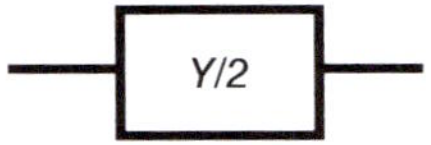	$Y/2 = \frac{1}{\sqrt{2}} \begin{pmatrix} 1 & -1 \\ 1 & 1 \end{pmatrix}$	$\|0\rangle \rightarrow \frac{\|0\rangle + \|1\rangle}{\sqrt{2}}$ $\|1\rangle \rightarrow \frac{\|1\rangle - \|0\rangle}{\sqrt{2}}$
Z gate: rotates the qubit state by π about the z-axis.	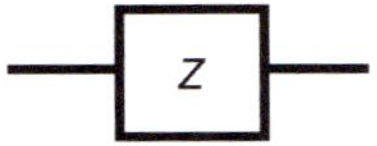	$Z = \begin{pmatrix} 1 & 0 \\ 0 & -1 \end{pmatrix}$	$\|0\rangle \rightarrow \|0\rangle$ $\|1\rangle \rightarrow -\|1\rangle$
S gate: rotates the qubit state by $\pi/2$ about the z-axis.	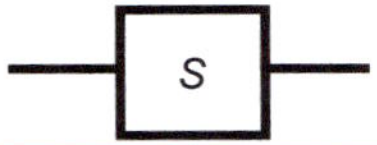	$S = \begin{pmatrix} 1 & 0 \\ 0 & e^{i\pi/2} \end{pmatrix}$	$\|0\rangle \rightarrow \|0\rangle$ $\|1\rangle \rightarrow i\|1\rangle$

(continued)

Table 2.3 (Continued)

Gates and rotations in Bloch sphere	Circuit representation	Matrix representation (up to a global phase)	Truth table
T gate: rotates the qubit state by $\pi/4$ about the z-axis.	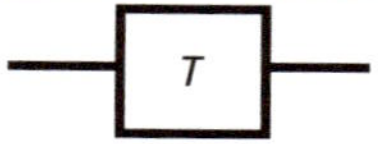	$T = \begin{pmatrix} 1 & 0 \\ 0 & e^{i\pi/4} \end{pmatrix}$	$\vert 0\rangle \to \vert 0\rangle$ $\vert 1\rangle \to e^{i\pi/4}\vert 1\rangle$
H gate: rotates the qubit state by π about an axis diagonal in the x–z plane. This is equivalent to an X-gate followed by a $\pi/2$ rotation about the y-axis.		$H = \frac{1}{\sqrt{2}}\begin{pmatrix} 1 & 1 \\ 1 & -1 \end{pmatrix}$	$\vert 0\rangle \to \frac{\vert 0\rangle + \vert 1\rangle}{\sqrt{2}}$ $\vert 1\rangle \to \frac{\vert 0\rangle - \vert 1\rangle}{\sqrt{2}}$
CNOT gate: applies an X-gate to the target qubit if the control qubit is in state $\vert 1\rangle$.	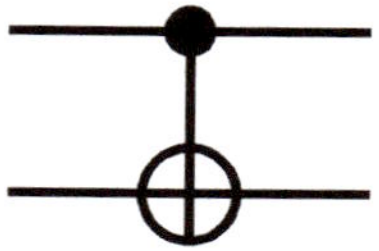	$\text{CNOT} = \begin{pmatrix} 1 & 0 & 0 & 0 \\ 0 & 1 & 0 & 0 \\ 0 & 0 & 0 & 1 \\ 0 & 0 & 1 & 0 \end{pmatrix}$	$\vert 00\rangle \to \vert 00\rangle$ $\vert 01\rangle \to \vert 01\rangle$ $\vert 10\rangle \to \vert 11\rangle$ $\vert 11\rangle \to \vert 10\rangle$
CZ or CPHASE gate: apply a Z gate or phase of -1 on target qubit if control is $\vert 1\rangle$. Alternatively, apply a phase of -1 on $\vert 11\rangle$	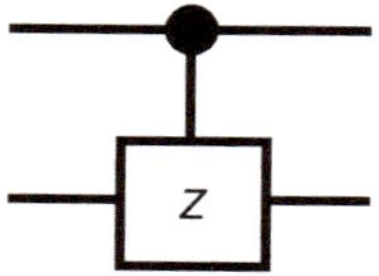	$\text{CPHASE} = \begin{pmatrix} 1 & 0 & 0 & 0 \\ 0 & 1 & 0 & 0 \\ 0 & 0 & 1 & 0 \\ 0 & 0 & 0 & -1 \end{pmatrix}$	$\vert 00\rangle \to \vert 00\rangle$ $\vert 01\rangle \to \vert 01\rangle$ $\vert 10\rangle \to \vert 10\rangle$ $\vert 11\rangle \to -\vert 11\rangle$

Source: With permission of American Chemical Society.

outcome of the measurement. The projector for measuring $|0\rangle$:

$$M_0 = |0\rangle\langle 0| = \begin{pmatrix} 1 & 0 \\ 0 & 0 \end{pmatrix}$$

And the projector for measuring $|1\rangle$:

$$M_1 = |1\rangle\langle 1| = \begin{pmatrix} 0 & 0 \\ 0 & 1 \end{pmatrix}$$

These operators satisfy the completeness relation: $M_0 + M_1 = I$, where I is the identity matrix.

Consider a qubit in the superposition state:

$$|\psi\rangle = \alpha|0\rangle + \beta|1\rangle$$

To determine the probability of measuring $|0\rangle$, we apply the operator M_0 to $|\psi\rangle$:
$M_0|\psi\rangle = \begin{pmatrix} 1 & 0 \\ 0 & 0 \end{pmatrix}\begin{pmatrix} \alpha \\ \beta \end{pmatrix} = \begin{pmatrix} \alpha \\ 0 \end{pmatrix}$, so that the probability of measuring $|0\rangle$ is given by

$$P(0) = \langle\psi|M_0|\psi\rangle = |\alpha|^2$$

Similarly, the probability of measuring $|1\rangle$ is

$$P(1) = \langle\psi|M_1|\psi\rangle = |\beta|^2$$

2.8 Example Quantum Algorithms–Deutsch and Deutsch–Jozsa Algorithms

We discuss two elementary quantum algorithms influential in the development of QC, the *Deutsch* [82] and the *Deutsch–Jozsa* algorithms, which use only the principles of superposition and interference to yield computational benefit. Although they do not solve any important problems, these were among the first to show that QC can be faster than any classical computer, by requiring substantially smaller number of operations.

Consider a function $f(x)$ that operates on a single bit and outputs a single bit. There are only four ways one could construct this function, as shown in Table 2.4.

The question for the computer is: "Is $f(x)$ going to output the same two numbers or opposite numbers?" A function is called *constant* if it always outputs the *same* result for all values of x and is called *balanced* if it outputs different results, i.e., 1 for half of all the possible values of x, and 0 for the other half. The functions f_1 and f_4 in Table 2.4 are constant, while f_2 and f_3 are balanced.

Here is the key question. Suppose you run a classical algorithm and see a result that $f(0) = 1$. Could you tell after that *one* application whether the function is

Table 2.4 Oracle for the Deutsch algorithm.

f_1	f_2	f_3	f_4
$f_1(0) = 0$	$f_2(0) = 0$	$f_3(0) = 1$	$f_4(0) = 1$
$f_1(1) = 0$	$f_2(1) = 1$	$f_3(1) = 0$	$f_4(1) = 1$

constant or balanced? For this case, it could be either the balanced function f_3 or the constant function f_4. Hence, it would be necessary for a classical computer to carry out two operations to evaluate both $f(0)$ and $f(1)$ to determine the answer.

With a quantum computer, the powers of superposition and interference will lead to the need for only *one* function application to determine the answer to the Deutsch problem.

When the algorithm is extended to test functions with any number of inputs, say n inputs, it is called the Deutsch–Jozsa algorithm. The corresponding classical algorithm will require up to $2^{n-1} + 1$ function evaluations in the worst case. The Deutsch–Jozsa algorithm leverages quantum parallelism to miraculously determine whether $f(x)$ is constant or balanced with just *one* function evaluation providing exponential speedup!

In the Deutsch algorithm, the *oracle* is a black-box function $f(x)$ that takes a single-input bit (0 or 1) and returns an output bit (0 or 1). You are not allowed to look inside the function or see how it works—you can only query it. As before, the goal of the algorithm is to determine whether the function is *constant* or *balanced* by using *just one* call to this "oracle." The word *oracle* is used because it behaves like a mysterious box that answers questions but reveals nothing about how the answer is produced.

In the Deutsch algorithm case, the quantum solution is based on the following procedure, first described in words before carrying out the math.

1) Put a qubit in a superposition of 0 and 1 using a Hadamard gate.
2) Operate on the qubit with the unknown function (oracle).
3) Apply another Hadamard gate.
4) Measure the qubit's state.

If the result 0, the function is constant. If the result is 1, the function $f(x)$ is balanced.

Now for the math. The oracle is rewritten as a phase oracle: $U_f \mid x\rangle = (-1)^{f(x)} \mid x\rangle$. This means: If $f(x) = 0$, the state $|x\rangle$ is unchanged. If $f(x) = 1$, the state $|x\rangle$ picks up a minus sign, which is equivalent to flipping the phase of the basis state $|x\rangle$ depending on the output of $f(x)$.

Start with a single qubit initialized in the $|0\rangle$ state, $|0\rangle = \begin{pmatrix} 1 \\ 0 \end{pmatrix}$.

Applying the Hadamard gate, we have

$$|\psi_1\rangle = H\,|0\rangle = \frac{1}{\sqrt{2}}\begin{pmatrix}1 & 1\\ 1 & -1\end{pmatrix}\begin{pmatrix}1\\ 0\end{pmatrix} = \frac{1}{\sqrt{2}}(|0\rangle + |1\rangle)$$

Constant Cases: Applying the Oracle function for the constant cases: f_1 and f_4:

$$f_1(0) = 0,\ f_1(1) = 0;\ \text{and}\ f_4(0) = 1,\ f_4(1) = 1.$$

Let's work out the case for f_1 to give ψ_2.

$$\psi_2 = U_f|\psi_1\rangle = (-1)^{f(x)}|\psi_1\rangle = (-1)^0\frac{|0\rangle + |1\rangle}{\sqrt{2}} = \frac{1}{\sqrt{2}}(|0\rangle + |1\rangle).$$

Applying a second H to $|\psi_2\rangle$ and working out the matrix algebra yield $H\left(\frac{1}{\sqrt{2}}(|0\rangle + |1\rangle)\right) = \frac{1}{\sqrt{2}}H|0\rangle + \frac{1}{\sqrt{2}}H|1\rangle$, using linearity and applying H to each component.

We already know the result for applying H to the $|0\rangle$ state from above. The result of applying the H gate to the $|1\rangle$ state is

$$H = \frac{1}{\sqrt{2}}\begin{pmatrix}1 & 1\\ 1 & -1\end{pmatrix}\begin{pmatrix}0\\ 1\end{pmatrix} = \frac{1}{\sqrt{2}}(|0\rangle - |1\rangle)$$

Putting the two results together, we get

$$\frac{1}{\sqrt{2}}\left(\frac{1}{\sqrt{2}}(|0\rangle + |1\rangle)\right) + \frac{1}{\sqrt{2}}\left(\frac{1}{\sqrt{2}}(|0\rangle - |1\rangle)\right)$$
$$= \frac{1}{2}(|0\rangle + |1\rangle + |0\rangle - |1\rangle) = \frac{1}{2}(2|0\rangle) = |0\rangle$$

After two applications of the Hadamard gate, the state $|0\rangle$ returns to itself.

$$H(H(|0\rangle)) = |0\rangle$$

We follow a similar process for f_4, and the result is also $|0\rangle$.

$$f_4(0) = 1, f_4(1) = 1.$$

$$U_{f_4}|0\rangle = (-1)|0\rangle, U_{f_4}|1\rangle = (-1)|1\rangle.$$

$$\left|\psi_2^{(4)}\right\rangle = U_{f_4}|\psi_1\rangle = U_{f_4}\frac{|0\rangle + |1\rangle}{\sqrt{2}}$$
$$= \frac{1}{\sqrt{2}}(-|0\rangle - |1\rangle)$$
$$= -\frac{1}{\sqrt{2}}(|0\rangle + |1\rangle).$$

Applying second H: $\left|\psi_3^{(4)}\right\rangle = H\left|\psi_2^{(4)}\right\rangle$:

$$\left|\psi_3^{(4)}\right\rangle = \frac{-\frac{1}{\sqrt{2}} + \left(-\frac{1}{\sqrt{2}}\right)}{\sqrt{2}}|0\rangle + \frac{-\frac{1}{\sqrt{2}} - \left(-\frac{1}{\sqrt{2}}\right)}{\sqrt{2}}|1\rangle$$
$$= \frac{-\frac{2}{\sqrt{2}}}{\sqrt{2}}|0\rangle + 0 \cdot |1\rangle$$
$$= -|0\rangle.$$

Balanced Cases: $f_2 : f_2(0) = 0, f_2(1) = 1$; and for $f_3 : f_3(0) = 1, f_3(1) = 0$.

Applying the Oracle function to the state after first Hadamard gives $\frac{1}{\sqrt{2}}(|0\rangle + |1\rangle)$.

$$U_{f_2}|0\rangle = (+1)|0\rangle, \text{ and } U_{f_2}|1\rangle = (-1)|1\rangle.$$
$$\left|\psi_2^{(2)}\right\rangle = U_{f_2}|\psi_1\rangle = U_{f_2}\frac{|0\rangle + |1\rangle}{\sqrt{2}}$$
$$= \frac{1}{\sqrt{2}}(U_{f_2}|0\rangle + U_{f_2}|1\rangle)$$
$$= \frac{1}{\sqrt{2}}(|0\rangle - |1\rangle).$$

Applying second H: $\left|\psi_3^{(2)}\right\rangle = H\left|\psi_2^{(2)}\right\rangle = \frac{\frac{1}{\sqrt{2}}+\left(-\frac{1}{\sqrt{2}}\right)}{\sqrt{2}}|0\rangle + \frac{\frac{1}{\sqrt{2}}-\left(-\frac{1}{\sqrt{2}}\right)}{\sqrt{2}}|1\rangle$

$$= \frac{0}{\sqrt{2}}|0\rangle + \frac{\frac{2}{\sqrt{2}}}{\sqrt{2}}|1\rangle$$
$$= |1\rangle.$$

After the second Hadamard gate, the state interferes to yield $|1\rangle$. And the same for f_3 aside from a phase. After the oracle and the Hadamaard gate the result is $-|1\rangle$, which is the same as $|1\rangle$ but with a phase shift.

Thus, the Deutsch algorithm determines the nature of the function *in just one application of the Oracle* versus two needed for the classical algorithm.

Figure 2.3 shows quantum circuit for the *Deutsch algorithm*. For the *constant* cases, since $f_1(0) = 0$ and $f_4(1) = 1$, you could sort of think of the oracle function as applying the *identity gate* in these cases. And since $f_1(1) = 0$ and $f_4(0) = 1$, you could think of the Oracle as applying an X gate to flip 0 to 1 and vice versa.

Similarly, for the balanced cases, $U_{f_2}|0\rangle = |0\rangle$ and $U_{f_2}|1\rangle = -|1\rangle$. Thus, in the computational basis $\{|0\rangle, |1\rangle\}$,

$$U_{f_2} = \begin{bmatrix} 1 & 0 \\ 0 & -1 \end{bmatrix} = Z.$$

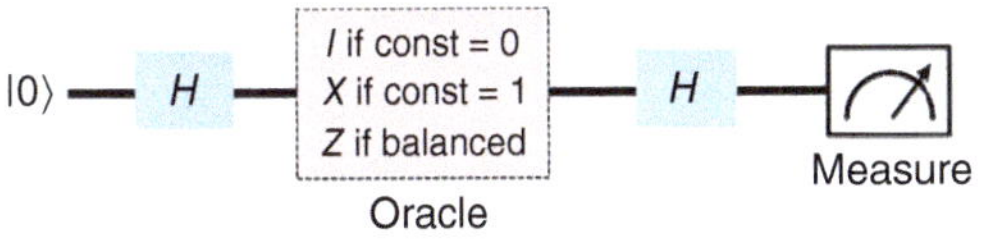

Figure 2.3 The input qubit, initialized in |0⟩, is first placed into a superposition by a Hadamard gate (H). The oracle then applies one of four possible operations—identity (I), Pauli-X, or Pauli-Z—depending on whether the function is constant or balanced. A final Hadamard transforms the state before measurement, allowing a single measurement to distinguish with certainty between constant and balanced functions.

So also $U_{f_3}|0\rangle = (-1)|0\rangle$, $U_{f_3}|1\rangle = (+1)|1\rangle$. Hence,

$$U_{f_3} = \begin{bmatrix} -1 & 0 \\ 0 & 1 \end{bmatrix} = -\begin{bmatrix} 1 & 0 \\ 0 & -1 \end{bmatrix} = -Z$$

Therefore, both balanced functions f_2 and f_3 correspond to the same effective operation: "apply Z" (up to a global phase).

In the Deutsch–Jozsa algorithm, we need n input qubits placed into superposition using Hadamard gates, producing a uniform superposition over all 2^n possible inputs. This allows the oracle to evaluate the function on all inputs simultaneously with one oracle call through quantum parallelism. Although preparing the superposition requires n Hadamards, the algorithm still needs only a single-oracle call. The Hadamard cost scales linearly with n, which is polynomial and negligible compared to the classical deterministic requirement of evaluating $f(x)$ on exponentially many inputs—up to $2^{n-1} + 1$ in the worst case.

2.9 Grover's Search Algorithm Example

Suppose there is a huge phone directory with millions of names. You are looking for one "Adam Sandler." The directory is *unsorted*, so you cannot use a binary search.

A classical program must look at the first name → compare; second → compare; and then third → compare. In the worst case, it may check almost every entry. If the database has N entries, a classical algorithm may need to look at half of them on average, and in the worst case, it must check them all. This takes time proportional to N.

Grover's algorithm climbs through the search space much faster. After only about $\sqrt{N}$ rounds (more precisely $\frac{\pi}{4}\sqrt{N}$), the right answer stands out sharply enough that a single measurement reveals it with high probability. The algorithm is based on the following steps:

Superposition, Oracle Application, Diffusion Operation, Repeat (Oracle + Diffusion) an optimum number of times, Finally Measurement.

For a database of N items, Grover's algorithm needs to place all N possibilities into a single-quantum state so it can search them in superposition. Hence, it chooses n qubits so that $2^n \geq N$. The first step is create a uniform superposition by applying the Hadamaard gate to all n qubits. At this stage, the system is everywhere at once, holding every possible answer with equal weight.

Next comes the *Oracle*, a special operation that marks the correct answer to a search problem. It acts on a state by flipping its phase only if it is the correct solution. In quantum form, it performs a phase flip:

$$|\text{name}\rangle \longrightarrow (-1)^{f\,(\text{name})}\ |\ \text{name}\rangle .$$

f (name) = 1 only if name = 'correct name' and 0 for all other names in this example.

Consequently, every wrong name stays the same while the correct name gets a minus sign that "marks" it. The oracle does not know the answer in advance. It is coded in the rule "match exactly this name." This phase flip later allows the algorithm's iterative amplification process to make the correct answer's amplitude grow, while all others shrink. The action of the Oracle is similar to placing a tiny tag on the right answer that quantum interference will detect.

Next comes the *diffusion operator*, a quantum operation that takes all amplitudes and reflects them about their *average* value. The diffusion operator reflects the amplitudes a_x about their average $\overline{a}$ to give values:

$$\overline{a} + (\overline{a} - a_x) = 2\overline{a} - a_x$$

The step amplifies the marked state's amplitude relative to the unmarked states (Figure 2.4). The phase flip introduced by the oracle turns into a real, measurable advantage. The correct answer becomes louder. The operator is called "diffusion" because it redistributes amplitudes globally in a way reminiscent of smoothing or averaging processes in physical diffusion.

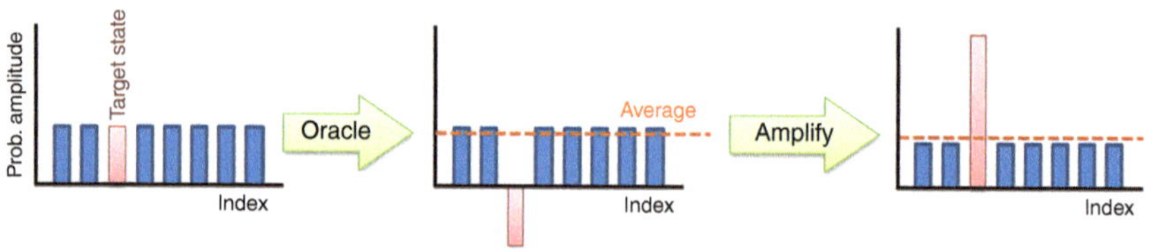

Figure 2.4 Illustration of the amplitude amplification process in Grover's search algorithm. An equal superposition state is prepared initially that contains the correct answer (target state). *Step 1:* the amplitude of the desired state is inverted (multiplied by −1) due to the application of the oracle. *Step 2:* a diffusion (inversion-about-the-mean) operation is applied, redistributing amplitudes so that the desired state's amplitude increases above the average, thereby boosting its probability of being measured.

The oracle/diffusion cycle is repeated several times. Each iteration nudges the marked state's amplitude higher, while the others shrink. With the right number of rounds—roughly the square root of the number of items—the marked state becomes overwhelmingly dominant. The probability of measuring the right name is near 1. The others have near *0* probability. Finally, there is a measurement to get a high probability for the target name: "Adam Sandler." Figure 2.4 illustrates the process and results [83].

Let us implement Grover's algorithm with three qubits, applicable to a database of eight items. We start from a very specific physical state:

$$|000\rangle$$

which is easy to prepare experimentally as the ground state of the system. We then apply the H gate to each qubit of the state $|000\rangle$ to produce the superposition:

$$H^{\otimes 3}|000\rangle = \frac{1}{\sqrt{8}}(|000\rangle + |001\rangle + |010\rangle + |011\rangle + |100\rangle + |101\rangle + |110\rangle + |111\rangle)$$

We created a quantum state that contains one amplitude for every possible index. Each computational basis state can serve as an address labeling a distinct database item. We pick a concrete target item = 101.

In the next step, the oracle flips the phase of the solution (target item) only. So, the state becomes

$$\frac{1}{\sqrt{8}}(|000\rangle + |001\rangle + |010\rangle + |011\rangle + |100\rangle - |101\rangle + |110\rangle + |111\rangle)$$

Unmarked state amplitudes are $+\frac{1}{\sqrt{8}} = 0.353$, and the marked state amplitude is $-\frac{1}{\sqrt{8}}$.

$$\text{The average amplitude is } \overline{a} = \frac{7 \cdot \frac{1}{\sqrt{8}} + \left(-\frac{1}{\sqrt{8}}\right)}{8} = \frac{6}{8\sqrt{8}} = \frac{3}{4\sqrt{8}} = 0.27.$$

As the next step, the diffusion operator reflects the amplitudes a_x about their average $\overline{a}$ to give

$$\overline{a} + (\overline{a} - a_x) = 2\overline{a} - a_x$$

The marked state amplitude becomes

$$2 \cdot \frac{3}{4\sqrt{8}} - \left(-\frac{1}{\sqrt{8}}\right) = \frac{6}{4\sqrt{8}} + \frac{1}{\sqrt{8}} = \frac{10}{4\sqrt{8}} = \frac{5}{2\sqrt{8}} = \frac{5}{4\sqrt{2}} \approx 0.884.$$

The unmarked state amplitudes become

$$2 \cdot \frac{3}{4\sqrt{8}} - \frac{1}{\sqrt{8}} = \frac{6}{4\sqrt{8}} - \frac{1}{\sqrt{8}} = \frac{2}{4\sqrt{8}} = \frac{1}{2\sqrt{8}} = \frac{1}{4\sqrt{2}} \approx 0.177.$$

Hence, unmarked state amplitudes drop and the marked state's amplitude grows.

The resulting state looks like

$$0.88|101\rangle + 0.18(|000\rangle + |001\rangle + |010\rangle + |011\rangle + |100\rangle + |110\rangle + |111\rangle)$$

The correct answer has already become much "louder" than the rest. For the marked state $|101\rangle$, the probability amplitude is 0.884. Any unmarked state amplitude becomes 0.177.

Repeating the (oracle + diffusion) once more, the result becomes

$$0.97|101\rangle - 0.09(|000\rangle + |001\rangle + |010\rangle + |011\rangle + |100\rangle + |110\rangle + |111\rangle)$$

At this point, the probability of measuring $|101\rangle$ is already $\approx 0.97^2 \approx 0.94$. This is essentially the optimal stopping point for $N = 8$. The wrong answers are diminished (small, negative amplitudes).

If one carries out (oracle + diffusion) a third time, it will lead to overshoot with a result like

$$0.57|101\rangle - 0.31(|000\rangle + |001\rangle + |010\rangle + |011\rangle + |100\rangle + |110\rangle + |111\rangle)$$

The procedure has rotated past the maximum—the marked state is no longer dominant. This is known as "overcooking" where the success probability drops again.

Note that the nearest integer of $\sim\frac{\pi}{4}\sqrt{N} = 2.2$ is two iterations, the state to be almost $|101\rangle$ and measurement yields the correct item with high probability.

Let us now discuss the sequence of gates needed to accomplish the first Oracle so that it flips the sign of $|101\rangle$ but leaves all the others the same. The circuit will be different if the marked item is different.

How to make a circuit that flips the sign of only $|101\rangle$ using standard gates?

First, the objective is to turn the target pattern of 101 into 111. The circuit is: apply X gate on qubit 2. This mapping turns

$$|101\rangle \longrightarrow |111\rangle.$$

Next apply the CCZ gate on qubits 1, 2, 3. The CCZ_{111} gate flips the phase *only* when all three qubits are 1. So $|111\rangle$ gets a minus sign. The selected circuit succeeds in applying the phase flip to the target.

$$|111\rangle \rightarrow -|111\rangle$$

Finally, apply X only on qubit 2 (reverse of step 1). That will give $-|101\rangle$, which is our intended goal, to reverse the sign of the target. The effective process can be thought to be a controlled-phase gate CCZ_{101}, which is the explicit form of oracle in this example.

When these same three circuits are applied to the other states, they will not produce any changes after all the gates. For example, take $|000\rangle$; apply X on qubit 2 to get $|010\rangle$; and then apply CCZ, which will do nothing, because all three qubits are not 111. Finally, apply X again to reverse the first operation to give $|000\rangle$.

The diffusion operator in gate form consists of five steps as a concrete circuit implementation of "inversion about the mean":

H on all qubits
X on all qubits
CCZ_{111}
X on all qubits
H on all qubits

The starting point now is post-oracle state:

$$|\psi\rangle = \frac{1}{\sqrt{8}}(|000\rangle + |001\rangle + |010\rangle + |011\rangle + |100\rangle - |101\rangle + |110\rangle + |111\rangle),$$

For all nontarget states (example $|000\rangle$), the amplitude $\alpha_i = a = 1/\sqrt{8}$.
For the target $|101\rangle$: $\alpha_{101} = a = -\frac{1}{\sqrt{8}}$.

The mean amplitude is $m = \dfrac{7a + (-a)}{8} = \dfrac{6a}{8} = \dfrac{3a}{4}$.

We apply the five steps to one nontarget state (choose $|000\rangle$) and to the target state $|101\rangle$.

Apply H on all qubits to the nontarget state: $|000\rangle$, recalling that $H|0\rangle = \frac{|0\rangle+|1\rangle}{\sqrt{2}}$.

$$H|0\rangle H|0\rangle H|0\rangle = \left(\frac{|0\rangle + |1\rangle}{\sqrt{2}}\right)\left(\frac{|0\rangle + |1\rangle}{\sqrt{2}}\right)\left(\frac{|0\rangle + |1\rangle}{\sqrt{2}}\right)$$

$$H^{(\mathrm{all})}|000\rangle = \frac{1}{\sqrt{8}}(|000\rangle + |001\rangle + |010\rangle + |011\rangle + |100\rangle + |101\rangle + |110\rangle + |111\rangle).$$

Now apply $X(\mathrm{all})$, which flips every bit:

$$|000\rangle \to |111\rangle,\quad |001\rangle \to |110\rangle,\quad |010\rangle \to |101\rangle,\quad |011\rangle \to |100\rangle,$$
$$|100\rangle \to |011\rangle,\quad |101\rangle \to |010\rangle,\quad |110\rangle \to |001\rangle,\quad |111\rangle \to |000\rangle.$$

Hence,

$$X(\mathrm{all})\ H(\mathrm{all})|000\rangle = \frac{1}{\sqrt{8}}(|111\rangle + |110\rangle + |101\rangle + |100\rangle + |011\rangle + |010\rangle + |001\rangle + |000\rangle).$$

Reorder to give the same uniform superposition:

$$X(\mathrm{all})\ H(\mathrm{all})|000\rangle = \frac{1}{\sqrt{8}}(|000\rangle + |001\rangle + |010\rangle + |011\rangle + |100\rangle + |101\rangle + |110\rangle + |111\rangle).$$

Next apply CCZ to X(all) H(all)$|000\rangle$; only the $|111\rangle$ term changes sign.

$$CCZX(\text{all})\ H(\text{all})|000\rangle = \frac{1}{\sqrt{8}}(|000\rangle + |001\rangle + |010\rangle + |011\rangle + |100\rangle + |101\rangle + |110\rangle - |111\rangle).$$

Next apply X (flip) with the result:

$$X(\text{all})\ CCZX(\text{all})\ H(\text{all})|000\rangle = \frac{1}{\sqrt{8}}(-|000\rangle + |001\rangle + |010\rangle + |011\rangle + |100\rangle + |101\rangle + |110\rangle + |111\rangle).$$

The single minus sign has moved from the $|111\rangle$ term to the $|000\rangle$ term.

Now apply the final H(all). This leads to a lot of messy algebra as H is applied to 0 and 1 qubits. But the final answer is

$$H(\text{all})\ X(\text{all})\ CCZX(\text{all})\ H(\text{all})\ |000\rangle = \frac{3}{4}|000\rangle - \frac{1}{4}(|001\rangle + |010\rangle + |011\rangle + |100\rangle + |101\rangle + |110\rangle + |111\rangle).$$

The above expression is the diffusion operator acting on the single basis state $|000\rangle$. It means that if the input is $|000\rangle$, then the output amplitude on $|000\rangle$ is $+\frac{3}{4}$. For all the other inputs, the output amplitude on $|000\rangle$ is $-\frac{1}{4}$. The "$\frac{3}{4}$" and "$-\frac{1}{4}$" are relative weights describing how diffusion maps one basis input into all outputs.

To get the Grover amplitude for $|000\rangle$, we sum those weights across all basis components present in the oracle superposition, including the special minus sign on $|101\rangle$.

Now recall that after the Oracle, the actual Grover input to the diffusion operator is

$$|\psi_{\text{oracle}}\rangle = \frac{1}{\sqrt{8}}(|000\rangle + |001\rangle + |010\rangle + |011\rangle + |100\rangle - |101\rangle + |110\rangle + |111\rangle).$$

So diffusion acts on a superposition:

$$|\psi_{\text{after}}\rangle = D|\psi_{\text{oracle}}\rangle.$$

Because D is linear,

$$D\left(\sum_x a_x|x\rangle\right) = \sum_x a_x D|x\rangle.$$

where

$$a_x = +\frac{1}{\sqrt{8}} \text{ for all } x \neq 101, \quad a_{101} = -\frac{1}{\sqrt{8}}.$$

Now we can compute the final amplitude for the nontarget state.

From $x = 000$: contribute $\frac{1}{\sqrt{8}} \cdot \frac{3}{4}$; from the six states $x \neq 000$, or 101: each contributes

$$\frac{1}{\sqrt{8}} \cdot \left(-\frac{1}{4}\right); \text{ from } x = 101\text{: contributes } \left(-\frac{1}{\sqrt{8}}\right) \cdot \left(-\frac{1}{4}\right)$$

So the Grover output amplitude for the nontarget $|000\rangle$ is

$$A_{000} = \frac{1}{\sqrt{8}}\left(\frac{3}{4} + 6\left(-\frac{1}{4}\right) + \left(+\frac{1}{4}\right)\right) = \frac{1}{\sqrt{8}}\left(\frac{-2}{4}\right) = -\frac{1}{2\sqrt{8}} \approx -0.1768$$

and the probability is

$$P(000) = |A_{000}|^2 = \frac{1}{32} \approx 0.03125.$$

Next we follow the same sequence of five gates on the target.

First, for H on all qubits of $|101\rangle$, recall that $H|1\rangle = \frac{|0\rangle - |1\rangle}{\sqrt{2}}$, $H|0\rangle = \frac{|0\rangle + |1\rangle}{\sqrt{2}}$.

Hence, $H^{(\text{all})}\ |101\rangle = \left(\frac{|0\rangle-|1\rangle}{\sqrt{2}}\right)\left(\frac{|0\rangle+|1\rangle}{\sqrt{2}}\right)\left(\frac{|0\rangle-|1\rangle}{\sqrt{2}}\right)$, giving

$$H^{(\text{all})}\ |101\rangle = \frac{1}{\sqrt{8}}(|000\rangle - |001\rangle + |010\rangle - |011\rangle - |100\rangle + |101\rangle - |110\rangle + |111\rangle).$$

$$X(\text{all})\ H(\text{all})\ |101\rangle = \frac{1}{\sqrt{8}}(|111\rangle - |110\rangle + |101\rangle - |100\rangle - |011\rangle + |010\rangle - |001\rangle + |000\rangle).$$

Reorder into standard basis order:

$$X(\text{all})\ H(\text{all})\ |101\rangle = \frac{1}{\sqrt{8}}(|000\rangle - |001\rangle + |010\rangle - |011\rangle - |100\rangle + |101\rangle - |110\rangle + |111\rangle)$$

Now for CCZ that flips only the $|111\rangle$term. Here $|111\rangle$ currently has a **plus** sign, so it becomes a **minus** sign:

$$CCZX(\text{all})\ H(\text{all})\ |101\rangle$$

$$= \frac{1}{\sqrt{8}}(|000\rangle - |001\rangle + |010\rangle - |011\rangle - |100\rangle + |101\rangle - |110\rangle - |111\rangle).$$

Apply $X(\text{all})$ to the $|101\rangle$ case

$$X(\text{all})\ CCZX(\text{all})\ H(\text{all})\ |101\rangle$$

$$= \frac{1}{\sqrt{8}}(|111\rangle - |110\rangle + |101\rangle - |100\rangle - |011\rangle + |010\rangle - |001\rangle - |000\rangle).$$

Reorder into standard basis order:

$$X(\text{all})\ CCZX(\text{all})\ H(\text{all})\ |101\rangle$$

$$= \frac{1}{\sqrt{8}}(-|000\rangle - |001\rangle + |010\rangle - |011\rangle - |100\rangle + |101\rangle - |110\rangle + |111\rangle).$$

Now apply the final $H(\text{all})$ which also needs a lot of algebra, with the final result

$$H(\text{all})\ X(\text{all})\ CCZX(\text{all})\ H(\text{all})\ |101\rangle = \frac{3}{4}|101\rangle - \frac{1}{4}(|000\rangle + |001\rangle + |010\rangle + |011\rangle + |100\rangle + |110\rangle + |111\rangle)$$

Similar to the analysis for the nontarget state, using the actual Grover input to the diffusion, we can compute the final amplitude for $|101\rangle$

$$A_{101} = \sum_x a_x \cdot (\text{coeff of } |101\rangle \text{ in } D|x\rangle).$$

Contribution from the $|101\rangle$ input component

$$a_{101} \cdot \frac{3}{4} = \left(-\frac{1}{\sqrt{8}}\right)\left(\frac{3}{4}\right) = -\frac{3}{4\sqrt{8}}.$$

Contribution from the other seven input components ($x \neq 101$). Each has $a_x = +\frac{1}{\sqrt{8}}$ and contributes $-\frac{1}{4}$, so

$$7\left(\frac{1}{\sqrt{8}}\right)\left(-\frac{1}{4}\right) = -\frac{7}{4\sqrt{8}}.$$

Adding, to finally get

$$A_{101} = -\frac{3}{4\sqrt{8}} - \frac{7}{4\sqrt{8}} = -\frac{10}{4\sqrt{8}} = -\frac{5}{2\sqrt{8}} \approx -0.8839$$

Probability of measuring the target $|101\rangle$ is

$$P(101) = |A_{101}|^2 = \left(\frac{5}{2\sqrt{8}}\right)^2 = \frac{25}{32} \approx 0.78125$$

Grover's algorithm is theoretically powerful, but practical quantum search with Grover is not yet used in real problems because current quantum devices are small and noisy. Its future impact is tied to progress in fault tolerance and error-corrected logical qubits.

It is clear that Grover's algorithm uses superposition via Hadamard gates. But Grover's also *does* use entanglement. It does so in a quiet way rather than with a single dramatic "entangle now" step. The entanglement emerges because the algorithm repeatedly applies global unitary operations to a superposition spread across many qubits. Grover's algorithm begins with a simple product state, but the moment the oracle flips the phase of the marked item, the qubits become entangled. The diffusion operator then acts globally on this entangled state, reinforcing the correlations and steering the amplitudes toward the solution. The speedup comes from this interplay: superposition provides many parallel paths, and entanglement ties those paths together so that interference can amplify the correct answer.

2.10 Two-Qubit States

When we have two qubits, each of which can be in the states $|0\rangle$ or $|1\rangle$, the combined system has a 4D state space with four possible basis states (see Section 2.11):

$$|0\rangle \otimes |0\rangle = |00\rangle$$
$$|0\rangle \otimes |1\rangle = |01\rangle$$
$$|1\rangle \otimes |0\rangle = |10\rangle$$
$$|1\rangle \otimes |1\rangle = |11\rangle$$

Here, the shorthand notation $|jk\rangle$ indicates that the first and second qubits are in the states $|j\rangle$ and $|k\rangle$, respectively. Any two-qubit state can be represented as a linear combination (or superposition) of these four basis states. A general two-qubit state, in bra–ket notation, is written as

$$|\psi\rangle = \alpha|00\rangle + \beta|01\rangle + \gamma|10\rangle + \delta|11\rangle$$

where α, β, γ, and δ are complex numbers that satisfy the normalization condition:

$$|\alpha|^2 + |\beta|^2 + |\gamma|^2 + |\delta|^2 = 1$$

which ensures that the total probability of measuring any of the four states is 1. The two-qubit state can also be represented as a 4D column vector:

$$|\psi\rangle = \begin{pmatrix} \alpha \\ \beta \\ \gamma \\ \delta \end{pmatrix}$$

Each of the four basis states has a unique vector representation. Using the Dirac notation basis as our vector basis:

$$|00\rangle = \begin{bmatrix} 1 \\ 0 \\ 0 \\ 0 \end{bmatrix}$$

$$|01\rangle = \begin{bmatrix} 0 \\ 1 \\ 0 \\ 0 \end{bmatrix}$$

$$|10\rangle = \begin{bmatrix} 0 \\ 0 \\ 1 \\ 0 \end{bmatrix}$$

$$|11\rangle = \begin{bmatrix} 0 \\ 0 \\ 0 \\ 1 \end{bmatrix}$$

2.11 Tensor Products

For a two-qubit system, the states $|00\rangle$, $|01\rangle$, $|10\rangle$, and $|11\rangle$ are represented as tensor products of the single-qubit states. Using the tensor product (denoted by $\otimes$), we find the vectors:

$$|00\rangle = |0\rangle \otimes |0\rangle = \begin{pmatrix} 1 \\ 0 \end{pmatrix} \otimes \begin{pmatrix} 1 \\ 0 \end{pmatrix} = \begin{pmatrix} 1 \\ 0 \\ 0 \\ 0 \end{pmatrix}$$

$$|01\rangle = |0\rangle \otimes |1\rangle = \begin{pmatrix} 1 \\ 0 \end{pmatrix} \otimes \begin{pmatrix} 0 \\ 1 \end{pmatrix} = \begin{pmatrix} 0 \\ 1 \\ 0 \\ 0 \end{pmatrix}$$

$$\vdots$$

$$|10\rangle = |1\rangle \otimes |0\rangle = \begin{pmatrix} 0 \\ 1 \end{pmatrix} \otimes \begin{pmatrix} 1 \\ 0 \end{pmatrix} = \begin{pmatrix} 0 \\ 0 \\ 1 \\ 0 \end{pmatrix}$$

$$|11\rangle = |1\rangle \otimes |1\rangle = \begin{pmatrix} 0 \\ 1 \end{pmatrix} \otimes \begin{pmatrix} 0 \\ 1 \end{pmatrix} = \begin{pmatrix} 0 \\ 0 \\ 0 \\ 1 \end{pmatrix}$$

Note that the tensor product of two kets or two vectors is given by

$$|\phi\rangle \otimes |\chi\rangle = \begin{pmatrix} a \\ b \end{pmatrix} \otimes \begin{pmatrix} c \\ d \end{pmatrix} = \begin{pmatrix} ac \\ ad \\ bc \\ bd \end{pmatrix}$$

2.12 Two-Qubit Gates

We now expand to two-qubit gates (Table 2.4). One class of such gates are *controlled* operations. One qubit is labeled the control qubit and the other the target qubit. Controlled operations involve an action on the target qubit that will change depending on the state of the control qubit.

Two-qubit gates are the basis of generating entanglement, such as the *CNOT* two-qubit entanglement gate. The action of the CNOT gate is to leave the target qubit alone if the control qubit is in state $|0\rangle$, and to flip the target qubit if the control qubit is in state $|1\rangle$. Therefore, CNOT can be written in a 4×4 matrix representation, with the columns and rows being the computational basis states of two qubits, $0, 0\rangle$, $|0, 1\rangle$, $|1, 0\rangle$, and $|1, 1\rangle$ as

$$U_{\text{CNOT}} = \begin{pmatrix} 1 & 0 & 0 & 0 \\ 0 & 1 & 0 & 0 \\ 0 & 0 & 0 & 1 \\ 0 & 0 & 1 & 0 \end{pmatrix}$$

The action of the CNOT gate on these states is

$$\begin{aligned} |00\rangle &\mapsto |00\rangle \\ |01\rangle &\mapsto |01\rangle \\ |10\rangle &\mapsto |11\rangle \\ |11\rangle &\mapsto |10\rangle \end{aligned}$$

The Hadamard gate, the phase gate, and the CNOT together generate the full Clifford group—the set of operations that map Pauli operators to other Pauli operators. Clifford circuits are powerful for stabilizer state preparation (Section 2.23), but they have a fundamental limitation: any computation built *only* from Clifford gates can be efficiently simulated on a classical computer. This is the content of the Gottesman–Knill theorem [84], and it means that Clifford gates alone cannot produce genuine quantum advantage. To cross that boundary, a circuit must include at least one non-Clifford operation. The standard choice is the T gate, a $\pi/8$ phase rotation, which cannot be expressed as a Clifford operation. Once a single non-Clifford gate is available—combined with the Clifford set—any unitary transformation can be approximated to arbitrary accuracy, the fully universal gate regime.

Another two-qubit gate that does a controlled operation is the *CPHASE, controlled phase gate* or *conditional phase shift*. It adds a phase (Z gate) to the target qubit if the control qubit is $|1\rangle$ so that $|11\rangle$ becomes $-|11\rangle$ and all other basis components remain the same. The matrix representation is given by

$$CPHASE = CZ = \begin{pmatrix} 1 & 0 & 0 & 0 \\ 0 & 1 & 0 & 0 \\ 0 & 0 & 1 & 0 \\ 0 & 0 & 0 & -1 \end{pmatrix}$$

There are other two-qubit gates where the distinction of control and target is not clear. These gates can be thought as simultaneously working on both qubits. Here are some examples.

The *SWAP* gate exchanges the states of two qubits without using one as a control for the other. It maps $|00\rangle \to |00\rangle, |01\rangle \to |10\rangle, |10\rangle \to |01\rangle$, and $|11\rangle \to |11\rangle$. Both qubits are swapped with each other regardless of their state. The gate's matrix representation is

$$SWAP = \begin{bmatrix} 1 & 0 & 0 & 0 \\ 0 & 0 & 1 & 0 \\ 0 & 1 & 0 & 0 \\ 0 & 0 & 0 & 1 \end{bmatrix}$$

$$SWAP = XX + YY + ZZ = \begin{pmatrix} 1 & 0 & 0 & 0 \\ 0 & 0 & 1 & 0 \\ 0 & 1 & 0 & 0 \\ 0 & 0 & 0 & 1 \end{pmatrix}$$

The *iSWAP gate* [85] is a variant of the SWAP gate that not only swaps the two qubits but also applies a phase shift: the iSWAP gate swaps the states of two qubits with an additional phase factor. It is also used for creating entanglement. For example, start with two qubits in the $|+\rangle\,|+\rangle$ state where

$$|+\rangle = \frac{1}{\sqrt{2}}(|0\rangle + |1\rangle)$$

is obtained by applying the Hadamard gate

$$|+\rangle = H|0\rangle = \frac{1}{\sqrt{2}}(\,|0\rangle + |1\rangle\,)$$

The initial state of the two qubits is

$$|+\rangle \otimes |+\rangle = \frac{1}{2}(|00\rangle + |01\rangle + |10\rangle + |11\rangle)$$

When we apply the *iSWAP*, we get

$$iSWAP\left(\frac{1}{2}(|00\rangle + |01\rangle + |10\rangle + |11\rangle)\right) = \frac{1}{2}(|00\rangle + i|10\rangle + i|01\rangle + |11\rangle)$$

This is an entangled state. The two qubits are no longer separable, as they now exist in a superposition where the state of each qubit is correlated with the state of the other.

The matrix representation is

$$iSWAP = XX + YY = \begin{pmatrix} 1 & 0 & 0 & 0 \\ 0 & 0 & i & 0 \\ 0 & i & 0 & 0 \\ 0 & 0 & 0 & 1 \end{pmatrix}$$

(See Section 2.14 on why entangled states are not separable.)

The *XX, YY, and ZZ* gates are generalizations of single-qubit Pauli operations extended to act on two qubits. They perform rotations in the two-qubit space based on interactions between the qubits.

The *bSWAP gate* or blue-sideband SWAP gate swaps $|00\rangle$ with $|11\rangle$ with added $\pi/2$ phase [86] [87].

$$bSWAP = XX - YY = \begin{pmatrix} 0 & 0 & 0 & i \\ 0 & 1 & 0 & 0 \\ 0 & 0 & 1 & 0 \\ i & 0 & 0 & 0 \end{pmatrix}$$

The bSWAP gate is called the bSWAP gate because of how it is driven and because the physical process it uses come directly from the *blue-sideband transition* in circuit (and cavity) QED (Section 3.30).

The *CZ (controlled-Z gate)* adds a phase of π to the $|11\rangle$ state, also for creating entangled states. There are different physical processes involved to implement these two-qubit operations. For example, the *CR (cross-resonance gate)* [88, 89] performs a CNOT operation [90]. The *RIP (resonance-induced phase) gate* is a flexible, conditional phase gate that can act on two or more qubits [91, 92].

As discussed earlier, the combination of single-qubit Clifford gates (such as H and S), an entangling two-qubit gate (such as CNOT or CZ), and at least one non-Clifford gate (commonly the T gate) is sufficient to achieve universal quantum computation. Universality refers to the ability of a QC system to approximate any arbitrary unitary operation—or equivalently simulate any quantum process—to arbitrary accuracy, given sufficient resources such as qubits, time, and gate depth. Often, more gates, e.g., $\{Y(\theta), Z(\theta)\}$, are added to the set of universal gates to reduce the number of operations or *circuit depth.*

Circuit depth is the number of sequential layers of gates. Since quantum states are fragile due to decoherence or noise, a deeper circuit means more error accumulation. Depth is especially important when scaling too many qubits because depth often grows faster than *circuit width*, which is the number of qubits. Reducing depth is a central goal in quantum compiler design.

Suppose we have three qubits and apply Hadamard gates to all of them at the same time. Even though there are three gates, they act in parallel on different qubits, so this counts as a single layer, giving the circuit a depth of 1. In contrast, if we apply an H on q_0, then a CNOT with q_0 as control and $q_1$as target, and then another CNOT from q_1 to q_2, these operations must occur sequentially. None of them can be parallelized, so the circuit depth is 3.

In the three-qubit Grover example, the algorithm applies three types of operations in sequence: the initial layer of Hadamards, followed by repeated rounds of the oracle and the diffusion operator. With two Grover iterations, the total depth is

$$1 \text{ (initialization)} + 2 \times \text{(oracle + diffusion)} = 5.$$

This illustrates how width (number of qubits) and depth (number of sequential layers) together determine the overall complexity of a quantum circuit.

In practice, hardware constraints—such as gate error rates, crosstalk, and calibration overhead—strongly influence which gates are used. For example, in superconducting qubit systems (Chapter 3), CR gates are widely adopted because they are relatively easy to implement, require no tunable couplers, and achieve lower error rates than many alternative two-qubit gate mechanisms. As a result, real-world circuit design often balances theoretical depth with the practical realities of the underlying hardware.

It is also possible to define gates involving more than two qubits. For three-qubit gates, the most well-known ones are the *Toffoli and Fredkin* gates. The Toffoli gate is a controlled-controlled-NOT (CCNOT), i.e., the state of the third qubit is flipped if and only if both the first two qubits are in state $|1\rangle$. The Fredkin gate is a controlled-SWAP (CSWAP) gate, swapping the states of the second and third qubits if and only if the state of the first qubit is $|1\rangle$. Most practical implementations of QC have limited connectivity between qubits, only allowing for pairwise interactions between nearest-neighbor qubits. This can prohibit direct implementations of multiqubit gates with three or more qubits, but it turns out that any multiqubit gate can be decomposed into a number of single- and two-qubit gates.

Table 2.5a and b shows the properties of the one-qubit, two-qubit, and three-qubit gates discussed.

While the T gate is the most commonly used non-Clifford resource for achieving universality, alternative non-Clifford gates such as the Toffoli or Fredkin gates can also promote the Clifford gate set to universality, since these gates cannot be generated within the Clifford group and enable operations beyond the stabilizer formalism. To become universal, a gate set must include something that breaks this symmetry—an operation that cannot be reduced to Clifford transformations. The Toffoli gate is not a Clifford gate. When you conjugate Pauli operators with Toffoli, they do *not* stay within the Pauli group. This immediately breaks the stabilizer structure and pushes you beyond the Clifford world. But Toffoli gates are typically *harder* to implement, and in many architectures, they are decomposed *into T*gates plus Cliffords.

Both the Toffoli and Fredkin gates precede QC. Tommaso Toffoli and Edward Fredkin asked deep questions: Can computation be done without destroying information? If the laws of physics are fundamentally reversible, computers should also be reversible. In classical computing, ordinary logic gates (AND, OR, NAND) erase information. Example: if the output of an AND gate is 0, you can't know whether the inputs were 00, 01, or 10. Toffoli asked: Can we build a universal computer using only reversible operations? He showed that is possible using a single three-bit gate: the one now called the *Toffoli gate*. Toffoli proved that classical computing can be done without erasing information. Toffoli gate allows classical logic

Table 2.5 Properties and representations for additional gates.

Gates and rotations in Bloch sphere	Circuit representation	Matrix representation	Truth table
SWAP gate: exchanges the states of two qubits (no "control/target") $\|ab\rangle \rightarrow \|ba\rangle$	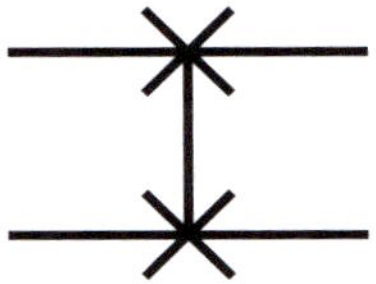	$\text{SWAP} = \begin{bmatrix} 1 & 0 & 0 & 0 \\ 0 & 0 & 1 & 0 \\ 0 & 1 & 0 & 0 \\ 0 & 0 & 0 & 1 \end{bmatrix}$	$\|00\rangle \rightarrow \|00\rangle$ $\|01\rangle \rightarrow \|10\rangle$ $\|10\rangle \rightarrow \|01\rangle$ $\|11\rangle \rightarrow \|11\rangle$
iSWAP gate: swaps the $\|01\rangle$ and $\|10\rangle$ states **and** adds a phase factor of i to both.	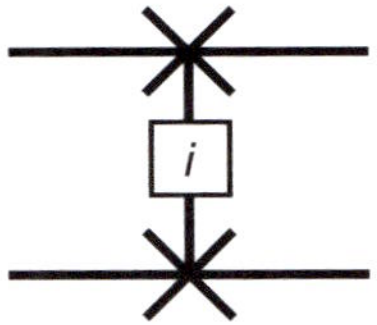 (no standard)	$i\text{SWAP} = \begin{bmatrix} 1 & 0 & 0 & 0 \\ 0 & 0 & i & 0 \\ 0 & i & 0 & 0 \\ 0 & 0 & 0 & 1 \end{bmatrix}$	$\|00\rangle \rightarrow \|00\rangle$ $\|01\rangle \rightarrow i\|10\rangle$ $\|10\rangle \rightarrow i\|01\rangle$ $\|11\rangle \rightarrow \|11\rangle$
bSWAP (blue-sideband SWAP): Swaps $00\rangle \leftrightarrow \|11\rangle$ while $\|01\rangle$ and $\|10\rangle$ are unchanged.	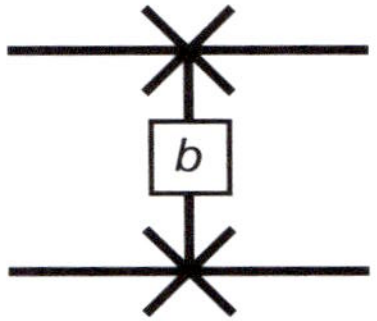 (no standard)	$b\text{SWAP} = \begin{bmatrix} 0 & 0 & 0 & 1 \\ 0 & 1 & 0 & 0 \\ 0 & 0 & 1 & 0 \\ 1 & 0 & 0 & 0 \end{bmatrix}$	$\|00\rangle \rightarrow \|11\rangle$ $\|01\rangle \rightarrow \|01\rangle$ $\|10\rangle \rightarrow \|10\rangle$ $\|11\rangle \rightarrow \|00\rangle$

(continued)

Table 2.5 (Continued)

Gates and rotations in Bloch sphere	Circuit representation	Matrix representation	Truth table
Toffoli (CCNOT): flips the third qubit iff the first two are both $\|1\rangle$ and all other three-qubit basis states unchanged.	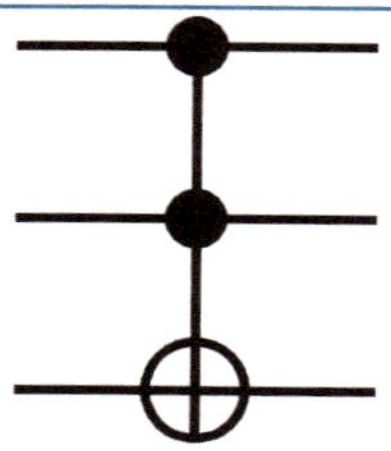	$\text{Toffoli} = \begin{pmatrix} 1 & 0 & 0 & 0 & 0 & 0 & 0 & 0 \\ 0 & 1 & 0 & 0 & 0 & 0 & 0 & 0 \\ 0 & 0 & 1 & 0 & 0 & 0 & 0 & 0 \\ 0 & 0 & 0 & 1 & 0 & 0 & 0 & 0 \\ 0 & 0 & 0 & 0 & 1 & 0 & 0 & 0 \\ 0 & 0 & 0 & 0 & 0 & 1 & 0 & 0 \\ 0 & 0 & 0 & 0 & 0 & 0 & 0 & 1 \\ 0 & 0 & 0 & 0 & 0 & 0 & 1 & 0 \end{pmatrix}$	$\|000\rangle \to \|00\rangle$ $\|001\rangle \to \|001\rangle$ $\|010\rangle \to \|010\rangle$ $\|011\rangle \to \|011\rangle$ $\|100\rangle \to \|100\rangle$ $\|101\rangle \to \|101\rangle$ $\|110\rangle \to \|111\rangle$ $\|111\rangle \to \|110\rangle$
Fredkin (CSWAP): swaps qubits 2 and 3 iff qubit 1 is $\|1\rangle$	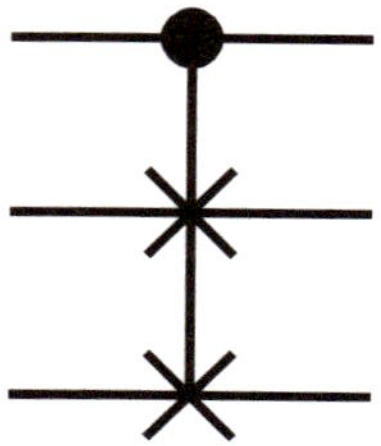	$\text{Fredkin} = \begin{pmatrix} 1 & 0 & 0 & 0 & 0 & 0 & 0 & 0 \\ 0 & 1 & 0 & 0 & 0 & 0 & 0 & 0 \\ 0 & 0 & 1 & 0 & 0 & 0 & 0 & 0 \\ 0 & 0 & 0 & 1 & 0 & 0 & 0 & 0 \\ 0 & 0 & 0 & 0 & 1 & 0 & 0 & 0 \\ 0 & 0 & 0 & 0 & 0 & 0 & 1 & 0 \\ 0 & 0 & 0 & 0 & 0 & 1 & 0 & 0 \\ 0 & 0 & 0 & 0 & 0 & 0 & 0 & 1 \end{pmatrix}$	$\|000\rangle \to \|00\rangle$ $\|001\rangle \to \|001\rangle$ $\|010\rangle \to \|010\rangle$ $\|011\rangle \to \|011\rangle$ $\|100\rangle \to \|100\rangle$ $\|101\rangle \to \|110\rangle$ $\|110\rangle \to \|101\rangle$ $\|111\rangle \to \|111\rangle$

to be embedded inside quantum circuits. Fredkin's gate conserves the number of ones (like a physical particle exchange). Fredkin viewed this as a model of physical interaction, like particles exchanging positions. It mirrors how particles collide and exchange momentum without being created or destroyed. QC now uses both their ideas.

2.13 Unitarity and Its Importance for QC

Unitarity is a fundamental property of quantum operations, particularly transformations that ensure conservation of probability and reversibility. Quantum states are described by probability amplitudes, and the square of the magnitude of these amplitudes gives the probability of measurement outcomes. Unitarity guarantees that the probabilities of all possible outcomes sum up to 1, regardless of the transformations applied. This is essential because in a physical system, total probability must always be conserved. Many quantum algorithms rely on interference effects that occur due to the coherent evolution of quantum states. Unitarity ensures that when states interfere, the total amplitude remains consistent, allowing constructive and destructive interference to be exploited in the correct way.

In quantum mechanics, unitary transformations are represented by unitary matrices. A matrix U is *unitary* if it satisfies the condition:

$$U^{\dagger}U = UU^{\dagger} = I, U^{-1} = U^{\dagger}$$

where $U^{\dagger}$ is the conjugate transpose (also called the Hermitian adjoint) of U, and I is the identity matrix. This condition implies that a unitary matrix has an inverse that is also its conjugate transpose, which is a key feature in quantum mechanics. If a system evolves forward in time by U, it can always be evolved backward by $U^{\dagger}$. No information is lost in the evolution. This reversibility is essential because quantum mechanics is fundamentally deterministic between measurements. Randomness enters only at measurement—not during the evolution.

As a result, in a quantum circuit, all operations are *reversible*, which is a fundamental difference from many classical operations. Because unitary transformations are invertible, it means that any quantum operation can, in principle, be undone or reversed leading to the initial state starting from the final state.

The Hadamard gate is reversible because applying it twice returns the original state. The H operation cancels itself out, and the qubit state is restored to its initial condition.

Applying H to $|0\rangle$ results in $\frac{1}{\sqrt{2}}(|0\rangle + |1\rangle)$.

Applying H again to $\frac{1}{\sqrt{2}}(|0\rangle + |1\rangle)$ returns

$$H\left(\frac{1}{\sqrt{2}}(|0\rangle + |1\rangle)\right) = |0\rangle$$

To show that the *Hadamard gate* is unitary, we need to confirm that the matrix representing the Hadamard gate satisfies the condition for unitarity:

$$H^{\dagger}H = HH^{\dagger} = I$$

Since the Hadamard matrix consists of real entries, its conjugate transpose $H^{\dagger}$ is simply the transpose of H. Therefore,

$$H^{\dagger} = H^{T} = \frac{1}{\sqrt{2}}\begin{pmatrix} 1 & 1 \\ 1 & -1 \end{pmatrix} = H$$

Now, let's compute $H^{\dagger}H$ which is $H \times H$ since $H^{\dagger} = H$.

$$\text{OHH} = \left(\frac{1}{\sqrt{2}}\right)^{2}\begin{pmatrix} 1 & 1 \\ 1 & -1 \end{pmatrix}\begin{pmatrix} 1 & 1 \\ 1 & -1 \end{pmatrix} = \frac{1}{2}\begin{pmatrix} 1 & 1 \\ 1 & -1 \end{pmatrix}\begin{pmatrix} 1 & 1 \\ 1 & -1 \end{pmatrix}$$

$$HH = \frac{1}{2}\begin{pmatrix} (1\cdot 1 + 1\cdot 1) & (1\cdot 1 + 1\cdot -1) \\ (1\cdot 1 + (-1)\cdot 1) & (1\cdot 1 + (-1)\cdot -1) \end{pmatrix} = \frac{1}{2}\begin{pmatrix} 2 & 0 \\ 0 & 2 \end{pmatrix} = \begin{pmatrix} 1 & 0 \\ 0 & 1 \end{pmatrix} = I$$

Similarly, we show that the CNOT gate is unitary,

$$U = \text{CNOT} = \begin{pmatrix} 1 & 0 & 0 & 0 \\ 0 & 1 & 0 & 0 \\ 0 & 0 & 0 & 1 \\ 0 & 0 & 1 & 0 \end{pmatrix}$$

Since the CNOT matrix consists only of real numbers, the conjugate transpose $U^{\dagger}$ is simply the transpose of U. Therefore,

$$U^{\dagger} = U^{T} = \begin{pmatrix} 1 & 0 & 0 & 0 \\ 0 & 1 & 0 & 0 \\ 0 & 0 & 0 & 1 \\ 0 & 0 & 1 & 0 \end{pmatrix} = U$$

Now, let's compute $U^{\dagger}U$, which is $U \times U$ since $U^{\dagger} = U$:

$$UU = \begin{pmatrix} 1 & 0 & 0 & 0 \\ 0 & 1 & 0 & 0 \\ 0 & 0 & 0 & 1 \\ 0 & 0 & 1 & 0 \end{pmatrix}\begin{pmatrix} 1 & 0 & 0 & 0 \\ 0 & 1 & 0 & 0 \\ 0 & 0 & 0 & 1 \\ 0 & 0 & 1 & 0 \end{pmatrix} = \begin{pmatrix} 1 & 0 & 0 & 0 \\ 0 & 1 & 0 & 0 \\ 0 & 0 & 1 & 0 \\ 0 & 0 & 0 & 1 \end{pmatrix} = I$$

2.14 Entanglement

We introduced this important concept in Chapter 1. Entanglement famously caused Einstein [93] great strife, and he derided it as “spukhafte Fernwirkung” or “spooky action at a distance.” His concern was that it seemingly violated special relativity, which embeds the idea that nothing, including information, can travel faster than the speed of light. If a maximally entangled pair (known as a Bell state [6]) is generated and the qubits are measured later after being separated by a very large distance, their outcomes are found to be perfectly correlated. Moreover, applying a local physical operation to one qubit—to flip its state with a magnetic interaction or any other unitary rotation—will instantaneously and coherently transform the joint two-qubit state, *as long as neither qubit is measured*. In this sense, the partner qubit's state is updated as well. This correlation *appears* to be sending information faster than the speed of light from one experimentalist to another, though of course no observer could causally confirm this. Quantum mechanics is nonlocal. The entangled pair is described by a single, inseparable quantum state. Nevertheless, Einstein is correct in that we cannot actually transmit preexisting information faster than light, only random signals.

To create entanglement, we typically start with two qubits in the $|0\rangle$ state:

$$|0\rangle \otimes |0\rangle = |00\rangle$$

First, we apply a Hadamard gate to the first qubit (the control qubit). As before, the Hadamard gate transforms $|0\rangle$ into a superposition of $|0\rangle$ and $|1\rangle$:

$$H|0\rangle = \frac{1}{\sqrt{2}}(|0\rangle + |1\rangle)$$

After applying the Hadamard gate to the first qubit, the state of the two-qubit system becomes

$$\frac{1}{\sqrt{2}}(|0\rangle + |1\rangle) \otimes |0\rangle = \frac{1}{\sqrt{2}}(|00\rangle + |10\rangle)$$

At this stage, the qubits are in a superposition state but are still not entangled. Now, we apply the CNOT gate with the first qubit as the control and the second qubit as the target. The CNOT gate will flip the target qubit's state if the control qubit is $|1\rangle$. After applying the CNOT gate, the state becomes

$$\frac{1}{\sqrt{2}}(|00\rangle + |11\rangle)$$

The two qubits are now entangled. This final entangled state, known as the Bell state, has several important properties. (We discuss later why it is called the Bell state.) The measurement outcomes of the two qubits become perfectly correlated. If you measure the first qubit and find it in the state $|0\rangle$, then the second qubit will also be $|0\rangle$. If the first qubit is measured as $|1\rangle$, the second qubit will also be $|1\rangle$. This correlation holds regardless of the physical distance between the qubits, a phenomenon known as quantum entanglement. Another manifestation of this correlation is that the qubits lose their individual identity and only exist as a whole.

Using the matrix representations above, the vector representation of the entangled state is

$$|\Phi^+\rangle = \frac{1}{\sqrt{2}}(|00\rangle + |11\rangle)$$

$$|\Phi^+\rangle = \frac{1}{\sqrt{2}}\left(\begin{pmatrix}1\\0\\0\\0\end{pmatrix} + \begin{pmatrix}0\\0\\0\\1\end{pmatrix}\right) = \frac{1}{\sqrt{2}}\begin{pmatrix}1\\0\\0\\1\end{pmatrix}$$

As we will see in the next section for two entangled qubits, when the first qubit is measured to be in the $|0\rangle$ state, the second qubit will also be in the $|0\rangle$ state. The vector representation of both qubits in the $|0\rangle$ state will be given by

First qubit $|0\rangle = \begin{bmatrix}1\\0\end{bmatrix}$, second qubit $|0\rangle = \begin{bmatrix}1\\0\end{bmatrix}$

Tensor product: $|0\rangle \otimes |0\rangle = \begin{bmatrix}1\\0\end{bmatrix} \otimes \begin{bmatrix}1\\0\end{bmatrix} = \begin{bmatrix}1\\0\\0\\0\end{bmatrix}$

Also written as

$$|00\rangle = \begin{bmatrix}1\\0\\0\\0\end{bmatrix}$$

If both qubits are in the $|1\rangle$ state, the tensor product is

$$|11\rangle = |1\rangle \otimes |1\rangle = \begin{bmatrix}0\\1\end{bmatrix} \otimes \begin{bmatrix}0\\1\end{bmatrix} = \begin{bmatrix}0\\0\\0\\1\end{bmatrix},$$

which is the vector representation of "both qubits in the $|1\rangle$ state."

Altogether, there are four maximally entangled two-qubit states called the Bell states or EPR (*Einstein–Podolsky–Rosen*) pairs. It refers to the famous 1935 paper by Albert Einstein, Boris Podolsky, and Nathan Rosen, where they introduced what is now called the *EPR paradox*—a challenge to the completeness of quantum mechanics. The Bell states are

$$|\Phi^+\rangle = \frac{|00\rangle + |11\rangle}{\sqrt{2}}$$

$$|\Phi^-\rangle = \frac{|00\rangle - |11\rangle}{\sqrt{2}}$$

$$|\Psi^+\rangle = \frac{|01\rangle + |10\rangle}{\sqrt{2}}$$

$$|\Psi^-\rangle = \frac{|01\rangle - |10\rangle}{\sqrt{2}}$$

In the $|\Psi^-\rangle$ states, measurements always give opposite results (01 or 10).

To obtain these other Bell states, apply Z or X operation to either qubit according to the following table, where subscripts stand for the qubit number.

$$|\Phi^+\rangle|H_1, \text{CNOT}|$$

$$|\Phi^-\rangle|H_1, \text{CNOT}, Z_1|$$

$$|\Psi^+\rangle|H_1, \text{CNOT}, X_2|$$

$$|\Psi^-\rangle|H_1, \text{CNOT}, X_2, Z_1$$

2.15 History of Bell Pairs

Imagine two particles as members of a pair are sent far apart. Before separating, the particles are allowed to agree on a plan. Each particle carries a set of hidden instructions, telling it how to respond if it is measured in any possible way. This is the classical idea called *local hidden variables*. *Local* means each particle only responds to what happens at its own location. *Hidden variables* mean the outcomes were already decided when the particles separated. Bell asked: If nature really works this way, what limits must the correlations between the two particles obey? He found that any such classical system must satisfy a strict mathematical limit, called *Bell's inequality*. It says that if particles carry preassigned answers, then the correlations between measurements can never exceed a certain strength. One common form is

$$|E(A, B) + E(A, B') + E(A', B) - E(A', B')| \leq 2$$

You can think of $E(A, B)$ as asking: "When A and B measure their particles this way, how often do their answers agree versus disagree?" No classical system based on local hidden instructions can produce correlations stronger than this bound.

Now take a pair of qubits in a Bell state, for example:

$$|\Phi^+\rangle = \frac{|00\rangle + |11\rangle}{\sqrt{2}}$$

These qubits have no definite individual values. Instead, they share a joint quantum state. When you measure one qubit, the other is instantly correlated with it—not because it knew the answer in advance, but because the state itself is one indivisible quantum object. If A and B measure their qubits in different directions, the correlations they observe are stronger than anything allowed by Bell's inequality. Quantum mechanics predicts, and experiments confirm

$$|E(A,B)+E(A,B')+E(A',B)-E(A',B')|=2\sqrt{2}$$

That is the violation. Entangled particles agree too well to be classical. Quantum mechanics creates correlations that are stronger than any classical model allows—without sending signals or breaking relativity.

2.16 Two-Qubit Measurements Applied to Entangled Qubits

To construct the measurement operators for a two-qubit system, we take the tensor product of a single-qubit measurement operator with the *identity* operator on the other qubit to ensure that the measurement affects the targeted qubit, while leaving the other qubit unchanged. Using the identity operator allows us to "select" the qubit we want to measure.

The identity operator, represented by $I=\begin{pmatrix}1 & 0\\ 0 & 1\end{pmatrix}$, acts as a "do-nothing" operation. When we apply it to a single qubit in a system, it ensures that the qubit's state remains unaffected.

Suppose we want to project the first qubit onto $|0\rangle$ in a two-qubit system. The measurement operator for the first qubit being $|0\rangle$ is

$$M_0=|0\rangle\langle 0|=\begin{pmatrix}1 & 0\\ 0 & 0\end{pmatrix}$$

To measure only the first qubit while preserving the second qubit, we take the tensor product of M_0 with I on the second qubit:

$$M_0^{(1)}=M_0\otimes I=\begin{pmatrix}1 & 0\\ 0 & 0\end{pmatrix}\otimes\begin{pmatrix}1 & 0\\ 0 & 1\end{pmatrix}=\begin{pmatrix}1 & 0 & 0 & 0\\ 0 & 1 & 0 & 0\\ 0 & 0 & 0 & 0\\ 0 & 0 & 0 & 0\end{pmatrix}$$

This resulting operator $M_0^{(1)}$will project the first qubit onto $|0\rangle$, while leaving the second qubit's state unchanged in case of a product or *separable state* (Section 2.17). However, for an entangled state, measuring one qubit affects the other.

Applying M_0 to the entangled state vector reveals both qubits to be in the $|0\rangle$ state.

$$M_0|\Phi^+\rangle=\begin{bmatrix}1 & 0 & 0 & 0\\ 0 & 1 & 0 & 0\\ 0 & 0 & 0 & 0\\ 0 & 0 & 0 & 0\end{bmatrix}\frac{1}{\sqrt{2}}\begin{bmatrix}1\\ 0\\ 0\\ 1\end{bmatrix}=\frac{1}{\sqrt{2}}\begin{bmatrix}1\\ 0\\ 0\\ 0\end{bmatrix}=\frac{1}{\sqrt{2}}|00\rangle$$

Similarly, the measurement operator for the second qubit being $|0\rangle$ is

$$I \otimes |0\rangle\langle 0| = \begin{bmatrix} 1 & 0 & 0 & 0 \\ 0 & 0 & 0 & 0 \\ 0 & 0 & 1 & 0 \\ 0 & 0 & 0 & 0 \end{bmatrix}$$

Apply the measurement operator to the state vector for the entangled state:

$$\begin{bmatrix} 1 & 0 & 0 & 0 \\ 0 & 0 & 0 & 0 \\ 0 & 0 & 1 & 0 \\ 0 & 0 & 0 & 0 \end{bmatrix} \cdot \frac{1}{\sqrt{2}} \begin{bmatrix} 1 \\ 0 \\ 0 \\ 1 \end{bmatrix} = \frac{1}{\sqrt{2}} \begin{bmatrix} 1 \\ 0 \\ 0 \\ 0 \end{bmatrix} = \frac{1}{\sqrt{2}} |00\rangle$$

Measuring the second qubit to be $|0\rangle$ collapses the entangled (Bell) state $|\Phi^+\rangle$ to $|00\rangle$. Both qubits are measured to be in the $|0\rangle$ state.

Next consider the measurement operator for the first qubit to be in the $|1\rangle$ state:

$$M_1 = |1\rangle\langle 1| \otimes I = \begin{bmatrix} 0 & 0 & 0 & 0 \\ 0 & 0 & 0 & 0 \\ 0 & 0 & 1 & 0 \\ 0 & 0 & 0 & 1 \end{bmatrix}$$

Apply M_1 to the entangled state:

$$M_1 |\Phi^+\rangle = \begin{bmatrix} 0 & 0 & 0 & 0 \\ 0 & 0 & 0 & 0 \\ 0 & 0 & 1 & 0 \\ 0 & 0 & 0 & 1 \end{bmatrix} \frac{1}{\sqrt{2}} \begin{bmatrix} 1 \\ 0 \\ 0 \\ 1 \end{bmatrix} = \frac{1}{\sqrt{2}} \begin{bmatrix} 0 \\ 0 \\ 0 \\ 1 \end{bmatrix} = \frac{1}{\sqrt{2}} |11\rangle$$

Both qubits are found to be in the $|1\rangle$ state. One can recheck this by applying the measurement operator N_1 for the second qubit to be in the $|1\rangle$ state

$$N_1 = I \otimes |1\rangle\langle 1| = \begin{bmatrix} 0 & 0 & 0 & 0 \\ 0 & 1 & 0 & 0 \\ 0 & 0 & 0 & 0 \\ 0 & 0 & 0 & 1 \end{bmatrix}$$

We find $N_1 |11\rangle = \begin{bmatrix} 0 \\ 0 \\ 0 \\ 1 \end{bmatrix} = |11\rangle$. So the second qubit is definitely $|1\rangle$. Both qubits are in the $|1\rangle$ state.

2.17 Entangled States and Separable States

An entangled state cannot be factored into a product of individual states for each qubit. Consider an entangled state:

$$|\Phi^+\rangle = \frac{1}{\sqrt{2}}(|00\rangle + |11\rangle)$$

For a state to be separable, it must be possible to write it as the tensor product of individual states for each qubit, say $|\psi\rangle = |\alpha\rangle \otimes |\beta\rangle|$ or simply $|\alpha\rangle|\beta\rangle$, where $|\alpha\rangle$ and $|\beta\rangle$ are states of the first and second qubits, respectively. Suppose, for contradiction, that this state could be factored as

$$|\Phi^+\rangle = |\alpha\rangle \otimes |\beta\rangle \text{ where } |\alpha\rangle = a|0\rangle + b|1\rangle \text{ and } |\beta\rangle = c|0\rangle + d|1\rangle$$

The product $|\alpha\rangle \otimes |\beta\rangle = ac|00\rangle + ad|01\rangle + bc|10\rangle + bd|11\rangle = ac|00\rangle + ad|01\rangle + bc|10\rangle + bd|11\rangle = \frac{1}{\sqrt{2}}(|00\rangle + |11\rangle)$.

Hence, $ad = 0$; $bc = 0$; $ac = \frac{1}{\sqrt{2}}$; and $bd = \frac{1}{\sqrt{2}}$.

If a = 0, then from normalization, $|b| = 1$. But if $a = 0$, then $ac = 0$, which contradicts the condition that $ac = \frac{1}{\sqrt{2}}$. Similarly for $bc = 0$.

In both cases, we reach a contradiction, meaning that it is impossible to satisfy all four conditions simultaneously. Therefore, there are no values of *a, b, c,* and *d* that can factor $|\Phi^+\rangle = \frac{1}{\sqrt{2}}(|00\rangle + |11\rangle)$ into a product of individual states for each qubit.

2.18 The Power of Entanglement for QC

A single qubit lives in a 2D complex Hilbert space. Four qubits live in 16D complex Hilbert space. Every pure four-qubit state can be written as $|\psi\rangle = \sum_{i=0}^{15} \alpha_i |i\rangle$. Each complex amplitude α_i has two real numbers (real + imaginary). On first glance, 32 numbers are required. But physics removes two degrees of freedom: one via *normalization:* $\sum_i |\alpha_i|^2 = 1$ and another due to the *global phase*, which is physically irrelevant. As a result, a generic (possibly entangled) four-qubit pure state needs

$$32 - 2 = 30 \text{ real parameters}$$

Now comes the crucial point. We see that the separable (nonentangled) states need far fewer numbers. A single-qubit state is $|\psi\rangle = \cos\frac{\theta}{2}|0\rangle + e^{i\phi}\sin\frac{\theta}{2}|1\rangle$.

That requires two real parameters: θ and ϕ. These are the Bloch sphere angles θ, ϕ. So four independent qubits require eight real parameters. Since the qubits are *not entangled*, their joint amplitudes must factorize: $\alpha_{ijkl} = a_i b_j c_k d_l$. Even though the state vector has 16 components, they are not independent, and only 8 real parameters control all of them. A classical computer trying to represent a separable four-qubit state needs just eight real numbers.

Entanglement removes the factorization constraint. Now α_{ijkl} *are independent*, though still subject to normalization and phase. Hence, the generic entangled four-qubit state needs all 30 parameters, which is nearly a four-fold increase, which grows *exponentially* with qubit number. This is how Quantum algorithms gain power with entanglement.

In summary, although both separable and entangled four-qubit states live in the same 16D Hilbert space; separable states are restricted to an eight-parameter submanifold, while generic entangled states explore the full 30-parameter space, giving entanglement its exponential power.

Four qubits in superposition already suggest parallel worlds, but when those qubits are entangled, the system becomes bound by invisible threads. It no longer behaves as four independent systems, but as a single-quantum tapestry whose pattern cannot be unraveled without destroying it. Information ceases to reside in any one qubit and instead lives in the correlations between them—measuring one qubit instantaneously reshapes the whole.

2.19 Quantum Errors

Quantum errors are ubiquitous in all hardware due to inevitable and unwanted interaction with the environment. Unlike classical errors, which generally involve flipping bits (from 0 to 1 or vice versa), quantum errors are more complex and require more sophisticated methods to detect and correct.

2.20 Nature of Quantum Errors

2.20.1 Bit-Flip Errors

A **bit-flip error** in a quantum system is similar to a classical bit flip: it changes the quantum state from $|0\rangle$ to $|1\rangle$ or vice versa. This type of error can be represented by the Pauli-X operation:

$$X = \begin{pmatrix} 0 & 1 \\ 1 & 0 \end{pmatrix}$$

A bit flip is equivalent to a 180° rotation about the x-axis. If a qubit starts at $|0\rangle$ (north pole), it moves to $|1\rangle$ (south pole), and vice versa. If the qubit is in a superposition, it remains in superposition, but the amplitudes swap.

2.20.2 Phase-Flip Errors

A phase-flip error alters the relative phase of a qubit's superposition state, but not its measurement outcome probabilities. If a qubit is in the state $|+\rangle = \frac{1}{\sqrt{2}}(|0\rangle + |1\rangle)$, a phase-flip error changes this to $|-\rangle = \frac{1}{\sqrt{2}}(|0\rangle - |1\rangle)$. This transformation is equivalent to changing $|1\rangle$ to $-|1\rangle$ while leaving $|0\rangle$ unchanged. Phase-flip errors are represented by the Pauli-Z operator:

$$Z = \begin{pmatrix} 1 & 0 \\ 0 & -1 \end{pmatrix}$$

The Z gate flips the qubit across the z-axis. A phase flip is equivalent to a 180° rotation about the z-axis.

2.20.3 Bit- and Phase-Flip Errors (Y Errors)

A bit- and phase-flip error is a combination of a bit flip and a phase flip. This error is represented by the Pauli-Y operator:

$$Y = iXZ = \begin{pmatrix} 0 & -i \\ i & 0 \end{pmatrix}$$

The Y operation changes a qubit's state as follows:

$$Y|0\rangle = i|1\rangle$$
$$Y|1\rangle = -i|0\rangle$$

Together, the three Pauli operators form a complete basis for describing all single-qubit errors (Figure 2.5), to understand noise, decoherence, and QEC. These Pauli-type errors are the simplest way a qubit can go wrong.

2.21 Amplitude Damping and Dephasing Errors

Amplitude damping errors occur because of energy loss, such as a photon escaping from a quantum system caused by interactions with the environment. This type of error tends to cause a qubit to relax toward the ground state $|0\rangle$, which is lower in energy. Amplitude damping is nonunitary and can be modeled by the following Kraus operators [1]:

$$K_0 = \begin{pmatrix} 1 & 0 \\ 0 & \sqrt{1-\gamma} \end{pmatrix}, \quad K_1 = \begin{pmatrix} 0 & \sqrt{\gamma} \\ 0 & 0 \end{pmatrix}$$

where γ represents the probability of amplitude damping.

The qubits also suffer from "dephasing" due to fluctuations in the qubit frequencies, which can be modeled using the following Kraus operators:

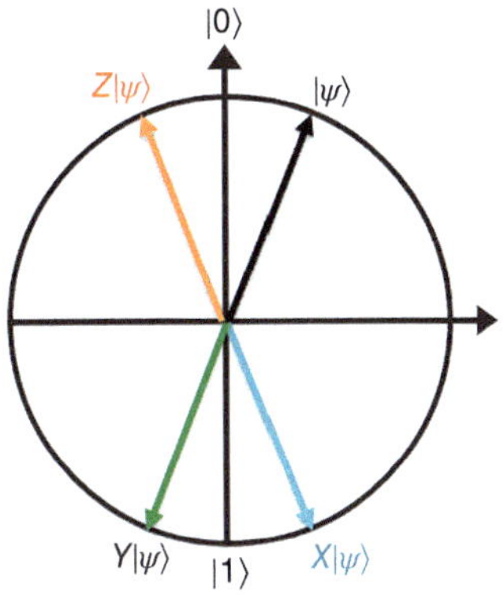

Figure 2.5 Visualizing different single-qubit errors on the initial state $|\psi\rangle$. The representation shows how the three fundamental Pauli errors—bit-flip (X), phase-flip (Z), and combined bit-and-phase-flip (Y)—act on an arbitrary qubit state $|\psi\rangle$. The black arrow indicates the original state $|\psi\rangle$, while the vertical axis marks the computational basis states $|0\rangle$ (north pole) and $|1\rangle$ (south pole). Applying a Z error (orange arrow) rotates the state by π about the vertical z-axis, flipping the phase between $|0\rangle$ and $|1\rangle$ without changing their populations. An X error (blue arrow) rotates the state by π about the x-axis, exchanging the amplitudes of $|0\rangle$ and $|1\rangle$ and producing a bit flip. A Y error (green arrow) combines both effects, corresponding to a π rotation about the y-axis, yielding simultaneous bit and phase flips.

$K_0 = \sqrt{1-\gamma_\phi} I, K_1 = \sqrt{\gamma_\phi} Z$. After the action of a noise channel on an initial density matrix ρ, the final state is given by $\rho_f = K_0 \rho K_0^\dagger + K_1 \rho K_1^\dagger$. For a discussion of density matrix, see Chapter 4.

2.22 Depolarizing Errors

Depolarizing errors occur when a qubit's state becomes a random mixture of the states $|0\rangle, |1\rangle, X|0\rangle, Z|0\rangle$, etc. These errors are modeled as the qubit's state being "scrambled" by applying one of the Pauli operators (X, Y, or Z) at random with certain probabilities. The effect of this error can be visualized as a shrinking Bloch vector toward the origin.

2.23 QEC Codes

The goal of a QEC code is to fix errors while preserving the quantum information. Extending classical error detection and correction codes into the quantum domain faces a unique challenge: directly measuring qubit states in the middle of a calculation will collapse superpositions, destroying the result. Even though the state of a qubit cannot be directly measured for checking errors, it is still possible to *measure relationships* between entangled qubits without measuring the values stored by the qubits themselves. By weaving qubits together in carefully designed patterns of entanglement, it allows us to measure not the qubits themselves, but the subtle relationships between them—hints of agreement and disagreement that betray the presence of errors. From these correlations, the code infers what went wrong and steers the system back on course. QEC repairs fragile quantum information without ever looking directly at it.

The strategy for error detection and correction is to add additional qubits and gates, as with classical computers. These logical qubits are encoded into multiple physical qubits by forming complex states. *Logical qubits* are protected by encoding information across many physical qubits or modes, so errors can be detected and corrected without destroying the quantum information. Examples discussed below are the Surface code and the Steane code. Bosonic logical codes (Binomial codes, Cat codes, GKP codes) will be discussed in Chapter 4.

There are two types of QEC protocols: (i) measurement-based ones that detect an error and then sends feedback signal (gate) to fix the error and (ii) autonomous QEC [94, 95] that utilizes *bath engineering* techniques to stabilize the logical states.

Measurement-based QEC (Figure 2.6) avoids directly measuring the qubits' states and instead measures *stabilizers* (properties of the encoded state) that reveal error *syndromes* (error patterns that indicate where errors occur) without collapsing superpositions. A stabilizer is an operator for which the ideal encoded state is an eigenstate with eigenvalue +1. In other words, stabilizers define what "being in the code space" looks like. As long as the system remains error free, every stabilizer measurement returns +1. When an error occurs, it flips the sign of one or more

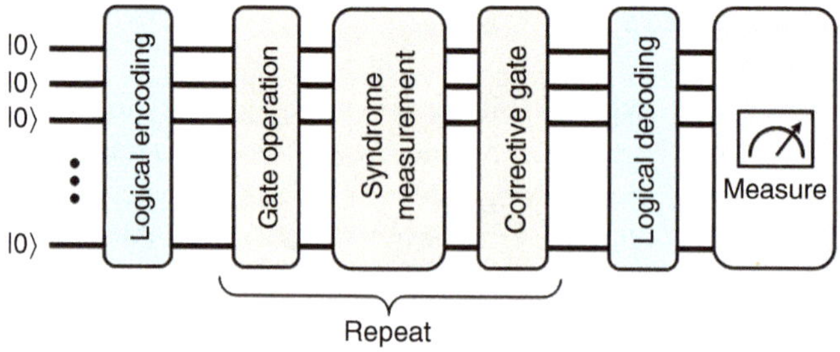

Figure 2.6 Measurement-based quantum-error correction (QEC) protocol. Multiple physical qubits are first logically encoded to represent a single logical qubit. During computation, gate operations are applied, followed by repeated cycles of syndrome measurement to detect errors. Based on the measured syndromes, appropriate corrective gates are applied to restore the logical state. Finally, the system undergoes logical decoding and is measured, ensuring reliable extraction of the logical qubit information despite the presence of physical errors.

stabilizers, producing a distinctive pattern of ± 1 outcomes called the error syndrome. This syndrome tells you *which* error happened without revealing anything about the actual quantum amplitudes or collapsing the encoded superposition.

The beauty of stabilizers is that they let you interrogate the system without touching the quantum information directly—which would destroy the superposition. Instead, you ask, "Is the global pattern of correlations still consistent with the code?" Errors disturb those correlations, and stabilizer measurements detect that disturbance while leaving the encoded logical state intact.

We will see a few examples below.

The *repetition code* [1] is one of the simplest QEC protocol and serves as an introductory concept in QEC, demonstrating the basics of protecting quantum information from errors—specifically, bit-flip errors, but it does not constitute a fully fault-tolerant logical code. It does NOT protect against phase-flip errors.

One logical qubit is encoded across multiple physical qubits by repeating the qubit state several times. Typically, we use a three-qubit repetition code as a basic example [96]. For a logical qubit $|0_L\rangle$ or $|1_L\rangle$, each physical qubit is set to $|0\rangle$ or to $|1\rangle$ as follows: $|0_L\rangle = |000\rangle$, $|1_L\rangle = |111\rangle$, so that all three physical qubits are in the same state.

If a bit-flip error occurs on one of the qubits, we can detect it by measuring the *parity* (the agreement or disagreement) between pairs of qubits. Parity measurements are also syndrome measurements. Measure if qubit 1 and qubit 2 are in the same state. Measure if qubit 2 and qubit 3 are in the same state. If qubits 1 and 2 disagree, but qubits 2 and 3 agree, an error likely occurred on qubit 1. If qubits 1 and 2 agree but qubits 2 and 3 disagree, an error likely occurred on qubit 3. If qubits 1 and 2 disagree and qubits 2 and 3 disagree, an error likely occurred on qubit 2. The correction involves flipping the erroneous qubit back, to match the majority state.

The repetition code is simplest example of encoding → syndrome → correction. But the basic repetition code only protects against bit-flip (Pauli-X) errors. It doesn't protect against phase-flip (Pauli-Z) errors. The repetition code *is, therefore,* not a fully fault-tolerant logical qubit.

Hence, more sophisticated QEC codes, such as the *Steane code* or *surface code*, are needed for scalable QC. Repetition codes require many physical qubits to encode a single logical qubit, leading to high overhead.

The Steane code is a QEC code designed to protect a single qubit of information against both bit-flip and phase-flip errors. It does this by encoding one logical qubit into a state spread across seven physical qubits, leveraging redundancy to detect and correct any errors that occur. The encoded logical qubits can be written as

$$
\begin{aligned}
|0_L\rangle = &\frac{1}{\sqrt{8}}(|0000000\rangle + |1010101\rangle + |0110011\rangle + |1100110\rangle \\
&+ |0001111\rangle + |1110000\rangle + |0011001\rangle + |1101001\rangle)
\end{aligned}
$$

$$|1_L\rangle = \frac{1}{\sqrt{8}}(|1111111\rangle + |0101010\rangle + |1001100\rangle + |0011001\rangle$$
$$+ |0001111\rangle + |1110000\rangle + |0110101\rangle + |1001110\rangle)$$

Any single-qubit error produces a unique syndrome that can be identified and corrected. Using the syndrome information, determine which qubit has experienced an error, and the type of error (bit-flip, phase-flip, or both). Apply the appropriate correction operator (Pauli X, Z, or Y) to the affected qubit to restore the correct encoded state.

The Surface code [97–99] encodes a single-logical qubit across a 2D array of physical qubits, arranged on a lattice (Figure 2.7). Logical qubits are represented by entangled states of these physical qubits, where the global structure (rather than individual qubits) holds the encoded information. This topological encoding makes the surface code resistant to localized errors because these errors need to act in a specific, extensive pattern to affect the logical qubit. A surface code is inherently robust to certain patterns of local errors so that error correction can focus on larger error "chains." For an error to affect the logical qubit, it needs to extend across a significant part of the lattice, which becomes highly improbable as the lattice size increases. Errors must form a continuous chain that spans the lattice, or forms a loop to impact the logical state, which makes it resistant to small, isolated errors. This topological aspect of the surface code provides a natural resistance to certain types of errors. However, since errors still accumulate

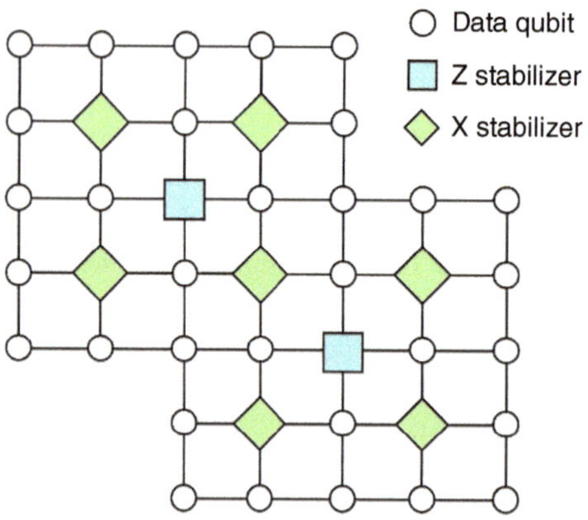

Figure 2.7 Schematic of the Surface code, which encodes a single-logical qubit using a two-dimensional lattice of physical qubits. Data qubits (white circles) occupy the vertices of the lattice, while ancillary qubits perform stabilizer measurements: Z stabilizers (blue squares) enforce parity checks of surrounding qubits in the computational basis, and X stabilizers (green diamonds) enforce parity checks in the Hadamard (phase) basis. Together, these stabilizers detect and correct errors while preserving the logical qubit state.

over time, active stabilizer measurements and corrections are necessary to sustain the code's integrity.

The surface code is a leading QEC code used to protect quantum information from errors, particularly in systems using transmon qubits (Chapter 3). It is typically implemented on a 2D grid of qubits arranged in a lattice structure with nearest-neighbor coupling. The qubits are divided into two types: data qubits, which store the quantum information, and ancillary qubits (also called syndrome or measurement qubits), which are used to detect errors. The surface code detects and corrects both bit-flip (X errors) and phase-flip (Z errors) errors by using stabilizer measurements. Each stabilizer is a combination of four neighboring qubits, and it measures whether the product of their states has changed. If an error occurs, the stabilizer measurement detects a deviation from the expected pattern. These measurements do not directly collapse the state of the data qubits but instead reveal information about the presence of errors. The process involves performing a series of CNOT gates between the data qubits and ancillary qubits, followed by a readout of the ancillary qubits. A stabilizer measurement produces a binary outcome (0 or 1) known as a syndrome. A syndrome of 0 indicates no error, while 1 indicates the presence of an error. The syndromes are fed into a classical decoder that processes them to identify the most likely pattern of errors. The decoder's output determines which qubits need corrective operations (e.g., applying X or Z gates) to restore the original quantum state. The last step is known as feedback, which needs to be performed in real time at a rate faster than the rate of errors in the qubits.

The surface code requires a large number of physical qubits to protect a single-logical qubit. Typically, hundreds or even thousands of physical qubits might be needed in the future for practical error correction.

2.24 Autonomous and Intrinsic Quantum Error Protection

Autonomous QEC embodies the idea of building a "self-healing environment" around the qubit so that the qubit continuously corrects itself without any measurement, or intervention. The key role is played by the *bath (or reservoir)*—a carefully engineered environment that the qubit is allowed to interact with. Normally, the environment damages coherence and injects noise. But in autonomous QEC, the bath is designed so that the only stable states of the combined system are the logical code states. Everything else—errors, stray excitations, unwanted phases—get drained away by *dissipative channels* to clean up the mistakes. For example, when errors show up as excitations, the bath is designed to act like a sponge offering a lower-energy pathway for the excitation to disappear. It is similar to having a room with perfect acoustics: any unwanted echo is instantly swallowed by the walls, leaving only the pure tones intended. Phase errors twist the state by rotating it into unwanted directions. The engineered dissipation creates a preferred orientation, and any deviation from that orientation is slowly undone. It is similar to a compass needle that relaxes back to North after being disturbed. With such built-in protection features, the bath becomes a guardian rather than a threat.

A simple illustration comes from the three-qubit bit-flip code. The logical information is stored in either $|000\rangle$ or $|111\rangle$. These two states are perfectly symmetric: all three qubits agree. Now imagine coupling this trio of qubits to an engineered reservoir that "rewards" agreement and "penalizes" disagreement. If all three qubits match, the reservoir leaves the system alone. But if one qubit flips—say the middle qubit becomes $|010\rangle$—the reservoir provides a gentle, energy-lowering pathway that nudges that qubit back into alignment with the other two. No controller decides what to do. The physics itself performs the correction.

Intrinsic QEC is a *subset* of autonomous QEC where the protection arises from the system's natural Hamiltonian (the fundamental energy structure) rather than engineered dissipation. Intrinsic quantum error protection refers to designing quantum systems or states that are naturally resilient to certain types of errors without needing frequent interventions or correction protocols. This intrinsic resilience is achieved by cleverly designing devices and interactions that are naturally not allowed (or has a very small probability of occurring) and encoding qubit states among those states. The system does not allow the errors to act on the logical information. The protection is passive, structural, and built into the physics using energy gaps or symmetries. The logical information is stored in global, nonlocal (spread across many qubits) degrees of freedom that local noise cannot access. The Hamiltonian erects an energy barrier around the code space, making it energetically costly for errors to disturb the logical information.

A good way to picture this is to imagine a system whose lowest-energy states form the logical qubit. As long as the system stays in its ground-state manifold, the logical information is safe. Any local error that tries to disturb the state must "climb" the energy barrier, which costs energy the environment typically cannot supply.

Consider a system of four physical qubits arranged so that the Hamiltonian strongly favors states where the qubits have even parity—either an even number of ones or an even number of zeros. The Hamiltonian penalizes any configuration with odd parity by giving it higher energy.

Now imagine a single-qubit bit flip. Bit flips change parity: flipping one qubit turns an even-parity state into an odd-parity one. But because odd-parity states sit at a higher energy, the system naturally avoids them. Unless the environment injects enough energy to overcome the gap, the bit flip cannot occur. The Hamiltonian forbids it energetically.

2.25 Short Introduction to Qudits

Qudits are generalizations of qubits [100]. While a qubit represents quantum information using a TLS (with states $|0\rangle$ and $|1\rangle$), a qudit operates in a higher-dimensional space, representing quantum states with d levels (where $d > 2$). For example, a *qutrit* is a qudit with $d = 3$ levels, meaning it can exist in states $|0\rangle$, $|1\rangle$, or $|2\rangle$. For every added state, the processing ability of a single qudit increases so that you need fewer qudits than qubits to encode and process the same amount of information.

For example, with trits, the number 65 can be written as $2.3^3 + 1.3^2 + 0.3^1 + 2.3^0$, as 2102, instead of (with bits) the longer $1000001 = 1.2^0 + 0.2^1 + 0.2^2 + 0.2^3 + 0.2^4 + 0.2^5 + 1.2^6$.

The principles in QC with qubits, such as superposition and entanglement, extend naturally to qudits, allowing similar, though sometimes more complex, computational operations with higher potential. In turn, shorter circuits reduce decoherence and error accumulation.

For a qutrit entangled state, we have

$$|\Phi^+\rangle_{\text{qutrit}} = \frac{1}{\sqrt{3}}\begin{pmatrix}1\\0\\0\\0\\1\\0\\0\\0\\1\end{pmatrix} = \frac{1}{\sqrt{3}}(|00\rangle + |11\rangle + |22\rangle)$$

Entanglement expands to include three possible outcomes: $|0\rangle$, $|1\rangle$, and $|2\rangle$. A common qutrit entangled state (often referred to as a qutrit Bell state or maximally entangled qutrit state) is

$$|\Phi^+\rangle_{\text{qutrit}} = \frac{1}{\sqrt{3}}(|00\rangle + |11\rangle + |22\rangle)$$

If one qutrit is measured and found to be $|0\rangle$, the other qutrit will also be $|0\rangle$. Similarly, if one is $|1\rangle$, the other will also be $|1\rangle$; and if one is $|2\rangle$, the other will also be $|2\rangle$. With qutrit entanglement, you have three correlated pairs, offering a higher potential for encoding information compactly. Multilevel systems like qutrits can be more resilient to certain types of noise, potentially enhancing fault-tolerant quantum computation.

A single entangled qutrit pair can hold more information than a single-entangled qubit pair. Two qutrits span a 9D Hilbert space, compared to the 4D space of two qubits. Three qutrits span 27 dimensions, while three qubits span only 8. The gap widens exponentially.

Implementing qudits also presents certain challenges. Manipulating and measuring qudits require more complex equipment and more precise control over multiple levels. QEC codes for qudits are less developed and require the development of new techniques.

2.26 Qudit Gates

Qudit gates generalize the idea of qubit gates to operate on d-dimensional systems. For qubits, the Pauli-X gate flips $|0\rangle$ to $|1\rangle$ and $|1\rangle$ to $|0\rangle$. For a qutrit (*three*-level qudit), this would shift $|0\rangle \rightarrow |1\rangle$, $|1\rangle \rightarrow |2\rangle$, and $|2\rangle \rightarrow |0\rangle$. For a d-level qudit, the generalized X_d gate is called a *shift operator* and cyclically shifts the state by one level. Consequently, X_d^2 gate (applying X_d twice) cyclically shifts each state by 2.

The Hadamard gate creates superposition in qubit systems. In qudit systems, the generalized Hadamard gate distributes amplitude over all d states. For a qutrit, this matrix would be a 3×3 matrix distributing the amplitude evenly across $|0\rangle$, $|1\rangle$, and $|2\rangle$ with appropriate phase factors. The first row maps each of the basis states to an equal superposition with equal phases. The second and third rows introduce phase differences via ω and ω^2, ensuring orthogonality and preserving the transformation's unitary nature.

$$H_3 = \frac{1}{\sqrt{3}} \begin{pmatrix} 1 & 1 & 1 \\ 1 & \omega & \omega^2 \\ 1 & \omega^2 & \omega \end{pmatrix}$$

where $\omega = e^{2\pi i/3} = -\frac{1}{2} + i\frac{\sqrt{3}}{2}$ is the complex third root of unity, and $\omega^2 = e^{4\pi i/3} = -\frac{1}{2} - i\frac{\sqrt{3}}{2}$.

For qutrits, the *controlled-shift gate* (sometimes called the CSUM gate) generalizes CNOT by conditionally shifting the target state in a cyclic manner, depending on the state of the control qutrit: $CSUM\,(|x\rangle \otimes |y\rangle) = |x\rangle \otimes |y + x\,mod\ 3\rangle$. The full matrix for the qutrit CNOT (CX_3) is

$$\text{CSUM3} = \begin{pmatrix} 1 & 0 & 0 & 0 & 0 & 0 & 0 & 0 & 0 \\ 0 & 1 & 0 & 0 & 0 & 0 & 0 & 0 & 0 \\ 0 & 0 & 1 & 0 & 0 & 0 & 0 & 0 & 0 \\ 0 & 0 & 0 & 0 & 0 & 1 & 0 & 0 & 0 \\ 0 & 0 & 0 & 1 & 0 & 0 & 0 & 0 & 0 \\ 0 & 0 & 0 & 0 & 1 & 0 & 0 & 0 & 0 \\ 0 & 0 & 0 & 0 & 0 & 0 & 0 & 1 & 0 \\ 0 & 0 & 0 & 0 & 0 & 0 & 0 & 0 & 1 \\ 0 & 0 & 0 & 0 & 0 & 0 & 1 & 0 & 0 \end{pmatrix}$$

We will return to the topic of Qudits after we discuss the architecture of Superconducting Qubits and Cavities in the following chapters.

2.27 Qutrit-Based Grover Search Algorithm

Given an unsorted list of N items, Grover's algorithm searches for a specific target item, or marked item, in $O(\sqrt{N})$ steps by amplifying the amplitude of the target state by leveraging quantum superposition and interference. Classically, searching an unsorted database with N items requires $O(N)$ steps on average.

Here is a simple example of qutrit-based Grover's algorithm to search for a particular state among three states: $|0\rangle$, $|1\rangle$, and $|2\rangle$. Our target could be any one of these, and we want our algorithm to amplify the probability of measuring that target state in a single step. Suppose our target state is $|1\rangle$. The algorithm consists of four steps: the Hadamard gate, the Oracle that chooses the target state, the diffusion operator, and, finally, the measurement to reveal the highest probability for the target state.

We have already discussed the Hadamard, Oracle, Diffusion, and Measurement operations. Apply the Hadamard gate to get an equal superposition $|\psi\rangle = \frac{1}{\sqrt{3}}(|0\rangle + |1\rangle + |2\rangle)$. Next, we need an oracle to mark the target state. The oracle applies a phase shift to the target state only. For instance, if the target state is $|1\rangle$, the oracle O would be

$$O = \begin{pmatrix} 1 & 0 & 0 \\ 0 & -1 & 0 \\ 0 & 0 & 1 \end{pmatrix}$$

This oracle flips the phase of $|1\rangle$ while leaving $|0\rangle$ and $|2\rangle$ unchanged. The state becomes

$$|\psi_2\rangle = \frac{1}{\sqrt{3}}(|0\rangle - |1\rangle + |2\rangle)$$

The mean amplitude $\overline{\alpha} = \frac{1}{3}\left(\frac{1}{\sqrt{3}} + \left(-\frac{1}{\sqrt{3}}\right) + \frac{1}{\sqrt{3}}\right) = \frac{1}{3} \cdot \frac{1}{\sqrt{3}}$.

Next we apply a qutrit version of the **diffusion operator**. The diffusion operator for qutrits can be written as

$$D_3 = 2|\psi\rangle\langle\psi| - I = \frac{2}{3}\begin{pmatrix} 1 & 1 & 1 \\ 1 & 1 & 1 \\ 1 & 1 & 1 \end{pmatrix} - \begin{pmatrix} 1 & 0 & 0 \\ 0 & 1 & 0 \\ 0 & 0 & 1 \end{pmatrix} = \begin{pmatrix} -\frac{1}{3} & \frac{2}{3} & \frac{2}{3} \\ \frac{2}{3} & -\frac{1}{3} & \frac{2}{3} \\ \frac{2}{3} & \frac{2}{3} & -\frac{1}{3} \end{pmatrix}.$$

The diffusion operator reflects the amplitudes around their mean value to amplify the amplitude of the target state.

For each state $|i\rangle$ with amplitude α_i, the updated amplitude is

$$\alpha_i \rightarrow 2\overline{\alpha} - \alpha_i$$

$$\text{For } |0\rangle: \text{ New amplitude} = 2 \cdot \frac{1}{3\sqrt{3}} - \frac{1}{\sqrt{3}} = \frac{-1}{3\sqrt{3}}$$

$$\text{For } |1\rangle: \text{ New amplitude} = 2 \cdot \frac{1}{3\sqrt{3}} + \frac{1}{\sqrt{3}} = \frac{5}{3\sqrt{3}}$$

$$\text{For } |2\rangle: \text{ New amplitude } = 2 \cdot \frac{1}{3\sqrt{3}} - \frac{1}{\sqrt{3}} = \frac{-1}{3\sqrt{3}}$$

After the diffusion step, our state becomes

$$|\psi_3\rangle = \frac{-1}{3\sqrt{3}}|0\rangle + \frac{5}{3\sqrt{3}}|1\rangle + \frac{-1}{3\sqrt{3}}|2\rangle$$

The amplitude of the target state $|1\rangle$ is significantly higher than the other states. Upon measurement, the probability of observing each state is proportional to the square of its amplitude.

$$\text{Probability of } |0\rangle: \left(\frac{-1}{3\sqrt{3}}\right)^2 = \frac{1}{27}$$

$$\text{Probability of } |1\rangle: \left(\frac{5}{3\sqrt{3}}\right)^2 = \frac{25}{27}$$

$$\text{Probability of } |2\rangle: \left(\frac{-1}{3\sqrt{3}}\right)^2 = \frac{1}{27}$$

With a probability of 25/27, we most likely measure the target state $|1\rangle$, successfully identifying it through Grover's search algorithm.

If the Grover algorithm was based on qubits, we would need two qubits to encode three states instead of the single qutrit. Because they can encode more possibilities, qutrits reduce the number required for certain quantum algorithms. For example, if we want to find a specific target state out of 81 possible states, it would require 7 qubits versus 4 qutrits or just 2 qudits with $d = 9$.

2.28 Conclusions

This chapter shows that the leap from classical to QC is a shift in the rules that information obeys. In classical machines, information is stored in bits that are always either 0 or 1. Quantum information, by contrast, lives in an abstract Hilbert space, where a qubit can exist in a superposition of possibilities, carrying not just values but also complex probability amplitudes that can interfere with one another.

These amplitudes are manipulated by quantum gates, which differ fundamentally from their classical counterparts. Classical logic gates discard information and are, therefore, irreversible, but quantum gates preserve information. They are described mathematically by unitary matrices that rotate and reshape quantum states without destroying them. By chaining such gates together into circuits, one can orchestrate patterns of interference that lie at the heart of quantum algorithms.

The real power of quantum computation emerges when multiple qubits become entangled. Entanglement creates correlations that cannot be reduced to independent states of individual qubits. Instead, the system behaves as a single, inseparable quantum object whose collective state encodes information in a way that has no classical analogue. This is what allows quantum computers to explore vast computational spaces in parallel and extract useful answers through interference.

At the same time, these fragile quantum states are threatened by noise and decoherence. To protect them, QEC codes such as repetition codes, Steane codes, and surface codes are used to spread logical information across many physical qubits, allowing errors to be detected and corrected without directly measuring and destroying the quantum state.

Finally, the chapter points beyond the standard qubit model. By extending quantum information into qudits—systems with more than two energy levels—one gains access to larger Hilbert spaces within a single physical device. These multi-level systems promise more compact, flexible, and potentially more noise-resilient architectures, pointing toward a future in which quantum hardware becomes both more powerful and more efficient.

Ultimately, the chapter positions these concepts as the essential toolkit for building quantum circuits, algorithms, and error-corrected architectures on superconducting platforms. It underscores that mastering these basics is a prerequisite for advancing into the practical and hardware-specific realms of QC explored in later chapters.

3

Superconducting Qubit

3.1 Overview Introduction

Qubits, gates, and algorithms discussed in Chapter 2 do not live in a vacuum—they must be embodied in real physical systems that can be built, controlled, and scaled. Every quantum circuit must ultimately be realized as voltages, currents, or photons (or electron spins) in the laboratory. At this point, our focus naturally shifts from quantum computation in theory to how it can be implemented in practice. The physical platform of superconducting circuits offers ways of turning mathematical qubits into tangible hardware. Superconducting circuits stand out because they combine the precision of microwave engineering with the rules of quantum mechanics, allowing the design of artificial atoms, connecting them with wires and cavities, and operating them with shaped electrical pulses. In this engineered, macroscopic quantum world, the abstract language of quantum algorithms takes physical form.

Chapter 3 introduces the physics that underpins *circuit quantum electrodynamics* (cQED [29, 31, 32, 44, 101–105], starting with superconducting qubits (such as the Cooper-pair box [CPB] and the transmon) and continuing on to how the qubits are coupled to microwave resonators. After a brief history of the development of the transmon, this chapter introduces the quantum framework of raising and lowering operators, which govern transitions between discrete energy states in the transmon. This leads naturally to the *Jaynes–Cummings Hamiltonian*, a central model describing how a TLS interacts coherently with a single-photon mode in a cavity.

It is a triumph of modern physics that we can now work at the level of a single photon—the smallest indivisible whisper of light—while the everyday world is flooded with torrents of illumination. A household light bulb releases trillions upon trillions of photons every second, an uncountable blaze; yet inside superconducting circuits and microwave cavities, we learn to summon just one photon, hold it still, shape its quantum phase (control the phase of the quantum field mode that contains the photon), and listen to its lifetime tick away (how long a photon remains stored in a particular physical system before it leaks out or is absorbed). We trap "light" not in mirrors of glass, but in resonant circuits and cavities fabricated from superconducting metals operated at temperatures colder than deep space. Here photons linger long enough to be counted, entangled (the quantum state of one is dependent on the other), and participate in light–matter interactions that cavities make possible. In circuit and cavity QED, light ceases to be a flood and becomes a single pulse—written, erased, and rewritten—revealing how one of the most abundant entities in the universe can be handled one quantum at a time.

From the Jaynes–Cummings Hamiltonian, the chapter explores critical concepts such as the rotating-wave approximation (RWA), which simplifies qubit–cavity dynamics by retaining only energy-conserving processes, and the physical consequences of strong coupling and dispersive (weak) coupling. Key quantum phenomena—such as Rabi oscillations, vacuum Rabi splitting, photon blockade, and entanglement between distant qubits via shared resonators—are introduced as the building blocks for quantum computation.

Finally, the discussion expands to measurement techniques, especially the dispersive regime that enables high-fidelity, quantum nondemolition (QND) readout of qubits where you can measure the qubit's state without kicking it out of that state. This chapter closes by linking theory to practice, describing how microwave components (Purcell filters, parametric amplifiers, attenuators, IQ mixers, etc.) integrate into the experimental setup of a dilution refrigerator (DR) to make real quantum processors possible.

3.2 Superconducting Qubit Platform

A superconducting qubit is one of the most remarkable inventions in modern physics because it is, literally, an artificial atom that can be built, designed, and controlled. In nature, atoms are tiny quantum systems with fixed energy levels set by the laws of physics. Instead of atoms given to us by nature, superconducting qubits take the same ideas of quantized energy and superposition engineered on a chip with wires, capacitors, and JJs cooled to near absolute zero.

In a superconducting qubit based on a JJ [106], electron pairs move collectively without resistance, forming a single coherent quantum object. The JJ [36, 107] interrupts the superconducting circuit with a thin insulating barrier through which electron pairs tunnel. Because of the nonlinear Josephson element, the electromagnetic energy in the circuit cannot vary continuously. Rather it is confined to a set of discrete quantum energy levels, directly analogous to the quantized levels of an atom. That is the remarkable feature: a device fabricated from patterned metal films on a chip behaves as a person-made quantum system.

By carefully engineering the junction along with its surrounding capacitance and inductance, we select two of these energy levels and define them as a qubit. The JJ lets us place this qubit into a superposition of its two energy states and control that superposition with precisely shaped microwave pulses. In doing so, we manipulate the quantum state of an artificial atom—using standard microfabrication techniques rather than calling on nature's chemistry.

Superconducting qubits do more than atoms. Their "atomic structure" is programmable. By changing the geometry of a capacitor plate or the thickness of an oxide barrier, the energy spectrum can be designed. By adding tunable junctions, the atomic properties can be adjusted in real time. We can couple many such artificial atoms together with microwave resonators, forming engineered molecules and crystals made of quantum circuits, some of which are impossible with ordinary matter.

Perhaps the most astonishing aspect is accessibility. Natural atoms live deep in the quantum world. Superconducting qubits, in contrast, can be lithographically drawn, wire-bonded, packaged, and plugged into instruments. Their wavefunctions live at microwave frequencies generated in the lab. We "talk" to them with coaxial cables.

More specifically, superconducting qubits are solid-state electrical circuits made from superconducting materials like aluminum, niobium, or tantalum, allowing control and measurement of two discrete quantum levels. The superconducting components enable resistance-free flow of current that minimize power dissipation and decoherence. Compared with the qubits based on other quantum systems (such as trapped ions, or optical systems), superconducting qubits offer the following advantages to become one of the leading candidates for scalable QC:

1) High designability. Different types of qubits such as charge qubits [108, 109], flux qubits [110], and phase qubits [111] can be designed by varying the connectivity and parameters of the three constituent components: capacitance, inductance, and Josephson energy.
2) Scalability. The preparation of superconducting qubits is based on the existing semiconductor microfabrication processes, leveraging advanced chip-making technologies suitable for manufacturing and scalability.
3) Coupling. The circuit nature and macroscopic size (millimeter scale) makes it relatively easy to couple multiple qubits together by capacitances or inductances or resonators.
4) Control. The operation and measurement of superconducting qubits are compatible with microwave control and operability so that commercial microwave instruments can be used for control and readout.

The primary disadvantage is their short coherence times. However, continued efforts in the past two decades have improved the relaxation times by six orders of magnitude from nanoseconds to milliseconds. In the past few years, superconducting qubits have also seen drastic improvements in gate operation fidelities, fast and high-fidelity qubit readout, demonstrations of error correction and quantum algorithms, all topics for this book.

Superconducting qubits require DRs to maintain low temperatures (10–20 mK) for acceptable coherence. Significant advances in the capacity of such cryostats and controls are needed before a device with millions of qubits can be operated to address real problems and demonstrate the advantage of quantum computers.

3.3 Brief History

3.3.1 From CPB to Transmon

The history of superconducting qubits [32] is a fascinating journey that began with the development of the CPB [108, 112] and evolved through various refinements, culminating in the invention of the *transmon* [113], which is the main subject of this chapter. Each design introduced improvements to suppress various sources of noise and increase coherence times. Before we discuss the several types of evolving qubits on the way to the transmon, we begin with an overview of this history.

The story of the transmon begins with the simplest superconducting qubit: the CPB (Figure 3.1a). In the late 1990s, researchers discovered that a tiny superconducting island connected to a reservoir through a JJ could behave like an artificial atom. In a superconductor, electrons bind together into Cooper pairs, and these pairs behave as a single-quantum object with charge $2e$. Note that at the 20-mK operating temperature of a qubit, almost all the electrons condense into these Cooper pairs. (For Al, the number of remaining normal electrons is astronomically small $\sim e^{-120}$.) Because the superconducting island is separated from the reservoir by a JJ, charge cannot flow freely across as in a normal wire. A Cooper pair on one side of the junction has a small but finite probability of appearing on the other side. Even though classically it does not have enough energy to cross the barrier, it can quantum mechanically tunnel through. Its wavefunction "leaks" through the barrier, giving the particle a finite probability of appearing on the other side. The tunneling process happens in whole units of $2e$. Each tunneling event shifts the island's charge by exactly one pair, producing discrete jumps between well-defined charge states. To be more accurate, Cooper pairs don't tunnel as identifiable particles. The Bose–Einstein condensate of Cooper pairs tunnels, but its quantized excitations carry charge $2e$, so the junction transfers charge in units of a Cooper pair.

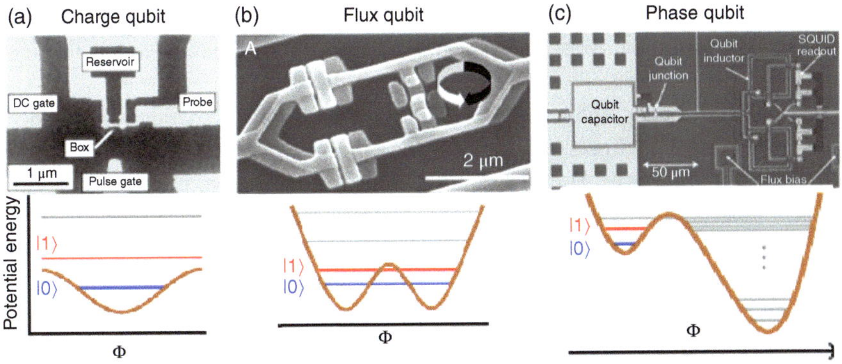

Figure 3.1 Early superconducting qubits—charge, flux, and phase implementations. (a) Charge qubit (Cooper-pair box). Electron micrograph of a small superconducting island ("box") coupled to a reservoir through a tunnel Josephson junction and capacitively biased by DC and pulse gates; a probe junction provides readout. The logical states are set by the number of Cooper pairs on the island, tuned by the gate-induced charge. Scale bar: 1 μm, with permission of American Physical Society. (b) Flux qubit. SEM of a micron-scale superconducting loop interrupted by Josephson junctions, supporting persistent currents circulating clockwise/counterclockwise (arrow), which form the qubit basis. The transition frequency is tuned with an external flux bias near $\Phi_0/2$; readout is commonly via an adjacent dc-SQUID. Scale bar: 2 μm, [118] / American Association for the Advancement of Science. (c) Phase qubit. Optical/SEM image of a large-area Josephson junction shunted by a capacitor and inductor, biased close to the critical current so the junction phase resides in a tilted potential; $|0\rangle$ and $|1\rangle$ are the lowest quantized levels in a single well. Readout uses SQUID switching to the voltage state; a flux-bias line provides control. Device dimensions are tens of micrometers (scale bar: 50 μm) [109]. Together, these archetypes established the core degrees of freedom—charge, flux, and phase—that underpinned subsequent transmon/Xmon and 3D cQED architectures. Adapted from Nadav Hatz (2007).

By applying a voltage to a nearby gate electrode, physicists can tune the energy landscape so that the island favors holding zero, one, or two extra pairs. The allowed charge states form a ladder of well-defined energy levels. Quantum mechanics then allows superpositions of these charge states, giving birth to the first superconducting qubit.

This early design was a conceptual breakthrough, but it came with a serious weakness: the qubit's energy levels depend sensitively on the exact electrostatic environment. Tiny fluctuations in background charge—stray electric fields, drifting impurities, or microscopic defects—shift the qubit frequency. The island's tiny capacitance makes its energy levels sensitive to offset charge. These shifts dephase the qubit's superpositions and collapse coherence within microseconds. The CPB proves that superconducting qubits are possible, but it also reveals the central challenge: the device is far too sensitive for real-world applications.

In addition, the small size of the island and JJ leads to a strong susceptibility to stray capacitance, local defects, and fabrication variation, leading to large device-to-device variability.

To discuss further, charge noise [78] originates from fluctuations in the local electromagnetic environment or charge environment. Mobile charges in nearby materials or interfaces, such as trapped electrons, can move unpredictably and create noise. Stray electric fields from nearby wiring, control lines, or other circuit elements can couple to the qubit. Fluctuations due to the external control electronics or thermal processes, excess or deficit charges in the local environment of the qubit gives rise to offset charge. Microscopic defects in the materials, such as dangling bonds or impurities, can behave as *two-level fluctuators* (Section 5.9), randomly changing their state and contributing to variations in offset charge. Imperfections in fabrication, such as nonuniformities in the dielectric or junction barriers, can leave behind residual charges that affect the qubit.

To solve this, researchers explored more noise-resistant forms. One path led to the *RF-SQUID qubit* and later the *flux qubit*, which stored quantum information in the direction of circulating supercurrents in a superconducting loop. Instead of counting Cooper pairs, these devices used magnetic flux, allowing the qubit states to be "current flowing clockwise" or "current flowing counterclockwise." These qubits were less sensitive to charge noise, but now they became sensitive to magnetic noise and imperfections in the loop.

The *RF-SQUID qubit* [107, 114, 115] was the first serious attempt to use magnetic flux as the basis of a quantum TLS. It was a beautifully simple device: a superconducting loop interrupted by a single JJ. By threading magnetic flux through the loop, researchers could shape its potential energy into a shallow double well, with the two lowest states—persistent currents circulating clockwise or counterclockwise—serving as the qubit states. The Yale group [116] explored variants with larger superconducting loops and adjusted control parameters in an effort to improve coherence. But the same simplicity that made the RF-SQUID appealing also made it fragile. With only one junction defining the entire potential landscape, the wells were soft and easily distorted, and even tiny fluctuations in magnetic flux shifted the qubit's transition frequency. Although the RF-SQUID is a flux-based device, it still retained some sensitivity to charge fluctuations as well. In the end, the RF-SQUID demonstrated that flux qubits were possible, but it also revealed just how delicate early designs were.

The *flux qubit* [114, 117, 118] uses a superconducting loop interrupted by two or multiple JJs (Figure 3.1b. The qubit states correspond to different circulating current directions, where the current creates a magnetic field. This setup allows the qubit to be controlled by magnetic flux rather than charge.

The modern flux qubit continued to grow out of the desire to tame the fragility. Instead of relying on a single junction, researchers redesigned the loop to contain three JJs—two identical ones and a third slightly smaller [110]. The multi-junction loop created a much deeper and more sharply defined double-well potential, with a barrier height and tunneling rate that could be engineered rather than tolerated. The qubit still lived in the space of circulating currents, but now the two logical states sat in a clean, well-controlled landscape instead of the soft, noisy wells of the RF-SQUID. Most importantly, the three-junction design allowed the qubit to operate at a special point—half a flux quantum—where its energy splitting became first-order insensitive to flux noise. Coherence improved dramatically, and the flux qubit became a practical, controllable quantum device.

Seen as a narrative, the RF-SQUID and the flux qubit are two chapters in the same story. The RF-SQUID was the pioneering idea: simple, intuitive, and limited. The flux qubit was the refinement: engineered, tunable, and more stable. Both rely on using circulating currents as quantum states—but only the later design had the robustness needed for real QC experiments.

Another branch produced the *phase qubit*, which used the quantum states of a JJ sitting in a tilted washboard-like potential. The qubit was encoded in different oscillation states of the junction's phase. The *phase qubit* [116] leverages the phase difference across a JJ as the qubit variable (Figure 3.1c). Here, the two energy levels of the qubit are represented by different phase states across the junction. Both designs aimed to reduce sensitivity to charge noise by focusing on flux or phase rather than charge. These devices were easier to control and measure. But they still suffered from material defects and microscopic noise that limited coherence times.

An important step forward came with the *quantronium*, developed in France. The *quantronium* [119] (Figure 3.2a) consists of a CPB with a loop structure. It was a clever hybrid that combined charge and phase degrees of freedom in a way that created a "sweet spot" where the qubit was simultaneously insensitive to both charge and flux noise than the standard CPB to first order. This showed that qubits could be designed to be protected by symmetry against their environment. The quantronium was the first hint that superconducting qubits could be made robust enough for large-scale QC.

Another offshoot of the CPB was the *Fluxonium* [120], a highly anharmonic device shunting a small JJ with the kinetic inductance of a series array of JJs having large capacitances (Figure 3.2b). It requires external flux bias to be operated. See also more information on energy landscape in the same figure. It improved protection against charge noise, though it required external flux bias and more complex fabrication.

The decisive breakthrough came with the *transmon* (transmission line shunted plasma oscillation qubit) [121], which kept the basic physics of the CPB but dramatically increased its shunt capacitance (Figures 3.3 and 3.4). This suppressed charge dispersion exponentially, making the qubit insensitive to charge

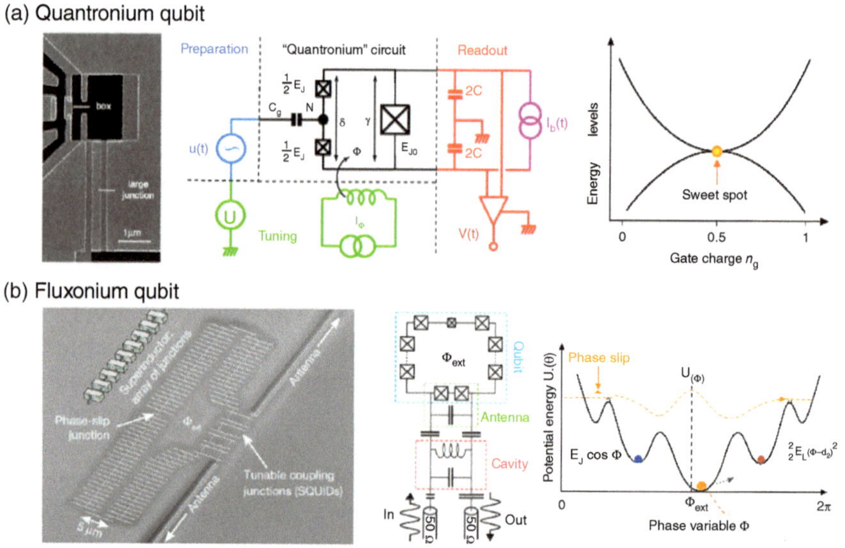

Figure 3.2 Quantronium and fluxonium superconducting qubits. (a) *Quantronium qubit (split Cooper-pair box with readout).* (Left) SEM of the device; scale bar 1 μm. (Middle) circuit showing a Cooper-pair box formed by two small Josephson junctions (energy E_J) in parallel, making a SQUID loop whose effective $E_J(\Phi)$ is tunable with an external flux bias I_Φ (green). The island is capacitively coupled to a gate C_g for microwave control $u(t)$ (blue). Readout (red) is performed through a separate, large Josephson junction pulsed by a bias current $I_B(t)$: the qubit state sets a circulating current in the loop that shifts the large junction's switching threshold (right) energy levels versus gate charge with a sweet spot, *Source:* with permission of American Association for the Advancement of Science. (b) *Fluxonium qubit* (small junction shunted by a super-inductance). Left: microscope image of a device in which a single "phase-slip" (small) junction is shunted by a super-inductance—a long array of Josephson junctions; scale bar 5 μm. Tunable coupling SQUIDs connect the qubit to an on-chip antenna (readout resonator). Right: lumped-element schematic: the qubit loop is threaded by an external flux Φ_{ext} that tunes the transition frequency; the antenna couples the qubit to a measurement cavity (in/out ports) (right) phase potential landscape; with permission of Springer Nature.

noise while preserving enough anharmonicity for microwave control. In effect, the transmon traded some anharmonicity for a huge gain in coherence. The transmon could be reliably fabricated, coupled to microwave resonators for high-fidelity readout with QND measurements, and integrated into large-scale circuits. Although its energy levels are not perfectly two-level—requiring care to avoid unwanted $|1\rangle \rightarrow |2\rangle$ transitions—the transmon's stability, high coherence, and ease of control made it the workhorse of modern superconducting quantum computing.

The transmon qubit can be of two types—fixed frequency, which consists of only one JJ, and tunable-frequency, which utilizes a SQUID loop. Another version

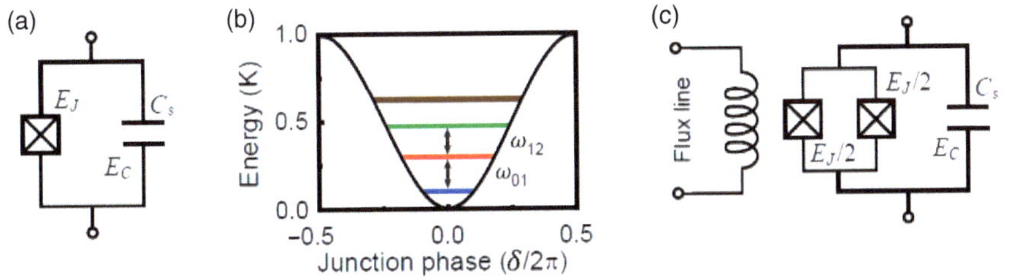

Figure 3.3 (a) Fixed-frequency transmon [121, 122]: a single Josephson junction with Josephson energy E_J shunted by a large capacitance C_s. The large shunt lowers the charging energy $E_C = e^2/2C_\Sigma$ (with $C_\Sigma \approx C_s$+ parasitics), yielding $E_J/E_C \gg 1$ to suppress charge dispersion. (b) Anharmonic spectrum: bound levels in the Josephson–cosine potential versus junction phase $\delta/2\pi$. The transition frequencies are unequally spaced—the qubit transition ω_{01} differs from $\omega_{12}$by the anharmonicity $\alpha = \omega_{12} - \omega_{01} \approx - E_C/\hbar$. This weak nonlinearity enables selective microwave control of $|0\rangle \leftrightarrow |1\rangle$ while suppressing leakage to $|2\rangle$. (Energy scale shown in kelvin for typical parameters.) (c) Tunable-frequency transmon: replacing the single junction by a DC SQUID (two junctions $E_J/2$ in parallel) makes an effective $E_J(\Phi) = E_{J,\Sigma} \cos(\pi\Phi \mid \Phi_0)$ set by a local flux line, allowing in-situ frequency tuning at the cost of added flux-noise sensitivity.

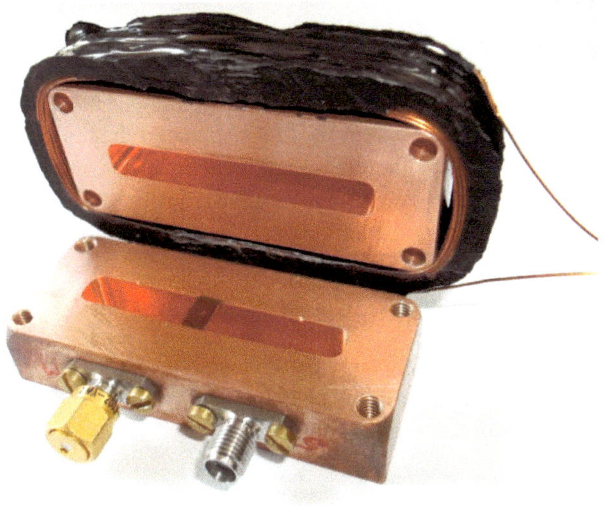

Figure 3.4 Image (Ref. [122] / Madhavi Chand / International Centre for Theoretical Sciences) of a 3D transmon sitting inside a rectangular cavity made of copper. On the outside, superconducting wire is wound and glued using stycast to apply magnetic flux to the tunable qubit (with permission).

of the transmon, the *3D Transmon*, places the transmon qubit inside a 3D cavity [123] to reduce electromagnetic coupling to the environment, further improving coherence times (Figure 3.4).

In the next parts of this chapter, we describe the basic theory of superconducting qubits. We begin with the theory of quantizing a quantum harmonic oscillator, such as the LC resonator. We then discuss further the JJ, which is the crucial component for realizing superconducting qubits. We subsequently describe the quantization theory of a transmon qubit, control of qubit states via gate operations, followed by a discussion of coupling transmons to resonators for qubit readout (measurement) and for multiqubit operations, the beginnings of circuit QED.

3.4 Digression: Quantum Mechanics Refresher

We discuss some of the basic quantum mechanics principles and techniques that are important to understand the operation of superconducting qubits, in particular operator methods that are useful for computation. A convenient reference for this refresher is [124].

3.5 Hamiltonian and Conjugate Variables

Determining the Hamiltonian of a system is the first step to deriving its dynamical behavior. The Hamiltonian governs the dynamics of canonically conjugate variables, such as position and momentum, x and p, through the Schrödinger equation in quantum mechanics. The conjugate relationship between x and p is expressed through commutators:

$$[\hat{x}, \hat{p}] = i\hbar.$$

(Note: We often drop the hat on top of the operator symbol.) This nonzero commutator leads to the Heisenberg uncertainty principle: $\Delta x \Delta p \geq \frac{\hbar}{2}$.

The *Hamiltonian* H is the total energy of a system expressed as a function of the conjugate variables x and p

$$H(x, p) = T(p) + V(x)$$

where $T(p)$ is the kinetic energy (momentum-dependent) and $V(x)$ is the potential energy (position-dependent). The conjugate nature of x and p means their time evolution is intertwined, with one influencing the other. The Hamiltonian dictates how the system evolves in time. The conjugate variables define a point in momentum space, and the Hamiltonian determines the trajectory of this point over time.

The Hamiltonian is an operator, and the time evolution of quantum states is governed by the Schrödinger equation.

$$i\hbar\frac{\partial\Psi}{\partial t} = \hat{H}\Psi$$

As the Schrödinger equation is a first-order linear differential equation, the temporal dynamics of the quantum system may be viewed as a straightforward example of a linear dynamic system with a formal solution

$$|\psi(t)\rangle = e^{-\frac{i\hat{H}t}{\hbar}}|\psi(0)\rangle.$$

The time-independent Hamiltonian governs the time evolution of the system through the operator $e^{-iHt/\hbar}$. Conjugate variables evolve over time under the influence of the Hamiltonian. $\partial H/\partial p$ relates the momentum to the velocity (rate of change of position), and $-\partial H/\partial x$ relates the position to the force (rate of change of momentum).

$$\frac{dx}{dt} = \frac{\partial H}{\partial p}, \quad \frac{dp}{dt} = -\frac{\partial H}{\partial x}$$

When the Hamiltonian H of a quantum system acts on the system's state vector, E is the energy eigenvalue

$$H|\psi\rangle = E|\psi\rangle$$

If the Hamiltonian has a set of n orthonormal eigenstates $\{|e_i\rangle\}$ with corresponding eigenvalues $\{E_i\}$, the eigenstates form a complete basis for the Hilbert space of the system.

$$H|e_i\rangle = E_i|e_i\rangle, \quad i = 1, 2, \ldots, n$$

In the energy eigenbasis, the action of the Hamiltonian is straightforward:

$$H|e_i\rangle = E_i|e_i\rangle$$

If we write the Hamiltonian as a matrix in this basis, the matrix will have eigenvalues E_i on its diagonal, and all off-diagonal elements will be zero (since the eigenstates are orthogonal):

$$H = \begin{pmatrix} E_1 & 0 & 0 & \cdots & 0 \\ 0 & E_2 & 0 & \cdots & 0 \\ 0 & 0 & E_3 & \cdots & 0 \\ \vdots & \vdots & \vdots & \ddots & \vdots \\ 0 & 0 & 0 & \cdots & E_n \end{pmatrix}.$$

Hence, the Hamiltonian can be written as the diagonal matrix on the basis of the energy states.

3.6 The Momentum and Energy Operators

We determine the momentum operator by requiring that, when we operate with $p_x^{(op)}$ on wavefunctions $u_{p0}(x, t)$, we get p_0 times the same wavefunction:

$$p^{(op)}u_{p0}(x,t) = p_0 u_{p0}(x,t)$$

where the wavefunction in position space is

$$u_{p0}(x,t) = \frac{1}{\sqrt{2\pi\hbar}} e^{i(p_0 x - E_0 t)/\hbar}$$

This is true for the following momentum operator: $p^{(op)} = \frac{\hbar}{i}\frac{\partial}{\partial x}$ as seen below when the momentum operator acts on the wavefunction.

$$p^{(op)}\frac{1}{\sqrt{2\pi h}}e^{\frac{i(p_0x-E_0t)}{h}} = \frac{\hbar}{i}\frac{\partial}{\partial x}\frac{1}{\sqrt{2\pi h}}e^{\frac{i(p_0x-E_0t)}{h}}$$

$$\frac{1}{\sqrt{2\pi h}}\frac{h}{i}\frac{ip_0}{\hbar}e^{\frac{i(p_0x-E_0t)}{h}} = p_0\frac{1}{\sqrt{2\pi h}}e^{\frac{i(p_0x-E_0t)}{h}}$$

In the same way, when the energy operator $E^{(op)}$ acts on the wavefunction $u_{p0}(x, t)$, we get E_0 times the same wavefunction

$$E^{(op)}\, u_{p0}(x,t) = E^{(op)}\, u_{p0}(x,t)$$

In this case, the energy operator is $E^{(op)} = i\hbar\frac{\partial}{\partial t}$, which we verify below:

$$\begin{aligned} E^{(op)}\frac{1}{\sqrt{2\pi h}}e^{i(p_0x-E_0t)/h} &= \frac{1}{\sqrt{2\pi h}}i\hbar\frac{-iE_0}{h}e^{i(p_0x-E_0t)/h} \\ &= E_0\frac{1}{\sqrt{2\pi h}}e^{i(p_0x-E_0t)/h} \end{aligned}$$

The position operator $x^{(op)}$ is simply given as

$$x^{(op)} = x$$

According to the above definitions, the Hamiltonian operator becomes

$$H = \frac{p^2}{2m} + V(x) = H^{(op)} = -\frac{\hbar^2}{2m}\frac{\partial^2}{\partial x^2} + V(x).$$

3.7 Commutators

If the Hamiltonian sets the energy landscape, the commutator tells how the system moves through that landscape, because the commutator between an observable and the Hamiltonian determines how that observable changes in time. The other important feature of the commutator is that when two operators commute, they can be known simultaneously; when they don't, the commutator measures the fundamental incompatibility between them. We discuss both of these features further.

In the Heisenberg picture, the time evolution of any observable A is given by

$$\frac{dA}{dt} = \frac{i}{\hbar}[H, A]$$

This is the quantum analog of Newton's laws. $\frac{i}{\hbar}[H, A]$ tells you how the system's energy structure pushes the observable around. The commutator encodes the equations of motion for quantum systems.

Note that a second term appears if the operator itself has explicit time dependence (e.g., a driven system).

$$\frac{dA}{dt} = \frac{i}{\hbar}[H, A] + \left(\frac{\partial A}{\partial t}\right)_{\text{explicit}}.$$

For a particle with Hamiltonian

$$H = \frac{p^2}{2m} + V(x),$$

the Heisenberg equation gives

$$\frac{dx}{dt} = \frac{i}{\hbar}[H, x] = \frac{p}{m},$$

$$\frac{dp}{dt} = \frac{i}{\hbar}[H, p] = -\frac{\partial V}{\partial x}.$$

These are analogous to the classical equations: velocity = momentum / mass; force = minus gradient of potential.

Quantum mechanics reproduces Newton's laws through commutators. Whenever you want to know how something evolves—spin, position, photon number, qubit state—you compute its commutator with the Hamiltonian.

Moving on to the second feature: the commutator $[A,B] = AB - BA$ determines whether two operators A and B can be measured simultaneously without uncertainty. Operators (or variables) do not necessarily commute. If $[A,B] = 0$, the operators commute, meaning they share a common set of eigenstates and can be measured simultaneously with precise values. If $[A,B] \neq 0$, the operators do not commute and are subject to the *uncertainty principle.*

Using p and x as examples, we define the commutator to be

$$[p, x] \equiv px - xp$$

We will now compute the commutator between the noncommuting p and x,

$$[p, x]\psi(x) = px\psi(x) - xp\psi(x) = \frac{\hbar}{i}\left(\frac{\partial}{\partial x}x\psi(x)\right) - x\frac{\hbar}{i}\frac{\partial}{\partial x}\psi(x)$$

$$[p, x]\psi(x) = \frac{\hbar}{i}\left(\psi(x) + x\frac{\partial\psi(x)}{\partial x} - x\frac{\partial\psi(x)}{\partial x}\right) = \frac{\hbar}{i}\psi(x)$$

Hence, for the commutator

$$[p, x] = \frac{\hbar}{i},$$

the position x and momentum p operators satisfy $[x,p] = i\hbar$, which gives rise to the position-momentum uncertainty relation: $\Delta x = \Delta p = \frac{1}{2}|\langle[x, p]\rangle| \geq \frac{\hbar}{2}$.

Similarly, for the commutator of E and t,

$$[E, t] = i\hbar$$

is verified as follows:

$$[E, t]\psi(x, t) = \left(i\hbar\frac{\partial}{\partial t}t - ti\hbar\frac{\partial}{\partial t}\right)\psi(x, t)$$

$$= i\hbar\psi(x, t) + \left(i\hbar t\frac{\partial}{\partial t} - i\hbar t\frac{\partial}{\partial t}\right)\psi(x, t)$$

$$= i\hbar\psi(x, t)$$

By contrast, the operators E and x do commute, so their commutator is

$$[E, x]\psi(x, t) = \left(i\hbar\frac{\partial}{\partial t}x - xi\hbar\frac{\partial}{\partial t}\right)\psi(x, t)$$

$$= \left(i\hbar x\frac{\partial}{\partial t} - i\hbar x\frac{\partial}{\partial t}\right)\psi(x, t) = 0$$

3.8 Quantum Mechanics for Simple Harmonic Oscillator

The quantum harmonic oscillator is one of the most elementary and fundamental systems in quantum mechanics, one of the first topics discussed in most introduction courses on quantum mechanics. Almost every physical system that vibrates, oscillates, or resonates can be understood, at least approximately, as a collection of quantum harmonic oscillators. A molecule's atoms vibrating in a bond, the electromagnetic field inside a cavity, a mode of light in an optical fiber, and a phonon in a crystal, all share the same mathematical structure. The equations for these systems reduce to the same quantum oscillator with its equally spaced energy levels and its creation and annihilation operators, making the harmonic oscillator the "alphabet" of quantum physics. The ladder of equally spaced energy levels gives us a natural language for describing how energy flows. The algebra of raising and lowering operators moves the system up or down this ladder. The quantum harmonic oscillator is the backbone of understanding and manipulating the quantum world.

The theory used for solving a simple quantum harmonic oscillator is essential for understanding the mechanisms of qubits. The Hamiltonian for the 1D harmonic oscillator, such as a mass on a spring, is the sum of the kinetic and potential energies:

$$H = \frac{p^2}{2m} + \frac{1}{2}kx^2$$

where k is the spring constant. The potential energy has a parabolic form in the position variable x. It is standard to remove the spring constant from the Hamiltonian, replacing it with the classical oscillator frequency $\omega = \sqrt{k/m}$. The Hamiltonian becomes

$$H = \frac{p^2}{2m} + \frac{1}{2}m\omega^2 x^2$$

The relationships of the Hamiltonian show that x and p are canonical conjugate variables as discussed above:

$$\frac{\partial H}{\partial x} = m\omega^2 x = kx = -\dot{p}$$

$$\frac{\partial H}{\partial p} = \frac{p}{m} = \dot{x}$$

with $[x, p] = i\hbar$.

We present two equivalent approaches to finding the energy levels. One is to solve the time-independent Schrödinger equation:

$$\left(-\frac{\hbar^2}{2m}\frac{d^2}{dx^2} + \frac{1}{2}m\omega^2 x^2\right)\psi(x) = E\psi(x)$$

This is a second-order differential equation describing a particle in a quadratic potential—a quantum harmonic oscillator. We solve for wavefunctions $\psi_n(x)$ and energy levels E_n explicitly. Quantization arises from requiring normalizable solutions. The solutions (Hermite polynomials multiplied by Gaussians) show nodes and symmetry clearly.

In the *raising and lowering ladder operator approach*, we factorize the differential H operator into a product of operators so that instead of solving a differential equation, we can use algebraic methods to build all energy eigenstates from the ground state. It shows the quantized nature of energy levels and transitions between them.

The Schrödinger approach goes through the nuts and bolts of quantum mechanics—how wavefunctions behave and how quantization emerges. The operator method reveals the deep structure of quantum systems—symmetries, algebraic relationships, and how quantum states evolve. We discuss the operator method first.

3.9 The Operator Method

The Hamiltonian can be rewritten in terms of an operator A and its Hermitian conjugate $A^{\dagger}$ (read as "A dagger"). As we will see shortly, these operators (also called creation and annihilation operators) have very special properties, and the operator methods developed are useful for solving many types of problems [125].

$$A \equiv \left(\sqrt{\frac{m\omega}{2\text{h}}} x + i \frac{p}{\sqrt{2m\hbar\omega}} \right), \quad A^{\dagger} = \left(\sqrt{\frac{m\omega}{2\text{h}}} x - i \frac{p}{\sqrt{2m\hbar\omega}} \right)$$

Here A is chosen so that $A^{\dagger}A$ is close to the Hamiltonian.

We can now rewrite x and p in terms the new operators.

$$A + A^{\dagger} = \frac{1}{\sqrt{2\hbar m\omega}} (2m\omega\hat{x}).$$

$$\hat{x} = \sqrt{\frac{\hbar}{2m\omega}} (A + A^{\dagger}).$$

Similarly, for the momentum operator,

$$A - A^{\dagger} = \frac{1}{\sqrt{2\hbar m\omega}} (2i\hat{p})$$

$$\hat{p} = i\sqrt{\frac{\hbar m\omega}{2}} (A^{\dagger} - A).$$

So, $a + a^{\dagger}$ is proportional to the displacement (position), and $a^{\dagger} - a$ is proportional to the conjugate momentum.

We now show that the Hamiltonian for the simple harmonic oscillator becomes

$$H = \hbar\omega \left(A^{\dagger}A + \frac{1}{2} \right)$$

First compute the quantity

$$A^{\dagger}A = \frac{m\omega}{2\hbar} x^2 + \frac{p^2}{2m\hbar\omega} + \frac{i}{2\hbar}(xp - px)$$

$$A^{\dagger}A = \frac{m\omega}{2\hbar} x^2 + \frac{p^2}{2m\hbar\omega} - \frac{i}{2\hbar}[p, x]$$

$$\hbar\omega(A^{\dagger}A) = \frac{p^2}{2m} + \frac{1}{2} m\omega^2 x^2 - \frac{1}{2}\hbar\omega.$$

Hence, the Hamiltonian can be written in terms of $A^{\dagger}A$ as

$$H = \hbar\omega \left(A^{\dagger}A + \frac{1}{2} \right)$$

Note that the commutator $[A, A^\dagger] = 1$ can be calculated as follows:

$$[A,A^\dagger] = \frac{m\omega}{2\hbar}[x,x] + \frac{1}{2m\hbar\omega}[p,p] - \frac{i}{2\hbar}[x,p] + \frac{i}{2\hbar}[p,x]$$
$$= \frac{i}{2\hbar}(-[x,p] + [p,x]) = \frac{i}{\hbar}[p,x] = 1.$$

Next, we compute the commutators of $A^\dagger$ *and* A with the Hamiltonian, which we will use to find the special properties of $A^\dagger$ *and* A operators.

$$[H,A] = \hbar\omega[A^\dagger A, A] = \hbar\omega[A^\dagger, A]A = -\hbar\omega A$$

$$[H,A^\dagger] = \hbar\omega[A^\dagger A, A^\dagger] = \hbar\omega A^\dagger[A,A^\dagger] = \hbar\omega A^\dagger$$

Hence,

$$[H,A] = -\hbar\omega A$$
$$[H,A^\dagger] = \hbar\omega A^\dagger$$

Let u_n be the n-th energy eigenstate of the Hamiltonian with energy E_n so that $Hu_n = E_n u_n$. We are now ready to show that $\mathrm{Au_n}$ is an eigenfunction of H with the eigenvalue

$$E_n - \hbar\omega$$

which implies that A lowers the energy by $\hbar\omega$.

Applying $[H, A]$ to the energy eigenfunction u_n gives

$$[H,A]u_n = -\hbar\omega A u_n$$
$$HAu_n - AHu_n = -\hbar\omega A u_n$$
$$H(Au_n) - E_n(Au_n) = -\hbar\omega A u_n$$
$$H(Au_n) = (E_n - \hbar\omega)(Au_n)$$

Similarly, $A^\dagger u_n$ is an eigenfunction of H with eigenvalue $E_n + \hbar\omega$ so that $A^\dagger$ raises the energy by $\hbar\omega$. Again, we show this by applying $[H, A^\dagger]$ to the energy eigenfunction u_n as

$$[H,A^\dagger]u_n = \hbar\omega A^\dagger u_n$$
$$HA^\dagger u_n - A^\dagger H u_n = \hbar\omega A^\dagger u_n$$
$$H(A^\dagger u_n) - E_n(A^\dagger u_n) = \hbar\omega(A^\dagger u_n)$$
$$H(A^\dagger u_n) = (E_n + \hbar\omega)(A^\dagger u_n)$$

The commutators $[H, A]$ and $[H, A^\dagger]$ reveal how A moves the system through the Hamiltonian's eigenstates. They expose the energy differences between states and determine whether A preserves energy, raises it, or lowers it. In this way, the commutators and the Hamiltonian work together to carve out the energy eigenstructure of the quantum system. The commutator *shifts* the energy by exactly one quantum.

$$[H, A] = -\hbar\omega A, [H, A^{\dagger}] = +\hbar\omega A^{\dagger}.$$

If u_0 is the ground state, then applying the lowering operator should give zero because we cannot go lower than the ground state, so that

$$Au_0 = 0$$

From the expression of the Hamiltonian in terms A and $A^{\dagger}$, we can deduce the ground-state energy.

$$Hu_0 = \hbar\omega\left(A^{\dagger}A + \frac{1}{2}\right)u_0 = \frac{1}{2}\hbar\omega u_0$$

The ground-state energy is $E_0 = \frac{1}{2}\hbar\omega$, and the states have energies

$$E = \left(n + \frac{1}{2}\right)\hbar\omega$$

We know that $A^{\dagger}$ raises the energy of an eigenstate

$$A^{\dagger}u_n = Cu_{n+1}$$

But we need to determine the coefficient C it produces in front of the new state.

$$|C|^2 = \langle A^{\dagger}u_n \mid A^{\dagger}u_n\rangle = \langle AA^{\dagger}u_n \mid u_n\rangle$$
$$= \langle (A^{\dagger}A + [A, A^{\dagger}])u_n \mid u_n\rangle = (n+1)\langle u_n \mid u_n\rangle = n+1$$

The effect of the raising operator is

$$A^{\dagger}u_n = \sqrt{n+1}u_{n+1}$$

Similarly, the effect of the lowering operator is

$$Au_n = \sqrt{n}u_{n-1}$$

The harmonic oscillator is a linear system with equal spacing of energy differences. Raising and lowering operators are also called *ladder operators* or *creation and annihilation* operators. These are powerful tools because they provide an algebraic approach to solving problems. The creation operator ($A\dagger$) is the "addition" tool. When it acts on a state, it increases the number of particles (such as photons in a field) by 1, essentially "creating" a new particle in the system. Annihilation operator (A) is a "removal" tool. When it acts on a state that describes a certain number of particles, it reduces the number of particles by 1. If there are no particles, using this operator gives zero.

Once the ground state $|0\rangle$ is found by solving $\hat{A}|0\rangle = 0$, higher-energy states are generated algebraically. Instead of solving the Schrödinger partial differential equation directly (which can be cumbersome), ladder operators exploit the symmetries and structure of quantum systems to obtain solutions more efficiently. Raising and lowering operators "step-up" or "step-down" the quantum state of a system, changing the eigenvalue of an observable. These operators express the dynamics and transitions concisely.

We also introduce the number operator as $N_{op} = A^{\dagger}A$. The number operator has eigenvalues $n = 0, 1, 2, \ldots$, and the energy eigenvalues are $E_n = \hbar\omega\left(n + \frac{1}{2}\right)$.

The wavefunctions of the harmonic oscillator can be understood as describing a single particle confined in a harmonic field, which has a parabolic or quadratic potential (Figure 3.5). The system Hamiltonian obtained from the position and momentum operators produces a spectrum of eigenstates, the Hilbert space, of equidistant energy (spaced by the natural frequency), also known as Fock states (see Section 4.4).

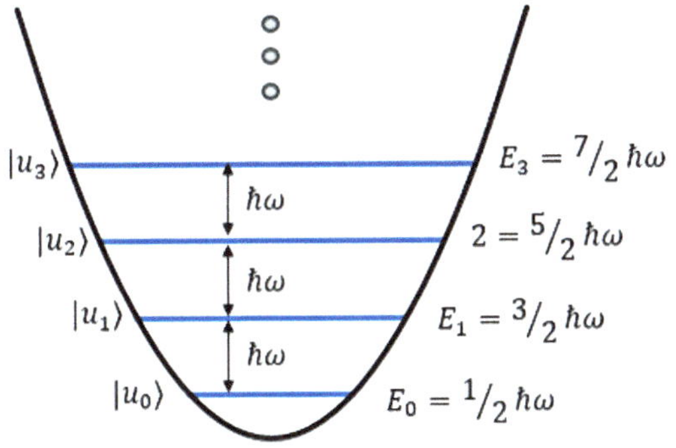

Figure 3.5 Energy eigenstates of a harmonic oscillator having a parabolic potential energy. Schematic of the equally spaced energy levels of a quantum particle confined in a parabolic (harmonic) potential $V(x) = \frac{1}{2}m\omega^2x^2$. The black curve represents the classical potential well, while the horizontal blue lines indicate the quantized stationary states $|u_n\rangle$ with energies

$$E_n = \left(n + \frac{1}{2}\right)\hbar\omega, \quad n = 0, 1, 2, 3, \ldots$$

shown explicitly for the ground state $E_0 = \frac{1}{2}\hbar\omega$ and the first few excited states $E_1 = \frac{3}{2}\hbar\omega$, $E_2 = \frac{5}{2}\hbar\omega$, and $E_3 = \frac{7}{2}\hbar\omega$. The constant spacing $\hbar\omega$ between adjacent levels is a defining feature of the harmonic oscillator, and it underlies the concept of equally spaced quanta (photons, phonons, or cavity excitations) in many physical systems. The gray dots above indicate higher excited states extending upward in the potential. Vertical arrows indicate the constant level spacing $\hbar\omega$.

3.10 LC Circuit Quantum Harmonic Oscillator

An exact analog of quantum mechanical oscillator is the ordinary LC oscillator (Figure 3.6), where the magnetic flux Φ of the inductor and the charge q on the capacitor plate follow the commutation relation $[\Phi, q] = i\hbar$. Here the flux and charge are analogous to the position and momentum of the quantum harmonic oscillator. The dynamics here is determined by the effective potential energy $\Phi^2/(2L)$ and kinetic energy $q^2/(2C)$, giving rise to an equally spaced multilevel system.

To quantize the LC circuit, we proceed in the canonical manner by finding a Hamiltonian for the system and its conjugate variables.

$$H = \frac{1}{2}CV^2 + \frac{1}{2}LI^2 = \frac{q^2}{2C} + \frac{\phi^2}{2L}$$

with $q = VC$ and $\phi = LI$. Here we see that the potential energy is parabolic in ϕ, which is the analogous to position (x) for the mechanical oscillator. The energy oscillates between electrical energy in the capacitor C and magnetic energy in the inductor L. The capacitor and inductor are assumed to be nondissipative elements, realized by superconductors.

The collective motion of electrons forms a single set of conjugate variables, obeying the necessary relationships to the Hamiltonian. The inductor and capacitor of

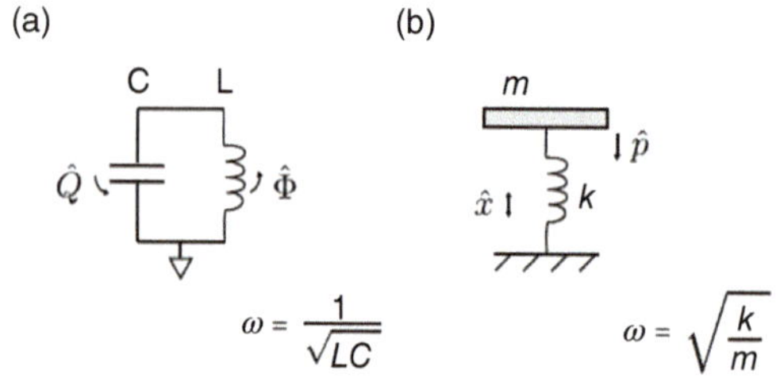

Figure 3.6 Quantum harmonic oscillators. (a) An electrical harmonic oscillator is constructed by placing a capacitor (C) in parallel with an inductor (L). The conjugate variables that describe the resulting oscillation are the charge on the capacitor q and flux through the inductor Φ. The natural frequency of this oscillator is related to these two circuit elements by $\omega = 1/\sqrt{LC}$. (b) The analogous mechanical circuit to the LC oscillator is a simple mass–spring system, with mass m and spring constant k. In this oscillator, the displacement of the spring x and the momentum of the mass p form the equivalent conjugate variables to the system as charge and flux in the electrical circuit. The mechanical oscillator has the resonant frequency $\omega = \sqrt{\frac{k}{m}}$.

an LC oscillator share the same voltage; therefore,

$$V = -L\dot{I} = \frac{q}{C}$$

From there, we obtain

$$\frac{\partial H}{\partial q} = \frac{q}{C} = -L\dot{I} = -\dot{\phi}$$
$$\frac{\partial H}{\partial \phi} = \frac{\phi}{L} = I = \dot{q}.$$

This shows that the variables q and ϕ are canonical conjugate variables, so they satisfy the commutative relation $[\phi, q] = i\hbar$.

Using a similar procedure as in the previous section, the annihilation and creation operators of a LC oscillator are chosen as

$$a = \frac{1}{\sqrt{2\hbar Z_c}}(\phi + iZ_c q)$$
$$a^{\dagger} = \frac{1}{\sqrt{2\hbar Z_c}}(\phi - iZ_c q)$$

Here $Z_c = \sqrt{L/C}$. From these, we obtain a set of expressions for the two conjugate variables in terms of creation and annihilation operators as

$$q = i\sqrt{\frac{\hbar}{2Z_c}}(a^{\dagger} - a)$$
$$\phi = \sqrt{\frac{\hbar Z_c}{2}}(a^{\dagger} + a),$$

These operators satisfy: $[a, a^{\dagger}] = 1$.

An LC circuit behaves like a single particle in a harmonic field because its dynamics are mathematically equivalent to those of a harmonic oscillator. This equivalence arises from the interplay of the inductive and capacitive energy storage, leading to quantized energy levels and collective behavior resembling a quantum particle.

It is customary (and useful for future analogies) to define the reduced flux and the reduced charge as $\phi \equiv 2\pi\Phi/\Phi_0$ and $n = Q/2e$, where $\Phi_0 = h/(2e)$ is the magnetic flux quantum. We also introduce the energy stored in the capacitor and the inductor as

$$E_C = \frac{e^2}{(2C)} \text{ and } E_L = \frac{(\Phi_0/2\pi)^2}{L}.$$

Finally, the Hamiltonian is similar to that of the mechanical harmonic oscillator:

$$H = \hbar\omega_r\left(a^{\dagger}a + \frac{1}{2}\right)$$

where $\omega_r = \sqrt{8E_L E_C}/\hbar = 1/\sqrt{LC}$ is the resonant frequency of the system.

The Hamiltonian can also be rewritten in terms of E_C E_L, and n, as is customary for the discussions of the transmon to come:

$$H = 4E_C n^2 + \frac{1}{2}E_L \phi^2$$

This form will be important to see the parallels for the upcoming CPB and transmon qubit.

The eigenstates of the Hamiltonian correspond to the number states $|n\rangle$, with eigenvalues:

$$E_n = \hbar\omega\left(n + \frac{1}{2}\right), \quad n = 0, 1, 2, \ldots$$

The transition energy between adjacent levels $|n\rangle \rightarrow |n+1\rangle$ is given by

$$\Delta E = E_{n+1} - E_n$$

$$\Delta E = \hbar\omega\left[(n+1) + \frac{1}{2}\right] - \hbar\omega\left[n + \frac{1}{2}\right]$$

$$\Delta E = \hbar\omega$$

The energy difference between adjacent levels of the quantized LC oscillator is

$$\Delta E = \hbar\omega = \hbar\frac{1}{\sqrt{LC}}.$$

Note that the transition energy is adjustable through modifying the capacitance and inductance of the LC circuit.

For an electromagnetic resonator (or microwave cavity), the solution is identical, but the position and momentum operators are replaced by the electric and magnetic fields.

3.11 Wavefunctions for the SHO via Schrödinger Equation

To derive the wavefunctions for the first few energy levels of a quantized LC simple harmonic oscillator (Figure 3.7), it is necessary to solve the Schrödinger equation. First, we recast operator $\hat{q} = -i\hbar\frac{\partial}{\partial\phi}$ (in analogy with the momentum operator) and ϕ (in analogy with the position operator) to rewrite the Hamiltonian as

$$H = \frac{\hat{q}^2}{2C} + \frac{\hat{\phi}^2}{2L} = -\frac{\hbar^2}{2C}\frac{\partial^2}{\partial\phi^2} + \frac{\phi^2}{2L}$$

The time-independent Schrödinger equation is $H\psi(\phi) = E\psi(\phi)$.

Substituting into the Hamiltonian, we have $\left(-\frac{\hbar^2}{2C}\frac{\partial^2}{\partial\phi^2} + \frac{\phi^2}{2L}\right)\psi(\phi) = E\psi(\phi)$.

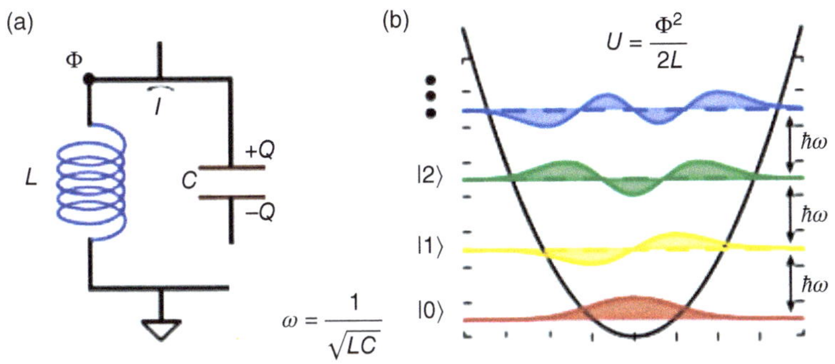

Figure 3.7 Quantized harmonic oscillator. (a) A harmonic oscillator in the form of a resonant LC circuit, where L is the inductance, C is the capacitance, and Φ a time-dependent flux that is analogous to the position variable in a classical oscillator, such as a pendulum. Similarly, just the momentum is conjugate variable to position, the charge, Q, here is the conjugate to Φ. (b) The equally spaced energy levels of a quantized LC oscillator. The wavefunctions corresponding to the first four energy levels are sketched in red, yellow, green, and blue, respectively. [126] / Yvonne Y. Gao.

Introducing the dimensionless variable $\chi = \sqrt{1/\hbar\sqrt{LC}\phi}$ and rewriting the Schrödinger equation by converting, the Hamiltonian becomes

$$\left(-\frac{\partial^2}{\partial\chi^2} + \chi^2\right)\psi(\chi) = \varepsilon\psi(\chi)$$

where $\varepsilon = 2E/\hbar\omega$.

This equation becomes the well-known quantum harmonic oscillator equation:

$$\frac{\partial^2\psi(\chi)}{\partial\chi^2} + (\varepsilon - \chi^2)\psi(\chi) = 0$$

The solutions are given by

$$\psi_n(\chi) = N_n e^{-\chi^2/2} H_n(\chi)$$

Here $H_n(\chi)$ represent Hermite polynomial of degree n and N_n is the normalization constant. The Hermite polynomials are defined as

$$H_n(x) = (-1)^n e^{x^2} \frac{d^n}{dx^n}(e^{-x^2})$$

The first few Hermite polynomials are

$$H_0(x) = 1$$

$$H_1(x) = 2x.$$

$$H_2(x) = 4x^2 - 2$$

$$H_3(x) = 8x^3 - 12x.$$

Hence, the wavefunctions for the first few levels are

$$\psi_0(\chi) = |0\rangle = N_0 e^{-\chi^2/2}$$

$$\psi_1(\chi) = |1\rangle = N_1 \chi e^{-\frac{\chi^2}{2}},$$

$$\psi_2(\chi) = |2\rangle = N_2(4\chi^2 - 2)e^{-\chi^2/2}$$

3.12 Josephson Junction

For a harmonic oscillator, the effective dimension of the Hilbert space (bounded by energy dissipation) could potentially provide an enormous computational space, as we will discuss in Chapter 4, assuming the essential ability to create and manipulate any arbitrary superposition of Fock states. This is where a qubit becomes indispensable. The homogeneity of the level spacing resulting from the linearity of the system makes controlling the photon occupation probability of each individual state impossible. This is because the application of any resonant drive at a fixed frequency will lead to transitions between *all* levels, resulting in a coherent state, i.e. a superposition of states. Introducing some nonlinearity into the system changes the response as needed. The energy levels become unevenly spaced and consequently the drive frequency is only resonant with a single pair of levels, resulting in oscillations between just those levels, e.g. between the ground and first excited state. The level selectivity allows for the preparation of useful superposition of states and ultimately QIP. We will return to role of Fock states in a harmonic oscillator for QC in Chapter 4.

For a quantized system to be used as a qubit, we need to be able to define only two energy states. These levels are known as computational subspace. We require that the target transition frequency is sufficiently distinct from other unwanted transitions connecting the levels in the computational subspace so that the qubit can be individually addressed without causing leakage to noncomputational states. The equally spaced energy levels of the LC oscillator make it impossible to isolate two levels for quantum computation without interference from higher levels. To mitigate the problem, we add anharmonicity (or nonlinearity), making the transition frequencies unequal. In general, the larger the anharmonicity, the better the transition. In practice, the amount of anharmonicity sets a limit on how short the pulses used to drive the qubit need to be.

The nonlinear element must also be nondissipative to preserve the coherence of quantum states. This motivates the introduction of the *JJ* [36, 127]. When we replace the linear inductor with the JJ in the *LC* circuit, it forms the simplest superconducting qubit, the CPB (Section 3.13).

When Brian Josephson, a graduate student at Cambridge, predicted his effect in 1962, he did so not from experiment but from theoretical audacity, daring to extend quantum mechanics across a gap where normally nothing should pass. He predicted that a supercurrent could flow between two superconductors separated by a thin insulating barrier, without any applied voltage—and that an applied voltage would produce an oscillating current with a precisely defined frequency. The Josephson effect was experimentally demonstrated in 1962, just one year later, by Philip W. Anderson and John M. Rowell. This rapid confirmation was extraordinary: a subtle quantum mechanical prediction, initially doubted by senior figures (including John Bardeen), was verified almost immediately. The experimental confirmation established the JJ as a macroscopic quantum object, a precise voltage-to-frequency standard, and the foundational element of SQUIDs, superconducting qubits, and cQED. Brian Josephson was awarded the Nobel Prize in Physics in 1973 for the prediction of the Josephson effect.

The JJ (Figures 3.8–3.11) is a circuit element that is nondissipative and nonlinear. It consists of two superconductors coupled by a weak link, typically a thin insulating barrier [35, 44]. When the insulator is thin enough, the supercurrent can tunnel from one superconductor to the other with zero voltage. Cooper pairs (Figure 3.10) coherently tunnel across this barrier, which leads to a phase difference ϕ between the macroscopic wavefunction on one side of the barrier compared with the other. In a superconductor, electrons pair up into Cooper pairs that behave like bosons. At low temperature, they all condense into one and the same quantum state described by a single macroscopic wavefunction. With a superconductor on either side of a JJ, each condensate has its own wavefunction:

$$\Psi_1 = \sqrt{n_1} e^{i\theta_1}, \quad \Psi_2 = \sqrt{n_2} e^{i\theta_2}$$

where n_s is the density of Cooper pairs, and θ is the quantum phase. Each is the collective quantum wavefunction of trillions of Cooper pairs acting as one object, a Bose–Einstein-like condensate. All the pairs share one phase, which is the physical

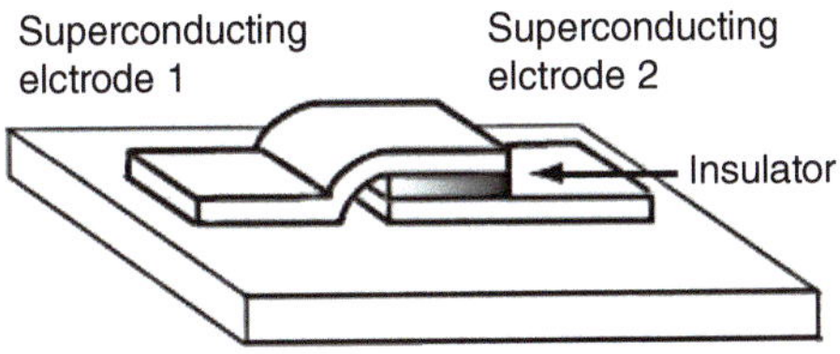

Figure 3.8 Josephson junction: Two superconducting electrodes (left/right) overlap across a nanometer thin insulating barrier (arrow), forming an SIS junction through which Cooper-pair tunnel without voltage (DC Josephson effect $I = I_c \sin \varphi$). The overlap also provides a junction capacitance C, so the junction is characterized by (I_c, C) and Josephson energy $E_J = \hbar I_c / 2e$. In practice, the barrier is typically AlOx $\sim 1-2$ nm between aluminum electrodes; this structure is the basic nonlinear element used in superconducting qubits and circuits.

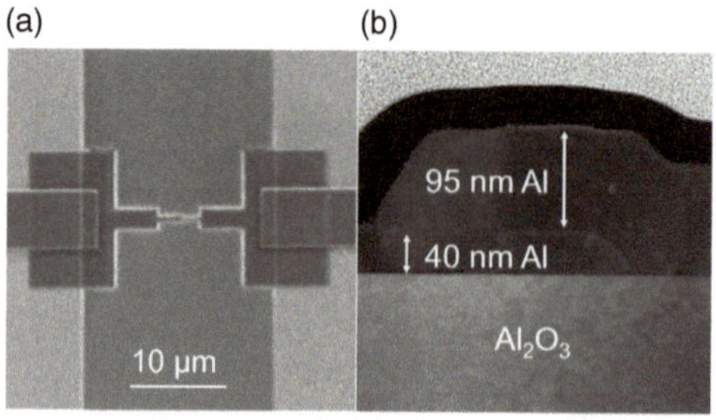

Figure 3.9 (a) Shadow-evaporated Al/AlO*x*/Al Josephson junction made with a Dolan bridge. *Source:* With permission of Superconducting Quantum Materials and Systems Center. A Dolan bridge is a nanoscale suspended mask that allows two angled metal depositions to overlap through a tiny window, creating the Josephson junction. Left: Top-view SEM of the completed overlap junction. Two aluminum electrodes (light gray) meet at the central constriction, where a suspended Dolan "bridge" in a bilayer resist defined the undercut aperture. The device was fabricated by double-angle evaporation: deposit first Al, perform an in-situ controlled oxidation to form a ~1–2 nm AlO*x* tunnel barrier, then deposit the second Al at the opposite angle to create the overlap junction. The larger rectangles on each side are wiring/lead pads. Scale bar: 10 μm. (b) Cross-sectional view of the metal stack showing the two Al films (~95 and ~40 nm) on an Al_2O_3 (sapphire) substrate; the ultrathin AlO*x* barrier lies at their interface (not resolved at this magnification). This Dolan bridge process yields sub-micron junction areas suitable for superconducting qubit circuits (with permission).

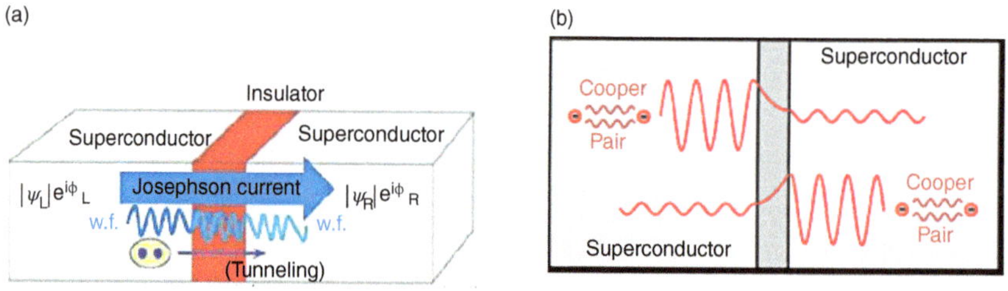

Figure 3.10 (a) Josephson tunneling across an SIS junction. Two superconductors separated by a nanometer thin insulating barrier form a superconductor–insulator–superconductor (SIS) junction. The condensate wavefunctions on the left and right, $\psi_L = |\psi_L|e^{i\phi_L}$ and $\psi_R = |\psi_R|e^{i\phi_R}$, overlap through the barrier so that Cooper-pair tunnel without dissipation. The phase difference $\phi = \phi_L - \phi_R$ drives a Josephson supercurrent $I = I_c \sin\phi$ at zero voltage; the blue arrow indicates this current. (b) Cooper-pair wavefunction overlap picture. Cartoon view of the same SIS junction emphasizing that each superconductor's pair wavefunction penetrates evanescently into the barrier; their overlap enables pair tunneling without breaking the pairs. With no applied voltage, a constant phase difference sustains a DC Josephson current; with a finite voltage, the phase advances as $\dot{\phi} = 2eV/\hbar$, yielding AC Josephson oscillations.

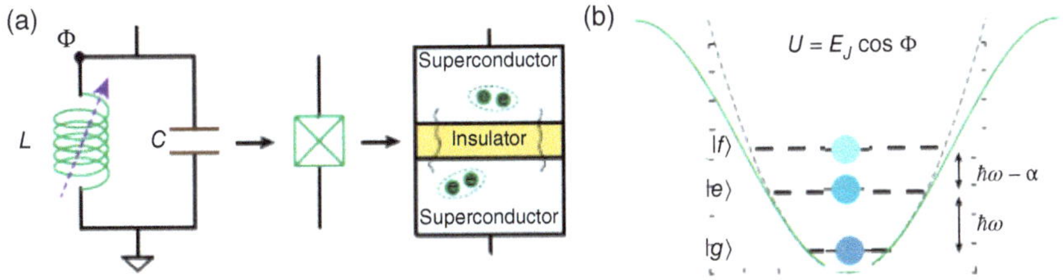

Figure 3.11 An anharmonic quantum oscillator. (a) A Josephson junction can be modeled as an oscillator with a nonlinear inductance. Physically, this element (represented by the green box (with a cross within) is composed of two superconductors separated by a thin insulating layer that allows the tunneling of electrons in pairs (Cooper pairs). When cooled far below the critical temperature of the superconductor, the tunneling events cause no significant dissipation, making the junction essentially a lossless nonlinear inductor. (b) Due to the presence of the junction, the oscillator's potential is no longer quadratic (gray-dashed line). Instead, it becomes a cosine function (green) of the node flux, Φ steeper near the bottom, flatter near the top. The nonlinearity breaks the equal spacing of the energy levels. As a result, the degeneracy is lifted and the energy eigenstates can now be addressed individually; [126] / Yvonne Y. Gao.

quantity that controls the supercurrent, the magnetic flux, and the voltage across the junction. Current flows because of phase difference across the junction.

The phase difference between the superconducting wavefunctions, $\phi = \theta_1 - \theta_2$, becomes the dynamical quantum variable of the JJ. A supercurrent flows because the Josephson relations tie the current directly to this phase difference, much like a pressure gradient drives fluid flow. When charge is added to or removed from the capacitor associated with the junction—even by a fraction of a Cooper pair—the conjugate variable ϕ shifts accordingly, producing a change in the current through the junction. If the junction is embedded in a superconducting loop, changes in the magnetic flux threading the loop impose a constraint on the total phase around the loop, thereby inducing a phase difference across the junction that drives current. In this way, both charge fluctuations and magnetic flux control the phase and, therefore, the current through the junction.

The phase difference determines the current as $I = I_c \sin\phi$, and the Josephson energy as $U(\phi) = -E_J \cos\phi$. Note that $U(\phi)$ determines how much potential energy is stored in the junction. It is the quantum potential well in which the phase of a superconducting circuit lives. Its slope determines the amount of current flow. The probability amplitude for Cooper pairs to tunnel depends on how well the two wavefunctions line up. When the phases line up ($\phi = 0$), the overlap is maximal, the energy is lowest, and slope of the potential is zero, so there is no current flow. At $\phi = \pi/2$, energy is not minimal, but slope is maximal, so the current is maximal. When ($\phi = \pi$), the overlap is minimal, the energy is highest, but the slope is zero again, and the current is zero again.

In a normal inductor, the inductance is linear with the magnetic flux ϕ with $I = \frac{\phi}{L}$. But as we will see below, for the JJ, the inductance is given by $L_J = \Phi_0/2\pi I_c \cos\phi$, where Φ_0 is the flux quantum. Inductance depends on ϕ, which depends on the current. That means the inductance changes as the current changes. This is what makes the junction a nonlinear inductance.

We move on to show more explicitly with math, how replacing the inductor in an LC oscillator (Figure 3.11a) with the JJ modifies the functional form of the potential energy from a parabolic form of the harmonic spectrum discussed above into a co-sinusoidal form (Figure 3.11b).

From $H = 4E_C n^2 + \frac{1}{2}E_L\phi^2$
into $H = 4E_C(n)^2 - E_J\cos(\phi)$.

First, we reiterate that, in contrast to the linear inductor current–flux relation,

$$I = \frac{\phi}{L},$$

the JJ has the nonlinear current–phase relation:

$$I_J = I_c \sin(\phi)$$

where I_c is the Josephson critical current (Figure 3.10), and I_J is the supercurrent through the junction [129]. This equation is known as the first Josephson relation.

The second Josephson relation connects phase to voltage V as

$$\frac{\partial\phi}{\partial t} = \frac{2eV}{\hbar}$$

We constrain ourselves to phenomena where the current is below the critical current. The critical current I_c is the maximum current that may flow through the junction before Cooper pairs are broken. Its magnitude is determined by the junction dimensions and material parameters. Once this current is exceeded, dissipation kicks in and a finite voltage develops across the junction accompanied by a resistive current. Operation in the quantum regime requires currents well below this critical current.

The inductance of the junction is related to the change in phase (instead of flux) with respect to the current:

$$L_J = \frac{\partial\Phi}{\partial I_J}$$

Let us take a partial derivative of the first Josephson relation with respect to I_J to obtain

$$1 = I_c \cos\phi \frac{\partial\phi}{\partial I_J}$$

Substituting $\Phi = \frac{\hbar}{2e}\phi$ gives $\frac{\partial\Phi}{\partial I} = \frac{\hbar}{2e}\frac{\partial\phi}{\partial I} = \frac{\hbar}{2e}\frac{1}{I_c\cos\phi}$.

Thus, the inductance is $L_J = \frac{\hbar}{2eI_c\cos\phi}$.

When no current flows through the junction, the value of $\phi = 0$, leading the maximum value of the inductance $L_{J0} = \hbar/2eI_c$.

Let us now compute the total energy stored. The energy E_C stored in the capacitor is given by

$$E_C = \frac{Q^2}{2C} = \frac{CV^2}{2}$$

The total energy associated with the JJ can be derived by integrating the voltage–current relationship.

$$E_J(\phi) = -\int I_J V dt = \int (I_c \sin\phi)\left(\frac{\hbar}{2e}\frac{d\phi}{dt}\right) dt = -\int \frac{\hbar I_c}{2e}\sin(\phi)d\phi$$
$$= -\frac{\hbar I_c}{2e}\cos(\phi) + \text{constant}$$

Choosing the constant such that $E_J(0) = -E_J$ (where $E_J = \frac{\hbar I_c}{2e}$), we get

$$E_J(\phi) = -E_J\cos(\phi)$$

Combining the capacitive and Josephson energies,

$$E = E_C + E_J = \frac{Q^2}{2C} - E_J\cos(\phi)$$

Since the system deals with Cooper pairs (each with charge 2e), it is customary to express charge in units of Cooper pairs; so, the number n of cooper pairs becomes, $n = Q/2e$.

Introducing the charging energy as $E_C = e^2/2C$, the total energy can now be expressed as

$$4E_C n^2 - E_J\cos(\phi).$$

The charging energy represents the electrostatic energy due to the number of Cooper pairs on the island, and the Josephson Energy $-E_J\cos(\phi)$ represents the energy associated with Cooper-pair tunneling across the junction. The charging energy of the capacitor plays the role of the kinetic term while the junction provides a nonlinear (cosine) potential, creating a periodic quantum well for the phase variable. Cooper pairs move back and forth across the barrier, which means the phase behaves like a particle in a periodic potential. Because it is a quantum particle, only certain standing-wave solutions of the Schrödinger equation are allowed that fit inside that potential. Each standing wave corresponds to a discrete energy level—quantized oscillations of the superconducting phase. In a natural atom, the superconducting phases are confined by the Coulomb potential and, therefore, occupy discrete orbitals. In a JJ, the superconducting phase is confined by the Josephson cosine potential and can occupy only discrete collective quantum states.

The Hamiltonian looks like a particle with "mass" $1/E_C$ moving in a cosine potential. Accordingly, the quantized Hamiltonian of the JJ and capacitor system becomes [121, 130]:

$$H = 4E_C(n)^2 - E_J \cos(\phi).$$

The anharmonicity arises from $-E_J \cos(\phi)$, which distinguishes it from a simple harmonic oscillator. In the harmonic approximation, expanding $-\cos(\phi) \approx -1 + \phi^2/2$, the Hamiltonian resembles that of an LC circuit:

$$H \approx 4E_C n^2 + \frac{1}{2} E_J \phi^2$$

The JJ quantizes the number of Cooper pairs n on the superconducting island and the superconducting phase ϕ across the junction. Thus, it turns *charge and phase* into a pair of quantum-mechanical conjugate variables, $[\phi, n] = i$, and by doing so, it forces a superconducting circuit to behave like a quantum system with discrete energy levels.

The key difference lies in the nonlinearity of the JJ, as seen by the expansion:

$$-E_J \cos\phi \approx -E_J \left(1 - \frac{\phi^2}{2} + \frac{\phi^4}{24}\right)$$

The term $\phi^4/24$ introduces anharmonicity. This anharmonicity breaks the harmonic oscillator approximation, resulting in unequally spaced energy levels. Including nonlinear aspects, the Hamiltonian is

$$H = -E_J + 4E_C n^2 + \frac{1}{2} E_J \phi^2 - \frac{E_J}{24} \phi^4 + \dots.$$

3.13 Cooper-Pair Box

We discussed this invention in Section 3.3.1. The CPB (Figure 3.12) consists of a JJ in parallel with its own capacitance C_J and connected to a gate voltage V_g through a gate capacitance C_g. The total capacitance of the system is

$$C_\Sigma = C_J + C_g$$

The charge Q stored on the capacitor contributes an electrostatic energy, or the charging energy.

$$E_C = \frac{Q^2}{2C_\Sigma},$$

The quantum Hamiltonian for the CPB is

$$H = 4E_C(n - n_g)^2 - E_J \cos(\phi)$$

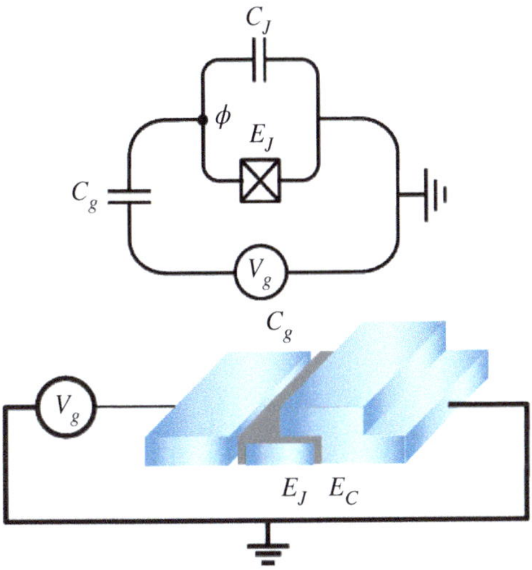

Figure 3.12 A schematic of the lumped-element electrical representation of a Cooper-pair box qubit. A small superconducting island (left node, labeled by the superconducting phase ϕ) is connected to a large superconducting reservoir (right node, ground) through a Josephson junction, indicated by the nonlinear element with Josephson energy E_J. In parallel with the Josephson junction is its intrinsic capacitance C_J, which arises from the oxide tunnel barrier separating the two superconducting electrodes. The island is also capacitively coupled to an external gate voltage V_g through the gate capacitor C_g. Together, C_J and $C_g$form the total island capacitance $C_\Sigma = C_J + C_g$, which sets the charging energy $E_C = e^2/(2C_\Sigma)$. The gate voltage V_g induces a controllable offset charge n_g on the island, allowing tuning of the qubit's electrostatic energy. The competition between the charging energy associated with C_Σ and the Josephson coupling E_J gives rise to discrete quantum energy levels, making the Cooper-pair box a tunable artificial atom. (b) A superconducting island connected to ground through a Josephson junction. The island is capacitively coupled to an external gate electrode via the gate capacitance C_g, which is biased by a gate voltage V_g [17].

where n_g is dimensionless gate charge, also called the offset charge, which is controlled by the gate voltage V_g. When voltage V_g is applied to the gate, it induces a charge on the island:

$$Q_g = C_g V_g$$

Since Cooper pairs carry charge $2e$, it is natural to express this in units of Cooper pairs:

$$n_g = \frac{C_g V_g}{2e}$$

Note that n_g is not an integer—it is a continuously tunable number that tells how many Cooper-pairs' worth of charge the gate is trying to place on the island. It is essentially a dial that shifts the preferred charge state of the island. n_g makes the system *tunable*. By changing the gate voltage V_g, we shift the electrostatic offset charge and reshape the potential felt by the Cooper pairs. Also, by fabricating the junction with different geometry, we change E_J. This lets us continuously tune the spacing between energy levels, selecting two of them to act as $|0\rangle$ and $|1\rangle$. So the CPB becomes an artificial atom with properties that can be engineered and tuned.

To quantize the Hamiltonian, we start by defining the ladder (creation and annihilation) operators $a^\dagger$ and a :

$$\phi = \phi_{\text{zpf}}(a + a^\dagger) = \left(\frac{2E_c}{E_J}\right)^{1/4}(a + a^\dagger)$$

$$n = -in_{\text{zpf}}(a - a^\dagger) = -i\left(\frac{E_J}{32E_c}\right)^{1/4}(a - a^\dagger)$$

where $\phi_{\text{zpf}} = (2E_C/E_J)^{1/4}$ is the zero-point fluctuation amplitude of the phase, and $n_{\text{zpf}} = (E_J/32E_C)^{1/4}$ is the zero-point fluctuation amplitude of the charge.

ϕ is the "position-like" variable, n is the "momentum-like" variable, so as in a harmonic oscillator we discussed the earlier section: $x \propto (a + a^\dagger)$, $p \propto i(a^\dagger - a)$.

Substituting the operators first into the harmonic Hamiltonian: $H_{\text{harmonic}} = 4E_C n^2 + E_J \frac{\phi^2}{2}$, we have

$$n^2 = n_{\text{zpf}}^2\left(\frac{a^\dagger - a}{i}\right)^2, \quad \phi^2 = \phi_{\text{zpf}}^2(a + a^\dagger)^2$$

And some simple algebra utilizing the commutation relation $[a, a^\dagger] = 1$ gives (as before)

$$H_{\text{harmonic}} = \hbar\omega_q\left(a^\dagger a + \frac{1}{2}\right)$$

$\hbar\omega_q$ is the harmonic oscillator energy quantum:

$$\omega_q = \frac{\sqrt{8E_C E_J}}{\hbar}$$

Next, adding the anharmonic terms that arise from higher-order terms in the cosine expansion:

$$\cos\phi \approx 1 - \frac{\phi^2}{2} + \frac{\phi^4}{24} - \dots,$$

the Hamiltonian becomes

$$H = \hbar\omega_q\left(a^\dagger a + \frac{1}{2}\right) - \frac{E_J\phi_{\text{zpf}}^4}{24}(a + a^\dagger)^4$$

Expanding $(a + a^\dagger)^4$ and keeping only relevant terms for low excitations:

$$(a + a^\dagger)^4 \approx 3 + 6a^\dagger a + 6(a^\dagger a)^2 + \text{other terms},$$

the Hamiltonian becomes

$$H = \hbar\omega_q\left(a^\dagger a + \frac{1}{2}\right) - \frac{\alpha}{2}(a^\dagger a) + \frac{\alpha}{2}(a^\dagger a)^2$$

where α is the anharmonicity defined as

$$\alpha = \frac{E_J \phi^4_{\text{zpf}}}{2} = -E_C$$

We often neglect the constant energy term $\frac{1}{2}\hbar\omega_q$ as it doesn't play any role in the dynamics of the system. Now, recall that the number operator (n) is the same as $n = a^\dagger a$:

$$E_n = n\hbar\omega_q + \frac{\alpha}{2}n(n-1)$$

where n is the energy-level index ($n = 0, 1, 2, \ldots$). The energy levels of the circuit are no longer equidistant.

Ground state ($n = 0$): $E_0 = 0$
First excited state ($n = 1$): $E_1 = \hbar\omega_q$
Second excited state ($n = 2$): $E_2 = 2\hbar\omega_q + \alpha$

The transition frequency between the ground and first excited state is approximately given by $\omega_{01} \approx \dfrac{(\sqrt{8E_J E_C} - E_C)}{\hbar}$.

We can interpret the above as describing a particle in a cosine potential (Figure 3.11b). The frequency $\omega_q = \sqrt{8E_C E_J}/\hbar$ corresponds to the frequency of small oscillations at the bottom of a well of the cosine potential.

The anharmonicity α of the CPB is a measure of the deviation of the energy spectrum from perfect harmonicity. It is defined as the difference between the frequency of successive transitions in the energy levels, specifically:

$$\alpha = (E_2 - E_1) - (E_1 - E_0) = E_{12} - E_{01}$$

The anharmonicity α is crucial for distinguishing the qubit's ground and first excited states, enabling its use as a qubit. The anharmonicity ensures that $\omega_{01} \neq \omega_{12}$, enabling selective addressing of qubit states.

3.14 Wavefunctions from Schrödinger Equation

Since the superconducting phase ϕ and the charge n are conjugate variables, it allows us to rewrite the Hamiltonian in the phase basis $|\phi\rangle$, using $\rightarrow -i\frac{\partial}{\partial\phi}$. The Hamiltonian becomes

$$H = 4E_C\left(-i\frac{\partial}{\partial\phi} - n_g\right)^2 - E_J\cos\phi$$

Reintroducing the gate (offset) charge, the wavefunction satisfies

$$\left[4E_C\left(\left(-i\frac{\partial}{\partial\phi}-n_g\right)-i\frac{\partial}{\partial\phi}-n_g\right)^2-E_J\cos\phi\right]\psi(\phi)=E\psi(\phi)$$

By introducing dimensionless variables

$$q=\frac{\phi}{\pi},\quad a=\frac{8E_C}{E_J},\quad b=\frac{16E_Cn_g}{E_J}$$

the Schrödinger equation becomes a *Mathieu equation*:

$$\frac{\partial^2\psi}{\partial q^2}+[a-2b\cos(2\pi q)]\psi=0$$

The solutions to the Mathieu equation are *Mathieu functions* [131], which describe the phase-dependent wavefunctions [44, 92]:

Even solutions: $\mathrm{ce}_n(q, a)$ (cosine-like)
Odd solutions: $\mathrm{se}_n(q, a)$ (sine-like)

Mathieu functions are implemented in many mathematical software libraries, such as Mathematica's MathieuC, and MathieuS. The Mathieu functions can be evaluated numerically. From the eigenenergies, we then define the charge dispersion for the mth energy band.

The energy levels of the charge qubit (Figure 3.13) are solutions to the Mathieu equation derived from the Hamiltonian [132]. The energy levels depend strongly on n_g as, for example, the ground-state energy can be approximated as

$$E_0(n_g)\approx-\frac{E_J}{2}\cos(2\pi n_g)$$

As stated earlier, the offset charge n_g arises from the external gate voltage $V_g = Q_g/C_g$. But it can also arise from spurious unwanted degrees of freedom in the environment. Figure 3.13 shows that the energy of the eigenstates varies, resulting in the variation of E_{01} known as the charge dispersion. Charge fluctuations in the circuit's environment lead to fluctuations in the qubit's energy levels and transition frequency and, consequently, to dephasing.

The CPB [40, 108, 119, 133] is thus susceptible to charge noise, which can be minimized by operation at a "magic point" where, to first order, the energy is independent of gate charge. Performing experiments at this charge "sweet spot" obtained by setting $n_g = \pm 0.5$ is crucial for obtaining longer coherence times [134]. This yields two states close to each other but with a large gap to higher states and with significant anharmonicity. But the qubit works only in this sweet spot as it is otherwise very sensitive to charge noise.

In this earlier superconducting qubit, the CPB was highly sensitive to charge fluctuations because the charging energy E_C was relatively large ($E_J \sim E_C$) as

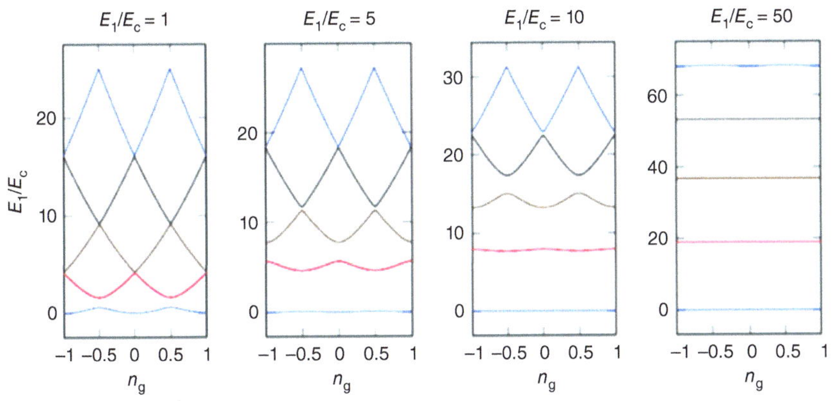

Figure 3.13 Charge qubit (Cooper-pair box) dispersion with gate charge and E_J/E_C. Dispersion of the lowest five eigenenergies of the Cooper-pair box Hamiltonian as a function of the reduced gate charge n_g for four different ratios of Josephson energy to charging energy: $E_J/E_C = 1, 5, 10, 50$ (left to right). The curves show how the quantized energy levels $E_m(n_g)$ evolve as the electrostatic offset charge is swept over one Cooper-pair period ($n_g \in [-1, 1]$). For small E_J/E_C (left panels), the spectrum exhibits strong charge dispersion with pronounced avoided crossings at half-integer n_g, reflecting the dominance of charging energy and the sensitivity to single Cooper-pair tunneling. As E_J/E_C increases, the Josephson coupling increasingly hybridizes neighboring charge states, smoothing the spectrum and reducing its dependence on n_g. In the large E_J/E_C (transmon) regime (right panel), the energy levels become nearly flat versus n_g, indicating strong suppression of charge noise while maintaining anharmonic level spacing suitable for qubit operation. The plotted energies are normalized to the charging energy E_C, and the colored curves correspond to the ground state and the first four excited states [132]. *Source:* [132] / Lev Samuel Bishop.

shown in panel a of Figure 3.13. As E_J/E_C increases, the charge dispersion decreases exponentially because the coupling between different charge states grows stronger, effectively delocalizing the wavefunction across many states. For large E_J/E_C, the ground-state energy dispersion ΔE_0 decays exponentially with $\sqrt{E_J/E_C}$.

$$\Delta E_0 \propto e^{-\sqrt{8E_J/E_C}}$$

This exponential suppression occurs because the wavefunctions become tightly localized due to the large Josephson energy E_J, making the system less sensitive to changes in n_g.

In summary, because the island is tiny and E_C is large, even tiny fluctuations of stray charges in the substrate or on nearby surfaces shift n_g. This causes large fluctuations in the qubit frequency, rapid dephasing, and the need to bias the device at a "sweet spot" just to get reasonable coherence. In other words, the CPB is an excellent artificial atom but a not a very good qubit because it is *too charge-sensitive*.

Increasing E_J/E_C reduces sensitivity to n_g, which is crucial for reducing the susceptibility of the qubit to charge noise and ensures better coherence properties. This suppression is the main motivation behind designing the transmon circuit.

3.15 Transmon Device

The *transmon regime* [121] is reached when $E_J \gg E_C$, with typical values of 20–80, meaning the Josephson energy dominates over the charging energy, making the qubit exponentially less sensitive to charge noise as the eigenstates spread over many charge states (Section 3.3.1). When E_J becomes large, the cosine potential dominates. The qubit's wavefunction is no longer sharply localized around a single charge state $|n\rangle$. Instead, it becomes a broad superposition over many charge states. A broad wavefunction in charge means the system is less sensitive to the value of n_g, and charge noise barely shifts the energy levels. The transmon (Figure 3.14) is the CPB redesigned for stability. The anharmonicity decreases algebraically as

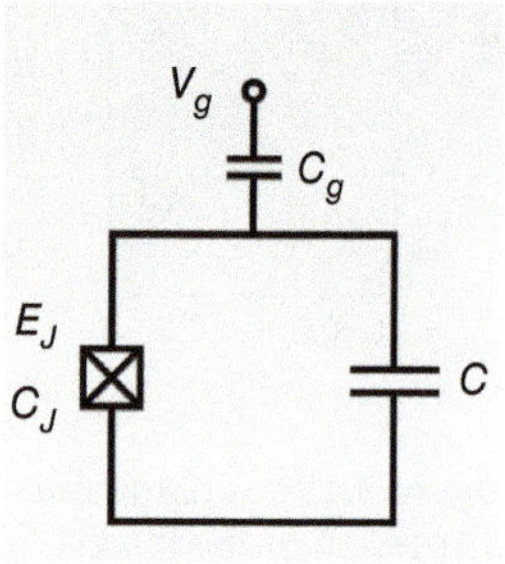

Figure 3.14 The circuit diagram of a transmon device where the crossed box represents a Josephson junction with its intrinsic capacitance. The junction is shunted by a large linear capacitor.

$$\alpha \sim \frac{E_C}{(E_J/E_C)^{3/4}}$$

But there remains sufficient anharmonicity for selective qubit control. For the case of the transmon, $\alpha = -E_C$ is usually designed to be 100–300 MHz, as required to maintain a desirable qubit frequency

$$\omega_q = \frac{(\sqrt{8E_JE_C} - E_C)}{\hbar} = 3 - 6\,\text{GHz}$$

Incidentally, if the resonance frequency of the oscillator $\omega/2\pi \gtrsim 5$ GHz when it is cooled to its ground state, at the operating temperatures of a commercial DR ($T \sim 10$ mK), the undesired population in the higher-energy levels is vanishingly small population suppressed by the Boltzmann factor

$$e^{-\hbar\omega/k_BT} \simeq e^{-20}$$

Taking an identical approach to quantize the transmon Hamiltonian, we find that the energy of the *nth* level can be approximated as

$$E_n \approx \hbar\omega_p\left(n + \frac{1}{2}\right) - \frac{E_C}{12}(6n^2 + 6n + 3).$$

$$E_n \approx \hbar\omega_p\left(n + \frac{1}{2}\right) - \frac{E_C}{2}n^2 - \frac{E_C}{2}n - \frac{E_C}{4}.$$

Here, $\omega_p = \sqrt{8E_J E_C}$ is known as the Josephson plasma frequency. The last term shifts all energy levels uniformly. Such a shift does not affect observable transition frequencies but reflects a difference in the zero-point energy reference and is ignored in many cases. Therefore,

$$E_n \approx \hbar\omega_p\left(n + \frac{1}{2}\right) - \frac{E_C}{2}n^2 - \frac{E_C}{2}n$$

$$E_n \approx \hbar\omega_p\left(n + \frac{1}{2}\right) - \frac{E_C}{2}n(n-1)$$

which essentially has the same form as for the CPB.

Ground-state energy ($n = 0$): $E_0 = \frac{\hbar\omega_p}{2}$

First excited state ($n = 1$): $E_1 = \frac{3\hbar\omega_p}{2}$

Second excited state ($n = 2$): $E_2 = \frac{5}{2}\hbar\omega_p$

We can calculate the transition energies of the adjacent levels as

$$\Delta E_{1-0} = E_1 - E_0 = \hbar\omega_p - E_C$$

$$\Delta E_{2-1} = E_2 - E_1 = \hbar\omega_p - 2E_C$$

The anharmonicity is the difference between the transition frequencies:

$$\alpha = \omega_{12} - \omega_{01} = -E_C$$

The last expression is valid when roughly $E_J/E_C > 20$. Even though one would like to increase this ratio arbitrarily to obtain even stronger suppression to charge noise, there are practical limits to this. First, the plasma frequency $\omega_p = \sqrt{8E_J E_C}$ is often designed to be in the microwave frequency range of 3–8 GHz. This means that one has to increase the ratio E_J/E_C while keeping the product $E_J E_C$ roughly fixed. In other words, E_J can be increased only at the expense of reducing E_C. Since E_C determines the anharmonicity or the ability of selective addressing of a transition, one cannot make it arbitrarily small. The magnitude of E_C sets a lower bound on the gate time for avoiding unwanted transitions. Due to these constraints, practical values for $|\alpha|$ lie in the range of 100–300 MHz while keeping the charge dispersion negligible (see right panel of Figure 3.13). In order to realize this kind of small charging energy, one needs to shunt the JJ with a large capacitor $C_s \gg C_J$, where C_J represents the capacitance intrinsic to the junction.

The transmon Hamiltonian, therefore, reduces charge noise in the $E_J \gg E_C$ regime while maintaining sufficient anharmonicity introduced by the ϕ^4 term. The resulting nonequidistant energy levels enable isolation of the lowest two eignestates, with transition frequency $\omega_q = \sqrt{8E_J E_C} - E_C$ allowing for robust qubit operation.

Among the many superconducting circuits employed as qubits, the transmon is one of the most widely adopted due to its simple geometry, reproducible performance, good coherence times, ease of control, and precise readout.

Another advantage of the transmon is its fabrication process, which involves well-established electron-beam and photolithographic procedures. This makes it possible for us to engineer and fabricate precise device parameters that are optimized for the particular study. These favorable traits make the transmon a highly robust and versatile resource that is becoming the staple of many cQED experiments. It has been used to demonstrate multiple high-fidelity logical operations and quantum simulations.

3.16 Two-Level Transmon Qubit

The superconducting circuit-based qubit with the transmon is called an "artificial atom" because it behaves like an atom with two quantum energy levels. Since we can restrict operations in the system to just these two lowest energy levels, it is useful to treat the transmon as a TLS, ignoring all higher levels. This view of the transmon will be helpful when we later discuss the interaction of the transmon with a resonator.

The lowest two eigenstates of the transmon, $|g\rangle$ and $|e\rangle$, are separated from the rest of the spectrum by an anharmonicity set by the JJ. By restricting ourselves to this pair, we isolate a clean, addressable quantum degree of freedom that behaves like an artificial atom. The higher levels are still there, but they are energetically far enough away that we can largely avoid populating them with carefully shaped microwave pulses. The two-level description also makes the interaction with microwave fields and resonators transparent. In this reduced space, the complicated charge and phase dynamics of the JJ collapse into a single electric dipole operator, written as $\sigma^+ + \sigma^-$. That lets us borrow the entire language of quantum optics.

We use the *ket* notations $|0\rangle$ and $|1\rangle$ for the ground and excited states, respectively. In this basis, the Hamiltonian can be written as $H = \frac{\hbar\omega_q}{2}\sigma_z$ as for a simple TLS, such as a spin-1/2 particle where σ_z is the Pauli z-operator in the qubit basis:

$$\sigma_z = \begin{pmatrix} 1 & 0 \\ 0 & -1 \end{pmatrix}$$

and ω_q is the angular frequency of the system's transition.

Alternatively, in terms of the number operators, $H = \hbar\omega_q a^\dagger a$. Here $\omega_q = \omega_{01}$ is the qubit transition frequency. One can move between these two operators using the relation $a^\dagger a = (1 - \sigma_Z)/2$.

To derive the Hamiltonian $H = \frac{\hbar\omega_q}{2}\sigma_z$ for a simple TLS, consider that a spin-1/2 particle has two eigenstates: spin-up or $|0\rangle$ with energy E_+ and spin-down or $|1\rangle$ with energy $|1\rangle$. The difference in energy between these states is

$$\hbar\omega_q = E_+ - E_-$$

The Hamiltonian of a system is diagonal on the basis of its energy eigenstates. For this system,

$$H = \begin{pmatrix} E_0 & 0 \\ 0 & E_1 \end{pmatrix}$$

if we express the energies of the two levels relative to their mean energy as

$$E_0 = -\frac{\hbar\omega_q}{2}, \quad E_+ = \frac{\hbar\omega_q}{2}$$

Substituting into the Hamiltonian, the matrix becomes

$$H = \begin{pmatrix} -\frac{\hbar\omega_q}{2} & 0 \\ 0 & +\frac{\hbar\omega_q}{2} \end{pmatrix} = -\frac{\hbar\omega_q}{2}\sigma_z$$

Note that this convention assumes that the energy of the ground state $|0\rangle$ is lower than that of the excited state $|1\rangle$. However, sometimes the opposite convention (coming from the spin-½ particles) $H = \frac{\hbar\omega_q}{2}\sigma_z$ is used, which produces the same dynamics.

When viewed as a TLS, the transmon is *not* operated as a harmonic oscillator. We then keep only the two ground and excited states $|g\rangle, |e\rangle$, and throw away all higher levels. This is a different 2D Hilbert space. All operators must now be written as 2×2 matrices. n is no longer a momentum. It is now an *electric dipole operator* that flips the qubit.

$\hat{\sigma}_+$: Raising operator for the qubit transition ($|0\rangle \rightarrow |1\rangle$), or $|g\rangle \rightarrow |e\rangle$
$\hat{\sigma}_-$: Lowering operator for the qubit transitions ($|1\rangle \rightarrow |0\rangle$), or $|e\rangle \rightarrow |g\rangle$

The operators σ^+ and σ^- are represented in matrix form if the qubit basis is taken as $\{|g\rangle, |e\rangle\}$:

$$|g\rangle\langle e| \equiv \sigma^- = \begin{pmatrix} 0 & 0 \\ 1 & 0 \end{pmatrix}, \quad |e\rangle\langle g| \equiv \sigma^+ = \begin{pmatrix} 0 & 1 \\ 0 & 0 \end{pmatrix}$$

$\sigma^+ = |e\rangle\langle g|$, $\sigma^- = |g\rangle\langle e|$ are the outer products that describe transformations on a quantum state as follows:

$\sigma^+|g\rangle = |e\rangle$; when applied to $|g\rangle$, it outputs $|e\rangle$ because $|e\rangle\langle g|g\rangle = |e\rangle$; and when applied to $|e\rangle$, it gives zero (no higher state to raise to): $\sigma^+|e\rangle = 0$. And similarly $\sigma^-|e\rangle = |g\rangle$ and $\sigma^-|g\rangle = 0$ (no lower state to drop to).

By defining these operators in terms of outer products, we capture their essential roles for transitioning between discrete quantum states.

The transmon's charge operator n is related to the qubit operators as $(\sigma^+ + \sigma^-)$ because the sum of the outer products satisfies both

$$n \propto |g\rangle\langle e| + |e\rangle\langle g| = \sigma^+ + \sigma^-$$

$$n \sim \sigma^+ + \sigma^-$$

Once we view the transmon *not* as a box that stores static charge but as an artificial atom with an electric dipole, the structure of the operator $\hat{n}$ becomes more clear. In the transmon regime, the qubit states $|g\rangle$ and $|e\rangle$ do not differ by having different amounts of charge sitting on the island; instead, they differ by the way charge oscillates across the JJ. Each state corresponds to a distinct pattern of dipole oscillation, much like the ground and excited states of a real atom correspond to different electron-cloud motions. Because the time-averaged charge in either stationary state is zero, the *dipole moment* has no permanent component within $|g\rangle$ or $|e\rangle$; its only role is to connect one oscillation pattern to the other. In quantum-mechanical terms, this means the dipole operator—and, therefore, the charge operator $\hat{n}$ that represents it is purely off-diagonal in the qubit basis: it takes $|g\rangle$ into $|e\rangle$ and $|e\rangle$ into $|g\rangle$. The operator is the sum of the raising and lowering operators. Thus, in the two-level transmon, $\hat{n}$ is simply the electric dipole moment of this artificial atom written in qubit language, and electric fields from a resonator couple to it by driving these dipole-mediated transitions.

3.17 SQUIDs

The *SQUID* (superconducting quantum interference device) architecture (Figure 3.15), achieved by connecting two JJs in parallel [135], enables enhanced sensitivity to magnetic fields and tunable inductance, which is essential for

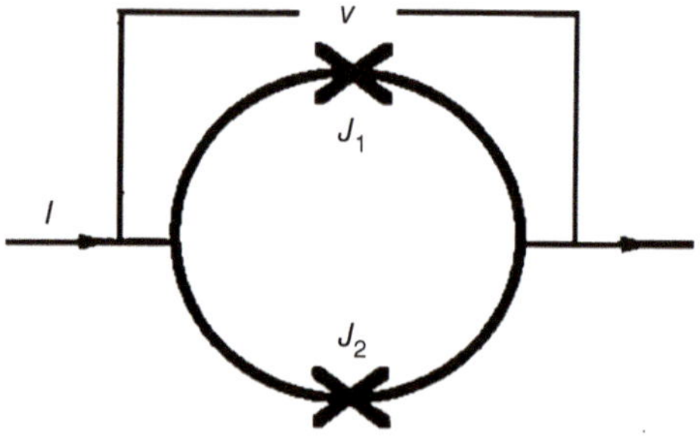

Figure 3.15 Schematic of a DC superconducting quantum interference device (SQUID). The figure shows a superconducting loop interrupted by two JJs J_1 and J_2, forming a closed superconducting ring. A bias current I enters from the left and exits on the right, splitting between the two junctions and recombining around the loop. The voltage V measured across the device reflects the collective Josephson dynamics of the two junctions. When an external magnetic flux threads the loop (not shown), it imposes a phase difference between the two junctions, causing their supercurrents to interfere. This quantum interference modulates the effective critical current of the SQUID in a periodic way, making the device an extremely sensitive detector of magnetic flux and a tunable nonlinear element in superconducting quantum circuits.

effectively controlling quantum states of qubits within a superconducting circuit [45, 46]. This configuration leverages the interference effects arising from the two junctions and contributes to improved coherence and stability in the quantum system, making SQUID-based designs a popular choice for various applications in QC, sensing, and metrology. However, the introduction of this control knob comes at a cost, as the qubit's susceptibility to magnetic flux noise increases significantly.

Because superconductivity is described by a single coherent wavefunction around the loop, the phase must "fit" continuously all the way around. When a magnetic field threads the loop, it adds a phase shift proportional to the magnetic flux, and this shift alters how supercurrents interfere with each other as they pass through the two junctions. The result is that the total current through the SQUID oscillates periodically as a function of magnetic flux, with a period equal to one flux quantum $\Phi_0 = h/2e \approx 2 \times 10^{-15}$ Wb. This makes the SQUID a sensitive detector for magnetic flux. The two paths around the loop form a quantum interferometer for Cooper pairs. Magnetic flux changes the phase difference accumulated along the two paths, just as a change in optical path length shifts the interference fringes in an optical interferometer. The supercurrent through the SQUID depends on that phase difference, so even a minute change in flux shifts the interference pattern and produces a measurable change in current or voltage. The entire condensate—trillions of Cooper pairs—participates in this interference.

Even a tiny fraction of a flux quantum causes a measurable change in the device's voltage or current. Modern SQUIDs can detect changes in magnetic flux as small as 10^{-6}–10^{-7} of a flux quantum, corresponding to magnetic fields below 10^{-11} gauss—orders of magnitude smaller than the Earth's magnetic field (about 0.5 gauss). In practical terms, this means a SQUID can sense the magnetic field produced by the electrical activity of the human brain or heart, or the tiny magnetic moment of a few million electron spins.

This extreme sensitivity opens remarkable scientific and technological possibilities. In medicine, SQUIDs enable magnetoencephalography (MEG) and magnetocardiography, allowing doctors and neuroscientists to map brain and heart activity noninvasively with millisecond time resolution. SQUIDs can ultimately detect magnetic fields comparable to those produced by a few thousand neuronal action potentials in the human brain, or even smaller signals in shielded environments. In geophysics and archaeology, SQUIDs detect subtle magnetic anomalies that reveal underground structures or mineral deposits. In fundamental physics, they are used to search for exotic phenomena such as axions, dark matter-induced magnetic signals, or tiny violations of fundamental symmetries. In quantum technology itself, SQUIDs serve as tunable inductors, sensitive amplifiers, and readout elements for superconducting qubits.

3.18 Rabi Oscillations

Rabi oscillations [45, 136] are a fundamental phenomena utilized in QIP. Rabi oscillations describe the coherent oscillations of a two-level quantum system between its ground state $|g\rangle$ [(equivalent to $|0\rangle$) and excited state $|e\rangle$ (equivalent to $|1\rangle$)] under the influence of a resonant or near-resonant driving field. The qubit doesn't just "jump" to the excited state. Instead, it oscillates back and forth between $|0\rangle$ and $|1\rangle$ at a frequency determined by the strength of the drive.

Rabi oscillations are the heartbeat of a working qubit. As the hallmark of quantum coherence, these oscillations arise from the interaction of the quantum system with an oscillating external field, such as a microwave field. The resonant case ($\omega = \omega_0$) leads to periodic oscillations of the population between $|g\rangle$ and $|e\rangle$ at the Rabi frequency Ω_R. Here ω_0 is the transition frequency between $|g\rangle$ and $|e\rangle$, and ω is the driving field frequency. The driving frequency ω should be close to the transition frequency ω_0. Off-resonance driving leads to generalized Rabi frequency $\tilde{\Omega}_R = \sqrt{\Omega_R^2 + (\omega - \omega_q)^2}$. The probability of the qubit being in the $|e\rangle$ state is an oscillatory function of the pulse duration, given by

$$P_e(t) = \frac{\Omega_R^2}{\tilde{\Omega}_R^2}\sin^2\left(\frac{\tilde{\Omega}_R t}{2}\right)$$

Note that off-resonant driving will lead to faster Rabi rate with a reduced contrast. In other words, the probability of finding the qubit in the excited state will not reach 1 unless driven on-resonance.

Rabi oscillations are the basis for implementing single-qubit rotations enabling pulse calibration, superposition creation, and coherence measurements. The simplicity of the on-resonance case makes it ideal for calibration and control, and it's the starting point for nearly all quantum operations in superconducting qubit systems.

An experimentally observed Rabi oscillation is shown in Figure. 3.16. It is obtained by applying an on-resonance drive on the qubit with varying time followed by a readout pulse. The decay of the oscillations gives an indication of the T_1 and T_ϕ parameters. By applying a resonant drive for a controlled duration, the system can be rotated between $|g\rangle$ and $|e\rangle$ or to any superposition state. A half-period drive ($t = \pi/\Omega_R$) creates a π pulse, flipping the qubit state $|g\rangle \rightarrow |e\rangle$. A quarter-period drive ($t = \pi/2\Omega_R$) creates a π/2 pulse, generating a superposition: $|g\rangle \rightarrow \frac{1}{\sqrt{2}}(|g\rangle + |e\rangle)$.

The transmon Hamiltonian under a time-dependent on-resonance microwave drive is

$$H = \frac{\hbar\omega_{01}}{2}\sigma_z + \hbar\Omega\cos(\omega_q t)\sigma_x$$

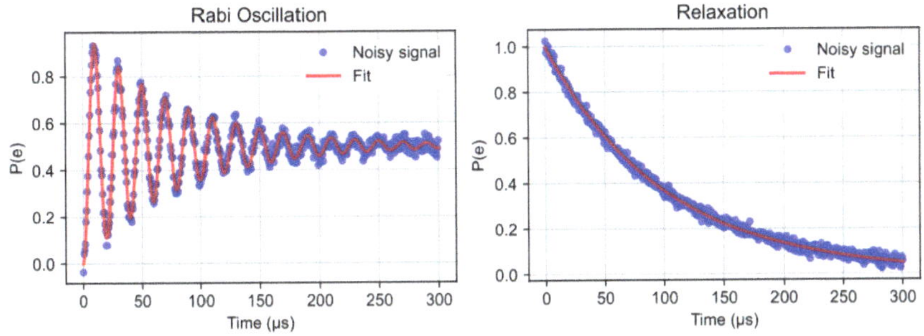

Figure 3.16 Standard pulse-sequence experiments used to extract key coherence and control parameters of a two-level quantum qubit system, plotted as the excited-state population P(e) versus time (μs). In all panels, blue markers represent experimentally sampled (or simulated noisy) data, while the red curves indicate fits to the expected theoretical models. **Left panel (Rabi oscillations [122]):** The qubit is driven resonantly, producing coherent oscillations between the ground and excited states. The gradual decay of the oscillation amplitude indicates decoherence processes, including energy relaxation and dephasing, which limit the fidelity of quantum control. The oscillation frequency corresponds to the Rabi frequency proportional to the drive strength. The time it takes to go from g to e (and reverse) is known as the π-pulse and half of that is called the $\pi/2$-pulse. **Right panel (Energy relaxation, T_1):** Following preparation in the excited state, the qubit relaxes exponentially toward the ground state. The monotonic decay of P(e) is well described by an exponential fit, from which the energy relaxation time T_1 is extracted.

ω_{01} is the qubit transition frequency.
σ_z acts on the qubit basis: it gives $+1$ for $|1\rangle$, -1 for $|0\rangle$.
Ω is the drive amplitude (Rabi frequency).
ω_q is the drive frequency, and on resonance: $\omega_q = \omega_{01}$.
σ_x flips the qubit: it couples $|0\rangle \leftrightarrow |1\rangle$.

The drive amplitude becomes the Rabi frequency because, in the rotating frame, it sets the rate at which the qubit state vector rotates on the Bloch sphere. Under an on-resonance drive, the qubit experiences an effective constant field along the x-axis of the Bloch sphere. This steady "push" causes the state vector to precess at a rate

$$\Omega_R = \Omega,$$

which is the Rabi frequency. In this picture, the amplitude of the microwave drive directly sets the angular velocity of the qubit's motion on the Bloch sphere.

The amplitude of the drive directly influences the Rabi oscillation frequency as the interaction strength depends on the *electric field strength and dipole coupling*. The applied microwave drive creates an oscillating electric field at the qubit location [92], and the qubit's electric dipole moment determines how strongly it couples to the field.

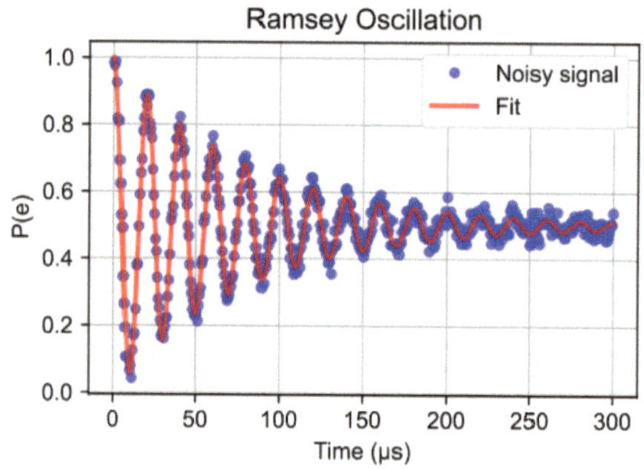

Figure 3.17 Ramsey Oscillation, T_{2R}: Two $\pi/2$ pulses separated by a variable delay create an interference pattern arising from phase evolution between the qubit states. The resulting oscillations (Ramsey fringes) decay in amplitude due to relaxation and dephasing caused by low-frequency noise (e.g., 1/f noise) resulting in fluctuations in the qubit transition frequency. The decay envelope yields the Ramsey dephasing time T_{2R}, while the oscillation frequency reflects detuning between the drive and qubit frequency.

The dipole moment of the transmon depends on several factors related to its physical and electrical properties. The electric dipole moment arises from the separation of charges across the JJ. A larger junction capacitance spreads the charge distribution, leading to a larger effective dipole moment. A sufficiently large coupling strength to external electric fields with a strong dipole moment is desirable for high controllability. While Figure 3.17a shows a single Ramsey trace used to extract coherence, Figure 3.17b aggregates such traces across a spatial grid to visualize how the qubit's frequency response varies with position. Together, they demonstrate how local conditions affect quantum coherence and control, enabling diagnostics of qubit performance.

3.19 Relaxation, Coherence, and Dephasing Times

The visibility of Rabi oscillations is limited by T_1 (energy relaxation time) and T_2 (coherence time) of the transmon. These characteristic times of a qubit must be long enough for the oscillations to be observed. Specifically, $\Omega \gg 1/T_1, 1/T_2$. These times are critical parameters that characterize qubit coherence and reliability in QC. People sometimes use "the coherence of the qubit" loosely to refer to both $T_1$and T_2.

T_1, the energy relaxation time (Figure 3.18b), measures how long a qubit remains in its excited state before decaying to its ground state (Figure 3.18a).

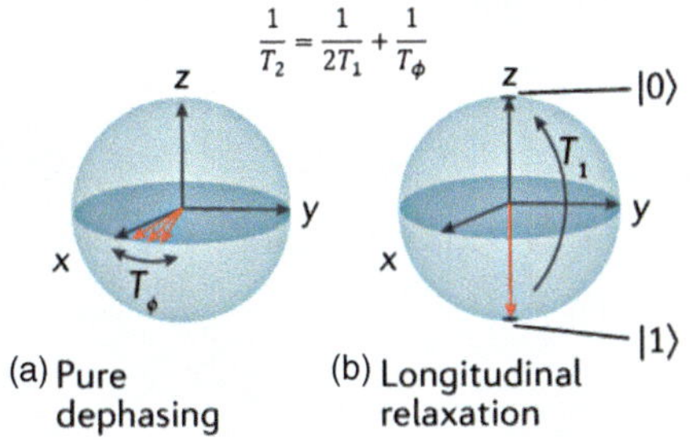

Figure 3.18 Decoherence pathways in a two-level quantum system. Schematic representation of the two primary decoherence mechanisms affecting a qubit, illustrated on the Bloch sphere. (a) Pure dephasing (T_ϕ) corresponds to random fluctuations in the qubit's transition frequency, leading to stochastic phase accumulation between the $|0\rangle$ and $|1\rangle$ states. This process preserves the population of the energy eigenstates but causes decay of the transverse coherence, represented by the shrinking and randomization of the Bloch vector within the equatorial (x–y) plane. (b) Longitudinal relaxation (T_1) describes irreversible energy loss from the excited state $|1\rangle$ to the ground state $|0\rangle$, typically mediated by coupling to environmental modes. This process drives the Bloch vector toward the south pole of the sphere and results in both population decay and loss of coherence. Together, these two channels determine the overall coherence time T_2, which combines contributions from energy relaxation and pure dephasing and sets the fundamental limit on qubit performance [46] / Springer Nature.

A long T_1 ensures the qubit can hold quantum information for extended periods, enabling more quantum operations before the state is lost.

The primary cause of low T_1 is energy dissipation to the environment, often through spontaneous emission or coupling to electromagnetic noise. As we discuss in Chapter 6, superconducting qubits a low T_1 can be caused by a variety of factors such as dielectric losses [138–140] in the substrate and energy loss due to imperfect materials in the qubit or its surrounding environment. Other loss channels are thermal noise arising from interactions with a higher-temperature environment, coupling to other modes, interaction with parasitic modes, resonators, or TLS [46].

T_2, the coherence time, measures how long a qubit maintains a coherent phase relationship in a superposition state (Figure 3.18a). A short T_2 leads to a rapid loss of quantum coherence, reducing gate fidelity and overall computation accuracy.

Dephasing occurs when the relative phase fluctuates due to interactions with the environment, even when no energy relaxation occurs. The main factors are low-frequency noise (e.g., 1/f-noise) and fluctuations in the qubit's transition frequency caused by environmental factors, such as magnetic flux, charge noise, or material imperfections. Residual thermal photons in the readout resonator, coupling between qubits or external signals, can also introduce phase noise.

The coherence time T_2 is given by the combined effects of energy relaxation (T_1) and pure dephasing (T_ϕ):

$$\frac{1}{T_2} = \frac{1}{2T_1} + \frac{1}{T_\phi}$$

If $T_\phi \to \infty$ (no pure dephasing), then $T_2 = 2T_1$. Environmental noise introduces pure dephasing such that

$$\frac{1}{T_2} > \frac{1}{2T_1} \text{ i.e. } T_2 < 2T_1.$$

Here, the factor 2 before accounts for the fact that it describes the decay of field energy rather than field amplitude, as is the case for T_2 and T_ϕ.

One also introduces the corresponding decay rates, $\Gamma_1 = 1/T_1$ $\Gamma_2 = 1/T_2 = \Gamma_1/(2 + \Gamma_\phi)$.

3.20 Implementing Gates with Transmons

3.20.1 Implementation of X and Y Rotations

Transmons are quantum systems whose states can be steered with carefully sculpted microwave pulses. When a microwave tone is applied at just the right frequency, the qubit responds like a spinning top nudged by a rhythmic push. By shaping the amplitude, phase, and duration of that push, it is possible to rotate the qubit's state around any chosen axis, tracing deliberate paths over the Bloch sphere. Quantum motion is achieved through pulse design. As the following discussion unfolds, the focus shifts from this intuitive picture of "steering" a single transmon to the richer landscape of multiqubit interactions, where precisely timed pulses and engineered couplings allow these artificial atoms to compute together.

To perform an X or Y rotation on a transmon qubit, a microwave pulse at the qubit's transition frequency (typically in the range of 4–8 GHz for superconducting qubits) is applied. This pulse couples to the qubit and induces transitions between the $|0\rangle$ and $|1\rangle$ states. The direction of the rotation (whether it's an X or Y rotation) depends on the phase of the microwave pulse relative to the qubit's reference frame. A pulse with a phase aligned with the x-axis of the qubit's Bloch sphere will result in an X rotation, meaning an anticlockwise rotation about the x-axis. Similarly, a pulse with a phase aligned with the y-axis will result in a Y rotation. Thus, a pulse with a phase of 90° relative to the x-axis will induce a Y rotation. It is common to use phase $\phi = 0$ to represent a rotation about the y-axis or in the zx plane. There are two ways to perform a reverse or clockwise rotation: either set the drive amplitude to negative or shift the phase by 180°.

The angle of rotation θ is controlled by the duration and amplitude of the microwave pulse: the amplitude of the microwave pulse determines the rate of rotation, and the length of the pulse determines how much rotation is accumulated. In other words, θ is equal to the area under the envelop of the on-resonance drive pulse. For example, a pulse that corresponds to the time needed to complete a half-cycle (π rotation) on the Bloch sphere will perform an X-gate or Y-gate (π-pulse) depending on the chosen phase.

A shorter pulse can achieve smaller rotations (e.g., π/2 rotation for a 90° rotation). By adjusting the amplitude, duration, and phase of the microwave pulse, arbitrary single-qubit rotations $R(\theta,\phi)$ can be performed where the rotation axis lies in the xy-plane, and the rotation angle is θ. To summarize:

X Gate (π-pulse about x-axis): A pulse with phase $-\pi/2$ and duration that corresponds to a π rotation will flip the qubit from $||0\rangle$ to $|1\rangle$ (or vice versa). Different notations are used to represent this gate, e.g., π_x, $R_X(\pi)$, X_π, or simply X.

Y Gate (π-pulse about y-axis): A pulse with phase $\phi = 0$ and the same duration will rotate the qubit around the y-axis by π, transforming the state in a similar manner but along the y-axis of the Bloch sphere. Different notations are used to represent this gate, e.g., π_y, $R_Y(\pi)$, Y_π, $Y(\pi)$, or simply Y.

The **Z gate** applies a phase shift of iπ to the qubit's state. In the computational basis, this operation can be represented as

$$Z = \begin{pmatrix} 1 & 0 \\ 0 & -1 \end{pmatrix}$$

This means that it leaves the $|0\rangle$ state unchanged but multiplies the $|1\rangle$ state by -1.

The Z gate for a transmon qubit is usually implemented as a virtual gate, meaning it is done by adjusting the phase in the control system rather than through a physical operation. This method is fast, precise, and avoids introducing additional noise or errors, making it highly efficient for quantum computation. Since no pulse is applied, there is no risk of introducing errors from imperfect pulses or additional decoherence.

There are different notations for other rotation angles that use θ_x, X_θ, or $X(\theta)$ and similarly for Y and Z.

$$X(\theta) = \begin{bmatrix} \cos\frac{\theta}{2} & -i\sin\frac{\theta}{2} \\ -i\sin\frac{\theta}{2} & \cos\frac{\theta}{2} \end{bmatrix}, \quad Y(\theta) = \begin{bmatrix} \cos\frac{\theta}{2} & -\sin\frac{\theta}{2} \\ \sin\frac{\theta}{2} & \cos\frac{\theta}{2} \end{bmatrix}, \quad Z(\theta) = \begin{bmatrix} 1 & 0 \\ 0 & e^{i\theta} \end{bmatrix}.$$

A generic rotation by angle θ about an axis, which makes an angle ϕ with respect to the y-axis, is given by

$$R(\phi, \theta) = \begin{bmatrix} \cos\frac{\theta}{2} & -e^{-i\phi}\sin\frac{\theta}{2} \\ e^{i\phi}\sin\frac{\theta}{2} & \cos\frac{\theta}{2} \end{bmatrix}$$

Note that this matrix will look slightly different with a different convention of ϕ.

Virtual Z gates can be easily combined with X and Y rotations to create arbitrary single-qubit rotations. For example,

$$X(\theta) = R\left(-\frac{\pi}{2}, \theta\right) = Y\left(\frac{\pi}{2}\right) \cdot Z(\theta) \cdot Y\left(-\frac{\pi}{2}\right)$$

$$Y(\theta) = R(0, \theta) = X\left(-\frac{\pi}{2}\right) \cdot Z(\theta) \cdot X\left(\frac{\pi}{2}\right)$$

As we discussed in Chapter 2, two-qubit gates (e.g., CNOT, iSWAP, CZ) are crucial for creating entanglement and enabling universal quantum computation. Coupling qubits allows them to interact in a controlled way, facilitating such gates. There are many ways to couple multiple qubits. A stronger coupling usually leads to a faster gate but often comes with some unwanted interactions known as crosstalk. A lot of research has been done to realize fast and high-fidelity gates while minimizing crosstalk. Below we discuss a few common ways of realizing coupling superconducting qubits.

3.21 Direct Capacitive Coupling

Real quantum computation only emerges when transmons can interact in a precise and programmable way. Bringing two transmons into communication means engineering a pathway—often a shared resonator, a direct capacitive link, or a tunable coupler—that allows excitations in one qubit to influence the state of the other without losing their individual identity. It is important to strike a balance: the qubits must remain well isolated most of the time to preserve coherence, yet become strongly connected when needed to generate entanglement. This interplay between isolation and interaction defines the heart of multiqubit transmon architectures, and understanding how these couplings are created, controlled, and exploited is essential for building scalable superconducting quantum processors.

A direct capacitive coupling [141] can be achieved by simply placing two qubits next to each other in close proximity (Figure 3.19). This type of coupling enables exchange interaction. Other ways to achieve similar exchange interaction are through coupling two qubits to a common resonator (known as bus resonator) or to a metallic strap. Coupling qubits via resonators is discussed in Section 4.18. One drawback of capacitive coupling is that it allows gates only between nearest neighbors. The capacitive coupling scheme has the topological advantage of requiring no additional control lines for the two-qubit coupling.

A common gate realized with this interaction between two transmons is the CR gate, in which the control qubit is driven at the resonant frequency of the target qubit. However, the frequency difference between the two transmons needs to be

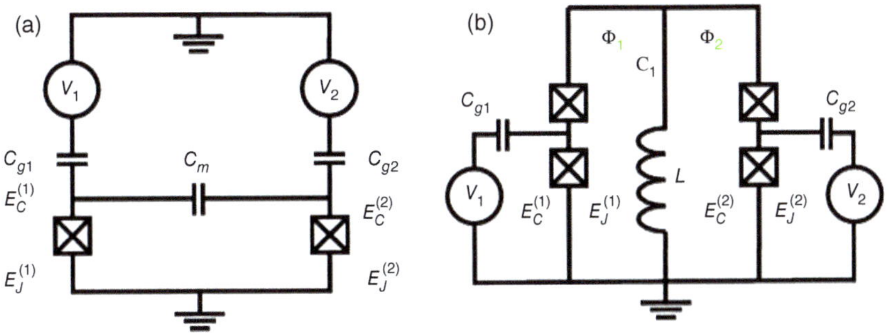

Figure 3.19 Capacitive and inductive coupling schemes for superconducting qubits. (a) Capacitive coupling. Two superconducting qubits (e.g., Cooper-pair boxes or transmons) are each defined by a Josephson junction with Josephson energy $E_J^{(1,2)}$ and charging energy $E_C^{(1,2)}$ and are biased by gate voltages V_1 and V_2 through gate capacitances C_{g1} and C_{g2}. The qubits are coupled via a mutual capacitance C_m, which mediates an electrostatic interaction between the island charges. (b) Inductive (flux-mediated) coupling. Two superconducting qubits are coupled through a shared inductive element L, such as a linear inductor or a tunable coupler. Each qubit is biased by a gate voltage (V_1, V_2) through capacitances C_{g1} and C_{g2} and includes Josephson junctions with energies $E_J^{(1,2)}$ and charging energies $E_C^{(1,2)}$. External magnetic fluxes Φ_1 and Φ_2 control the effective Josephson energies and hence the qubit frequencies and coupling [17] / Jerry Moy Chow.

smaller than the anharmonicity. Because the two transmons are coupled, the drive on the control qubit produces a small, conditional response in the target qubit that depends on the state of the control. This conditional rotation can be used to build a CNOT-like entangling gate, and it can be implemented using only microwave control lines, with no need to tune qubit frequencies dynamically.

In practice, however, the raw CR interaction is accompanied by unwanted terms, such as single-qubit $X$and Z rotations on both qubits and small ZZ interactions leading to crosstalk—one qubit's frequency depends on what its neighbor is doing. The ZZ interaction is the background of a coupled-qubit system—a conditional energy shift that is always there because two transmons are always connected. Qubits can pick up conditional phases even when no gate is applied. A modified version known as echoed cross-resonance (ECR) gate enables speeding up the gate and cancelling some effects of the crosstalk due to its constant always-on nature. The ECR gate is a pulse-sequence refinement that cancels the spurious terms. By inserting a π pulse (an "echo") on the control qubit halfway through the CR drive, the undesired single-qubit effects average to zero, while the desired two-qubit ZX interaction adds constructively. What remains is a clean, robust entangling operation that can be directly compiled into a CNOT or controlled-phase gate.

3.22 Backaction Between Coupled Qubits

Backaction between qubits is the unavoidable consequence of allowing quantum systems to interact. Whenever two qubits are coupled—whether directly through capacitors or inductors, or indirectly through a resonator or tunable coupler—the state of one qubit subtly influences the dynamics of the other. This influence is also what makes entanglement possible, but it also means that qubits are never perfectly independent. Even when no gate is being applied, the mere presence of a coupling allows information about one qubit's state to leak into another. The ZZ interaction is one of the manifestation of backaction.

Physically, backaction most often appears as a state-dependent shift of energy levels. The transition frequency of one qubit changes depending on whether a neighboring qubit is in its ground or excited state, leading to the accumulation of unwanted conditional phases during idle periods. In the case of the weak anharmonicity of transmons, higher excited states can reshape the energy of two-qubit states.

There can be several consequences of uncontrolled backaction. Qubits pick up extra phases while idling, coherence times are reduced by correlated noise, and gates become less accurate. There are several ways to control backaction. As discussed in Section 3.21, echo techniques symmetrize pulse sequences so that unwanted conditional shifts cancel out, leaving only the desired entangling terms. Tunable couplers allow interactions to be switched on during a gate and effectively turned off during idle periods. Careful frequency allocation, pulse shaping, and isolation of measurement circuitry further confine backaction to where it is intended.

3.23 Tunable Coupling Using a Coupler

A tunable coupler [43] is realized by placing a frequency tunable resonator in between two qubits. Often, a tunable transmon is used as the coupler where an external flux is needed to tune the inductance of a SQUID loop. A tunable coupler enables the interaction to be turned on and off. The resonator is detuned to a far frequency when no interaction is needed, and it is quickly brought into some resonance condition when application of a gate is needed. By biasing the coupler into a regime where its effective interaction vanishes, the qubits can be made nearly independent during idle periods, suppressing backaction. In this architecture, backaction is actively gated.

Tunable couplers can provide very strong coupling allowing fast gates. Typically, iSWAP, $\sqrt{i\text{SWAP}}$, or CZ type of gates (Chapter 2) are realized with tunable couplers. However, due to the requirements of additional control lines, this type of coupler increases device complexity with added noise channels. It also requires careful calibration to avoid crosstalks.

3.24 Implementation of a Hadamard Gate on a Transmon Qubit

Recall that the Hadamard gate (H) is a fundamental single-qubit quantum gate that creates superposition states. It maps the computational basis states as follows:

$$|0\rangle \xrightarrow{H} \frac{1}{\sqrt{2}}(|0\rangle + |1\rangle), \quad |1\rangle \xrightarrow{H} \frac{1}{\sqrt{2}}(|0\rangle - |1\rangle).$$

It forms a cornerstone for creating superposition states. The Hadamard gate is implemented using microwave pulses to rotate the qubit state on the Bloch sphere. The Hadamard gate can be decomposed into a sequence of rotations on the Bloch sphere:

$$H = R_X(\pi)R_Z(\pi/2)$$

where $R_X(\theta)$ and $R_Z(\theta)$ represent rotations about the x- and z-axes, respectively. A resonant microwave pulse is applied at the qubit's transition frequency to induce a rotation about the x-axis. The pulse amplitude and duration are calibrated to achieve a π rotation. $R_Z(\theta)$ rotations can be achieved "virtually" by adjusting the phase of subsequent microwave pulses rather than applying an actual physical pulse. This is fast and introduces no additional decoherence. The gate operation must be faster than the qubit's coherence time to minimize errors. Precise amplitude, frequency, and duration calibration of microwave pulses is essential for accurate implementation of gates.

3.25 Entangling Coupled Transmons

Let us consider the case of direct capacitive coupling for realizing entangling operations. In this case, each transmon has dedicated control lines for applying microwave pulses to manipulate the qubit state and is further coupled to its own readout resonator (Section 3.26), to measure the state of the qubit. For a CR gate [88, 89], a microwave drive at the frequency of one qubit (e.g., qubit A) is applied to the other qubit (qubit B). This interaction causes the target qubit to "flip" (change its state) depending on the state of the control qubit. The result is a special kind of interaction where the control qubit determines how the target qubit behaves. Additional control pulses are used to tune this interaction into a CNOT or CZ gate.

Refer to the control qubit as Q_c and the target qubit as Q_t. Say the two qubits start in the state $|0\rangle_c\,|0\rangle_t$. Applying a Hadamard gate to Q_c creates a superposition:

$$|Q_c\rangle = \frac{1}{\sqrt{2}}(|0\rangle + |1\rangle).$$

While Q_t is still $|0\rangle$, the state of the system becomes $\frac{1}{\sqrt{2}}(|0\rangle_c|0\rangle_t + |1\rangle_c|0\rangle_t)$.

Now the CNOT gate is implemented using the CR interaction: it acts only when $Q_c = |1\rangle$. The state becomes

$$\frac{1}{\sqrt{2}}(|0\rangle_c|0\rangle_t + |1\rangle_c|1\rangle_t)$$

This is a Bell state, the entangled state (Section 2.15). The two qubits are now correlated as follows:

$$\text{If } Q_c = |0\rangle, \quad Q_t = |0\rangle.$$
$$\text{If } Q_c = |1\rangle, \quad Q_t = |1\rangle.$$

3.26 Coupling Transmons to Resonators

Besides operating gates on qubits, and coupling qubits together, it is important to measure the state of the qubit as well as protect it from the environment [142]. In 2004, a new quantum circuit paradigm was introduced [6, 31], which is called cQED that enables efficient qubit control and readout by coupling a transmon to a linear resonator [45, 143]. The coupling of a transmon to a microwave resonator [coplanar waveguide (CPW), planar 2D circuits, or a bulk cavity 3D] opens up significant applications for quantum information [45, 85]. It forms the backbone of QC architectures as well as for scaling up to larger quantum processors.

At the heart of this architecture is the *Jaynes–Cummings Hamiltonian*, which describes a nonlinear transmon mode interacting with a harmonic cavity through a controllable light–matter coupling. In the *dispersive regime*, this interaction shifts the resonator frequency in a qubit state-dependent way, enabling quantum nondemolition measurement by probing the cavity without destroying the qubit's energy. In the *strong coupling regime*, excitations coherently swap between qubit and resonator, forming hybrid states that allow quantum information to be written into, stored in, and retrieved from the cavity. A shared resonator can also serve as a *hub* that mediates entanglement and two-qubit gates between qubits. A high-Q resonator can also act as a *quantum memory*, holding photonic states far longer than the transmon. In the following, we discuss cQED in the light of these important developments.

Resonators can be made in a 2D or 3D geometry using superconducting material like aluminum (Al), niobium (Nb), or tantalum (Ta). The resonator can

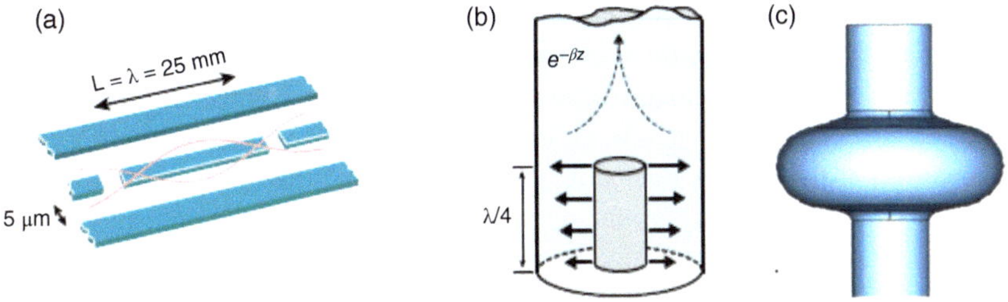

Figure 3.20 Three complementary realizations of superconducting microwave resonators used in circuit quantum electrodynamics. (a) A planar coplanar waveguide (CPW) or microstrip resonator fabricated on a chip, with $L \simeq \lambda \approx 25$ mm for a few gigahertz fundamental mode. The narrow center conductor and gaps confine the microwave electric field (red field lines) to the surface dielectric and vacuum, creating a standing-wave voltage pattern along the resonator that can be capacitively coupled to transmon qubits. [144] / American Physical Society / CC BY 4.0. (b) A three-dimensional quarter-wave ($\lambda/4$) coaxial cavity resonator, in which the electromagnetic field is confined between the outer superconducting can and an inner center post. The field amplitude decays evanescently ($e^{-\beta z}$) away toward the open end, while strong transverse electric fields near the post enable efficient coupling to qubits or antennas inserted into the cavity [144] / American Physical Society / CC BY 4.0. (c) The smooth, elliptical cell shape supports low-loss standing microwave modes with extremely long photon lifetimes, making such cavities ideal as quantum memories or as ultra-coherent buses for mediating interactions between transmon qubits. Together, these panels show the progression from lithographically defined 2D resonators to macroscopic 3D cavities, all governed by the same cQED Hamiltonian but offering different trade-offs in coupling strength, coherence, and scalability [144] / American Physical Society / CC BY 4.0.

be a superconducting LC circuit or a CPW resonator or a 3D microwave cavity. In this chapter, we discuss mainly the CPW resonator, leaving the microwave cavity approaches to Chapter 4.

The CPW resonator [143, 145] consists of a conductor of width w, thickness t, and finite length d, placed between two ground planes at a distance s (Figure 3.20a). This structure is usually deposited on a low-loss dielectric substrate whose thickness is much greater than w, d, and s. The dimensions are engineered to concentrate the field between the central conductor and the ground ($s \ll d$) to achieve a 1D mode with strong electrostatic coupling to a transmon qubit. The essential parameters that describe the characteristic of the resonator are the characteristic impedance $Z_r = \sqrt{L_0/C_0}$ and the phase velocity $v_0 = 1/\sqrt{L_0 C_0}$, where C_0 and L_0 are the capacitance to ground and the inductance per unit length, respectively. Typical values of these parameters are $Z_r = 50\Omega$ and $v_0 = 1.3 \times 10^8$ m/s. We generally want large w at the edges of the chip to make connections to external cables and small w at the center for miniaturization purposes.

One can obtain a resonator imposing zero-current or zero-voltage boundary conditions on the CPW. For a $\lambda/4$ resonator, one end of the central conductor is left open (point of zero-current), while the other end is shorted to the ground. Both ends of the central conductor are left open for a $\lambda/2$ resonator. The gaps in the center-pin transmission line serve as the mirrors of the cavity, with microwave radiation forming standing waves within the center strip line. Through careful engineering of these capacitive gaps and the length of the center strip line, the resonant frequency and quality factor of the resonators can be designed.

In the 3D quarter-wave coaxial resonator cavity (Figure 3.20b), the inner and outer conductors of a coaxial transmission line are short-circuited at the bottom of the cylindrical structure and open-circuited at a distance $\lambda/4$ away. For quantum oscillators using 3D structures, the electric field resides primarily in vacuum, which is lossless. Figure 3.20b shows a coaxial cavity design that has demonstrated high-coherence properties [146] with $Q \sim 10^8$. The quality factor is enhanced by the lack of seams if the entire structure is machined out of a single block of high-purity (4N) aluminum. This type of 3D stub cavity is a popular candidate for microwave quantum memories with storage times exceeding milliseconds. Figure 3.20c shows a superconducting radio-frequency (RF) elliptical cavity cell, illustrating a high-Q 3D resonator geometry commonly used in accelerator and quantum hardware. We will discuss 3D cavity for QC more completely in Chapter 4.

In the CPW case, for a transmission line of length d, capacitance per unit length c, and inductance per unit length l, $V(x,t)$ and $I(x,t)$ denote the voltage and current at position x along the transmission line at time t, respectively. The equations of motion are derived from Maxwell's equations, and the lumped-element model

$$\frac{\partial^2 V(x,t)}{\partial x^2} = lc\frac{\partial^2 V(x,t)}{\partial t^2}$$

is a wave equation with solutions corresponding to standing waves. The standing-wave solutions can be expressed as

$$V(x,t) = \sum_n V_n(t)\cos\left(\frac{n\pi x}{d}\right).$$

The current $I(x,t)$ is related to the spatial derivative of the voltage:

$$I(x,t) = \sum_n I_n(t)\sin\left(\frac{n\pi x}{d}\right)$$

where $n = 1, 2, 3, \ldots$ represents the mode index, and $\frac{n\pi}{d}$ is the wavenumber.

The transmon qubit is generally capacitively coupled to a resonator line as shown in Figure 3.21a, along with the corresponding lumped-element circuit model shown in Figure 3.21b.

We now turn to the development of the interaction between the resonator and the transmon.

The CPW resonator plays the role of the harmonic oscillator that traps microwave photons whose electric and magnetic fields slosh back and forth like

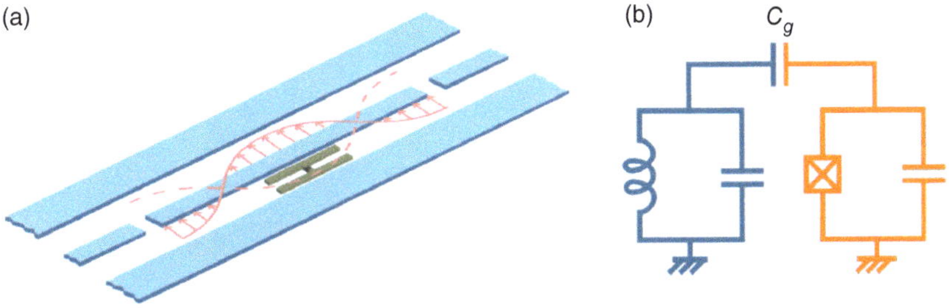

Figure 3.21 Qubit–resonator coupling. (a) Schematic representation of a transmon qubit embedded in a resonator. (b) Qubit–resonator lumped-element circuit [105] / American Physical Society.

an idealized mass on a spring. Into this evenly spaced ladder of photon states sits the transmon qubit—a weakly anharmonic artificial atom whose nonlinear JJ carves out a usable TLS from the oscillator-like spectrum. When the two are brought into proximity, their electric fields overlap, and a dipole-like interaction allows excitations to be exchanged between qubit and cavity, a process captured by the *Jaynes–Cummings Hamiltonian*. In the strong coupling regime and within the *RWA* (Section 3.28), this Hamiltonian predicts coherent oscillations of a single quantum of energy between light and matter, hybridizing them into dressed states that form the backbone of resonator-mediated entanglement and two-qubit gates. By tuning frequencies and couplings, the same framework also describes multiqubit interactions through a shared cavity, turning the resonator into a quantum bus.

When the qubit and cavity are detuned into the *dispersive regime*, energy exchange is frozen out, but their mutual influence remains: the cavity frequency shifts depending on the qubit state, enabling high-fidelity, nondestructive readout. At the same time, the resonator can play a protective role—by engineering its impedance and spectral density, one can suppress qubit decay through the *inverse Purcell effect* (Section 3.35), letting the cavity both listen to the qubit and shield it from the electromagnetic environment, a dual function that lies at the heart of cQED architectures.

As before with the simple harmonic oscillator, the transmission line resonator circuit is quantized starting from the Hamiltonian:

$$H = \sum_n \left[\frac{Q_n^2}{2C_n} + \frac{V_n^2}{2L_n} \right]$$

where the conjugate momentum for each mode n is $Q_n = C_n \frac{dV_n}{dt}$ and $C_n = \frac{cd}{2}$, $L_n = \frac{2ld}{n^2\pi^2}$.

To quantize the system, promote V_n and Q_n to operators with the commutation relation: $[\hat{V}_n, \hat{Q}_n] = i\hbar$

We define ladder operators:

$$\hat{a}_n = \sqrt{\frac{C_n\omega_n}{2\hbar}}\hat{V}_n + i\sqrt{\frac{1}{2\hbar C_n\omega_n}}\hat{Q}_n,$$

$$\hat{a}_n^\dagger = \sqrt{\frac{C_n\omega_n}{2\hbar}}\hat{V}_n - i\sqrt{\frac{1}{2\hbar C_n\omega_n}}\hat{Q}_n,$$

where $\omega_n = \frac{1}{\sqrt{L_n C_n}}$.

Rewriting the Hamiltonian in terms of ladder operators gives $\hat{H} = \sum_n \hbar\omega_n \left(\hat{a}_n^\dagger\hat{a}_n + \frac{1}{2}\right)$.

Usually, the fundamental mode, with $n = 1$, is used, and as a result, one can write the resonator Hamiltonian without the sum and subscripts and with frequency $\omega_{\mathrm{r}} = \pi/d\sqrt{lc}$, and thus $\hat{H} \approx \hbar\omega_r a^\dagger a$.

The transmon is treated as a TLS, described by the Pauli operators as discussed before (Section 3.16):

$$\hat{H}_{\text{transmon}} \approx \frac{\hbar\omega_q}{2}\hat{\sigma}_z,$$

where ω_q is the qubit frequency, and $\hat{\sigma}_z$ is the Pauli z-operator for the qubit.

$\hat{\sigma}_+$: Raising operator for the qubit transition $(|0\rangle \rightarrow |1\rangle)$, or $|g\rangle \rightarrow |e\rangle$

$\hat{\sigma}_-$: Lowering operator for the qubit transitions $(|1\rangle \rightarrow |0\rangle)$, or $|e\rangle \rightarrow |g\rangle$. As we showed earlier in the treatment of the transmon as a TLS, the transmon's charge operator n is related to the qubit operators as $(\sigma^+ + \sigma^-)$.

$$n \propto |g\rangle\langle e| + |e\rangle\langle g| = \sigma^+ + \sigma^-.$$

The charge operator induces transitions between energy levels because a qubit excitation corresponds to an oscillation of charge across the JJ. When the transmon is placed inside a resonator, the cavity's electric field couples directly to this charge degree of freedom, giving an interaction Hamiltonian of the form

$$\hat{n}\,\hat{V} \propto (\sigma^+ + \sigma^-)(a + a^\dagger),$$

which allows photons in the resonator to be absorbed or emitted as the qubit flips. This charge-mediated coupling enables a single microwave photon in the CPW to drive a qubit transition and likewise allows an excited qubit to release its energy back into the cavity. The result is a coherent exchange of excitations between the transmon and the resonator—the hallmark of the light–matter interaction captured by the Jaynes–Cummings Hamiltonian.

The total Hamiltonian of the transmon–resonator system is

$$\hat{H} = \hat{H}_{\text{transmon}} + \hat{H}_{\text{resonator}} + \hat{H}_{int}.$$

$$H = \hbar\omega_r a^\dagger a + \frac{\hbar\omega_q}{2}\sigma_z + \hat{H}_{int}.$$

The interaction is expressed as

$$\hat{H}_{int} = \hbar g(\hat{a}^\dagger + \hat{a})(\hat{\sigma}_+ + \hat{\sigma}_-).$$

Here g is the coupling strength.

On expansion of the interaction term, we keep the terms $a^\dagger\sigma^-$ for which the oscillator loses one quantum of energy, and the qubit gains energy (a transition from $|g\rangle$ to $|e\rangle$), and the term $a\sigma^+$ for which the oscillator gains one quantum of energy, and the qubit loses energy (a transition from $|e\rangle$ to $|g\rangle$. These terms are known as energy-conserving terms. The other terms $a^\dagger\sigma^+$ and $a\sigma^-$ represent non-resonant processes that are negligible in the RWA (Section 3.28). This simplifies the interaction part of the Hamiltonian [105, 147–150] to

$$H_{\text{int}} = \hbar g(\sigma^+ a + \sigma^- a^\dagger).$$

3.27 Jaynes–Cummings Hamiltonian

The full Jaynes–Cummings Hamiltonian is

$$H = \hbar\omega_r a^\dagger a + \frac{\hbar\omega_q}{2}\sigma_z + \hbar g(\sigma^+ a + \sigma^- a^\dagger)$$

The Jaynes–Cummings Hamiltonian provides a framework for studying qubit–resonator interactions. It is a solvable model that accurately describes a TLS in interaction with a single mode of the electromagnetic field. Only the first two levels of the transom play an important role. It is often the case, however, that quantitative agreement between theoretical predictions and experiments is obtained only when accounting for higher transmon energy levels and the multimode nature of the field.

The eigenstates of the Jaynes–Cummings Hamiltonian form a ladder of hybridized states $|n, g\rangle$, $|n-1, e\rangle$, where n is photon number in the resonator, g is ground state of the qubit, and e is the excited state of the qubit. The energy levels are given by the general expression:

$$E_n^{\pm} = n\hbar\omega_r \pm \frac{\hbar}{2}\sqrt{\Delta^2 + 4g^2 n}$$

where $\Delta = \omega_q - \omega_r$. Figure 3.22 shows the ladder of energy levels illustrating how bare atomic states $|g\rangle, |\,e\rangle$ and cavity photon number states $|n\rangle$ reorganize into Jaynes–Cummings dressed states when the systems are coupled.

The Jaynes–Cummings Hamiltonian turns a superconducting circuit into a quantum optical system. It is the mathematical statement that the qubit and the cavity are no longer separate actors but partners in a coherent interchange, exchanging energy and information in quantized steps. A photon in the cavity can be absorbed by the qubit; an excited qubit can emit a photon into the cavity. This reversible exchange is the engine behind *vacuum Rabi oscillations*, dressed

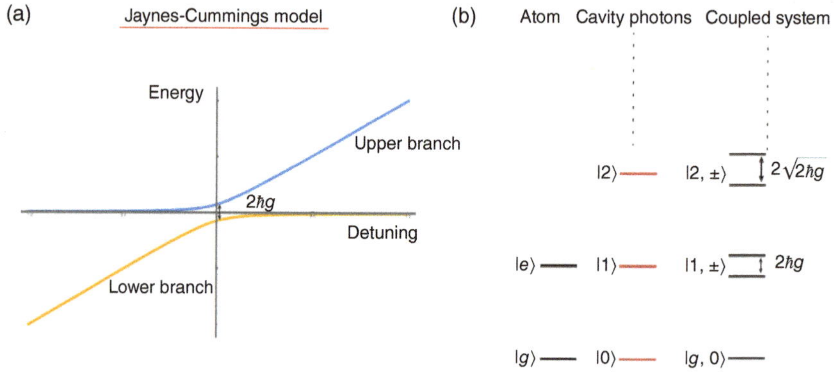

Figure 3.22 Jaynes–Cummings energy spectrum: from bare states to dressed light–matter eigenstates. (a) Eigenenergies of a transmon qubit coupled to a single microwave cavity mode plotted versus detuning $\Delta = \omega_q - \omega_r$. In the absence of coupling, the qubit excitation $|e, 0\rangle$ and the one-photon cavity state $|g, 1\rangle$ would cross at $\Delta = 0$. The light–matter interaction hybridizes these states, producing an avoided crossing and two "dressed" polariton branches (upper and lower). At exact resonance, the splitting between the branches is $2\hbar g$, the vacuum Rabi splitting, which directly measures the coherent coupling rate g. For large detuning, the branches asymptotically recover the bare qubit- and photon-like dispersions, illustrating the crossover from strongly mixed light–matter states to weakly perturbed uncoupled modes. (b) Ladder of energy levels showing how bare atomic states $|g\rangle, |e\rangle$ and cavity photon number states $|n\rangle$ reorganize into Jaynes–Cummings dressed states when the systems are coupled. The uncoupled basis consists of tensor-product states $|g, n\rangle$ and $|e, n\rangle$. Within each excitation manifold $n+1$, the degenerate pair $|e, n\rangle$ and $|g, n+1\rangle$ mix to form symmetric and antisymmetric superpositions $|n, +\rangle$ and $|n, -\rangle$. These dressed states are split by $2\hbar g\sqrt{n+1}$, producing the characteristic $\sqrt{n}$ scaling of the Jaynes–Cummings spectrum (e.g., $2\hbar g$ for the first doublet, $2\sqrt{2}\hbar g$ for the second). The ground state $|g, 0\rangle$ remains uncoupled, while higher rungs form a nonlinear ladder of eigenstates that underlie vacuum Rabi oscillations, photon number-dependent shifts, and resonator-mediated qubit entanglement in cQED, (adapted from Ref. [105]).

states, and the entire language of strong coupling. From this, Hamiltonian flows nearly everything that makes cQED powerful: dispersive readout, photon number-dependent frequency shifts, qubit-mediated interactions between resonators, resonator-mediated interactions between qubits, and the ability to engineer gates by shaping the exchange of excitations. The Jaynes–Cummings model is the conceptual backbone that explains why superconducting qubits can be measured, entangled, and controlled with precision.

3.28 The Rotating-Wave Approximation

This approximation [151, 152], which we just used in the previous section, is a crucial concept in cQED to simplify the interactions between qubits and electromagnetic fields. The RWA simplifies the Hamiltonian by eliminating the nonresonant terms, which oscillate at high frequencies and average out over time, leaving only the resonant interaction terms. The underlying idea is that in a system where the qubit and cavity are close to resonance (*i.e.*, $\omega_q \approx \omega_r$), the terms $\hat{\sigma}_+\hat{a}$ and $\hat{\sigma}_-\hat{a}$ oscillate slowly, while the terms $\hat{\sigma}_+\hat{a}^\dagger$ and $\hat{\sigma}_-\hat{a}$ oscillate rapidly at frequencies around $2\omega_q$ or $2\omega_c$ The RWA makes the problem analytically tractable by reducing the complexity of the interaction Hamiltonian.

In most practical scenarios, the qubit and cavity are designed to be near resonance, so the nonresonant terms indeed contribute very little to the dynamics of the system. The RWA respects energy conservation in the sense that it only includes interactions where the resonator emits a photon and excites the qubit, or absorbs a photon and de-excites the qubit. The nonresonant terms would imply processes that violate this conservation, such as the simultaneous creation or annihilation of a photon and a qubit excitation. Many quantum gates in cQED are designed under the RWA.

The RWA is not always valid. Its accuracy depends on the relative magnitudes of the qubit-cavity detuning ($\Delta = \omega_q - \omega_c$) and the coupling strength g. When the detuning becomes large ($g \sim \Delta$), or when the system operates at very high frequencies, the neglected terms can start to have a noticeable effect, and the full Hamiltonian needs to be considered.

3.29 Strong Coupling Regime

On-resonance ($\omega_q = \omega_r$)

The strong coupling regime is reached when the interaction rate of the atom and a single photon (g) is large. The qubit and resonator strongly exchange energy, leading to *vacuum Rabi oscillations*. The energy exchange is coherent and fast. In the strong coupling regime, the coupling strength g overwhelms their decay rates, so $g \gg \kappa, \gamma$, where κ is the oscillator decay rate and γ is the qubit dephasing rate. These oscillations are not the forced ones by a microwave drive as for a single transmon qubit undergoing ordinary Rabi oscillations; they are the natural "motion" of a hybrid light–matter system. The oscillation frequency is set by the coupling strength g, not by any applied power, and halfway through each cycle the qubit and the cavity are entangled, sharing a single excitation in a superposition of $|e, 0\rangle$ and $|g, 1\rangle$.

The cavity is no longer a passive container; it becomes part of the qubit's identity. The transmon's excitation is shared with the electromagnetic mode, and the cavity's photons inherit a trace of the qubit's nonlinearity. Everything is hybridized and intertwined. The coupled system forms hybridized eigenstates (dressed states) with energies split by 2g. The dressed states are linear combinations of the original quantum states and the electromagnetic field states.

In this resonant regime, the eigenstates of the system are no longer pure qubit or photon states but superpositions. For a single photon, the eigenstates are

$$|+\rangle = \frac{1}{\sqrt{2}}(|e,0\rangle + |g,1\rangle), \quad |-\rangle = \frac{1}{\sqrt{2}}(|e,0\rangle - |g,1\rangle).$$

The energy splitting is called the *vacuum Rabi splitting*. The split peaks correspond to the transitions between the hybridized states of the qubit–resonator system.

This is an essential signature of strong coupling between the qubit and the cavity. Observing the splitting confirms that a qubit and a resonator interact coherently. It enables the study of system properties, such as anharmonicity of the qubit and the resonator quality factor by observing the dependence of splitting on frequency and driving strength. Measuring the linewidths of the split peaks reveals information about qubit and resonator decay rates.

Rabi splitting enables the study of anharmonicity of the qubit and the resonator quality factor by observing the dependence of splitting on frequency and driving strength. The splitting between these states corresponds to the Rabi frequency $\Omega_R = 2g$.

When the qubit and cavity are resonant ($\omega_q = \omega_c$), the energy levels split into a ladder. The energy levels of these dressed states in this special case ($\Delta = 0$) are from the general eigenstates above:

$$E_{n,\pm} = n\hbar\omega_c \pm \hbar g\sqrt{n}$$

The strong coupling regime [38, 39, 43, 45] facilitates entanglement and coherent energy exchange between the qubits and the resonators. This interaction is foundational for quantum gates, quantum memory, and a quantum bus. The strong coupling between the transmon and the resonator ensures that operations such as controlled gates can be performed fast and efficiently. When multiple qubits are involved, the resonator can allow for fast entanglement generation between qubits. A higher coupling strength facilitates faster operations, though it must be balanced to avoid decoherence.

3.29.1 Photon Blockade

In this quantum phenomenon, a single photon effectively blocks the transmission of additional photons through a system when the qubit and cavity are strongly coupled, because the effect relies on the anharmonic energy structure of the Jaynes–Cummings ladder, resolvable in the strong-coupling regime. It occurs when the presence of one photon shifts the energy levels of a system, making it energetically unfavorable for subsequent photons to enter the system.

Here's the basic mechanism: when a photon enters the cavity, it causes the energy levels of the combined system (qubit + cavity) to shift due to the strong nonlinearity of the interaction. The energy required a second photon to enter the cavity changes due to the presence of the first photon. This shift can detune the system from the resonance condition. Because the system is no longer resonant for the second photon, it is effectively blocked from entering the cavity. The result is that the system only allows one photon to pass through or be absorbed at a time, creating a single-photon stream. The interaction between the photon and the qubit must be strong enough to shift the energy levels significantly. This is often achieved in systems operating in the strong coupling regime.

Photon blockade gives a way to make light behave as if it were made of individual, well-spaced particles rather than a continuous wave (CW). This makes it possible to build highly reliable single-photon sources, the kind needed for quantum communication and photonic QC, where information is carried one photon at a time.

3.30 Dispersive Regime

When the qubit and cavity are detuned by an amount $\Delta = \omega_q - \omega_r$ that is large compared to the coupling strength g, the two systems can no longer exchange real excitations, and the dynamics enter the dispersive regime [43, 45, 153, 154]. The name *dispersive regime* comes from the language of wave physics. A medium is called dispersive when waves do not exchange energy with it but instead experience frequency-dependent phase shifts. In cQED, the qubit plays the role of such a medium for the cavity field, and vice versa.

In this regime, a qubit and a resonator are kept far apart in frequency, but the qubit is still connected quantum mechanically to the cavity field, and it is constantly "trying" to interact with it. The qubit is essentially an electric dipole embedded in the resonator's electromagnetic mode. For extremely brief moments allowed by quantum uncertainty, the qubit can partially excite the cavity field and then immediately reabsorb that excitation. No real photon is created, but the electromagnetic field inside the resonator is momentarily distorted. The qubit partially borrows the energy needed to excite the cavity field, then immediately gives it back. These brief, forbidden excursions never produce a real photon, but they leave a trace: the field configuration is momentarily distorted before snapping back. Recall that a cavity mode is not an empty container, even when it holds no photons. It is filled with fluctuating electric and magnetic fields—the zero-point field of the electromagnetic mode. The qubits' electric dipole, i.e., its charge distribution responds to those fluctuations.

These fleeting, unobservable distortions are called *virtual photons*. If a second qubit is coupled to the same resonator, it interacts with this modified electromagnetic environment. The energy levels of the second qubit shift in a way that depends on what the first qubit is doing. Both qubits are responding to the same altered vacuum field of the resonator.

In this limit, the interaction term of the Jaynes–Cummings Hamiltonian produces a qubit-state-dependent shift of the resonator frequency. The cavity resonance is displaced by $\pm\chi$, where $\chi = g^2/\Delta$, depending on whether the qubit is in $|g\rangle$ or $|e\rangle$. Equivalently, the resonator frequency becomes $\omega_r - \chi$ when the qubit is in $|e\rangle$ and $\omega_r + \chi$ when it is in $|g\rangle$. At the same time, the qubit's own transition frequency acquires a photon number-dependent shift—the AC Stark shift. This is analogous to how an external electric field perturbs atomic energy levels. The interaction modifies the qubit's phase evolution without changing its population. The large detuning also suppresses unwanted energy exchange with the environment, since the resonator acts as a frequency filter that protects the qubit from noise at its transition frequency.

Starting from the Jaynes–Cummings Hamiltonian and applying second-order perturbation theory in the limit $|\Delta| \gg g$, one obtains the dispersive Hamiltonian

$$H_{\text{disp}} = \hbar(\omega_r + \chi\sigma_z)a^\dagger a + \frac{\hbar}{2}(\omega_q + \chi)\sigma_z.$$

$\chi = g^2/\Delta$ is the dispersive shift. This Hamiltonian is diagonal in the qubit and photon number basis because real photon exchange is suppressed. The coefficient of the term $a^\dagger a$ determines the oscillator frequency. Since the eigenvalue of σ_z is ±1, cavity frequency takes one of two values:

$$\omega_r^{(g)} = \omega_r + \chi, \quad \omega_r^{(e)} = \omega_r - \chi,$$

depending on whether the qubit is in $|g\rangle$ or $|e\rangle$. Thus, the cavity exhibits two distinct resonance responses separated by 2χ. This is the basis for the *dispersive readout* method [43, 45], where the qubit state is inferred by measuring the shift in the cavity's resonance frequency (Figure 3.23 [137]).

At the same time, the qubit's own transition frequency acquires a photon number-dependent shift—the AC Stark shift—analogous to how an external electric field perturbs atomic energy levels. This modifies the qubit's phase evolution without changing its population. The large detuning also protects the qubit from environmental noise, since the resonator acts as a frequency filter that suppresses unwanted energy exchange at the qubit frequency.

Dispersive readout is a *QND* measurement of σ_z. A microwave probe is sent into the resonator, and the phase or amplitude of the transmitted or reflected signal is measured. Because the dispersive Hamiltonian commutes with σ_z, the measurement does not induce transitions between $|g\rangle$ and $|e\rangle$. The qubit's energy eigenstates remain intact, but any superposition collapses into one of them, fully consistent with quantum measurement theory. If the qubit is found in $|g\rangle$ on the first measurement, all subsequent measurements will return $|g\rangle$ as well.

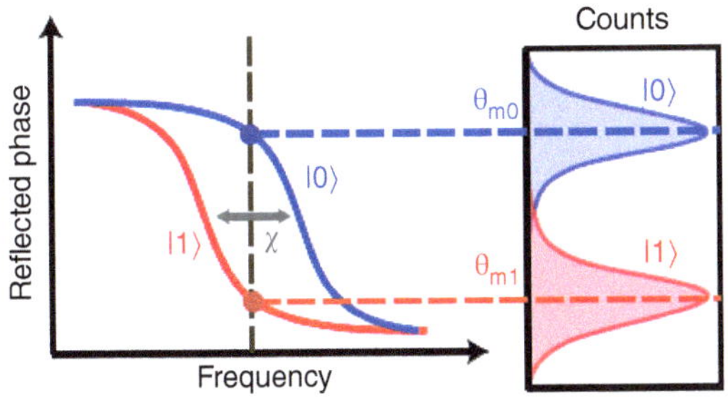

Figure 3.23 Dispersive qubit readout using a superconducting readout cavity. Due to the qubit–cavity dispersive interaction, the cavity's resonance frequency exhibits two distinct values separated by χ, corresponding to the qubit being in the $|0\rangle$ (blue) or $|1\rangle$ (red) state. Consequently, the reflected phase response of the cavity depends on the qubit state, resulting in two distinguishable cavity outputs. This property is used for qubit state measurement, and the cavity used for this purpose is referred to as the readout cavity or readout resonator.

High-fidelity readout requires that the dispersive shift be comparable to or larger than the resonator linewidth, $\chi \gtrsim \kappa$. Under this condition, the two cavity responses are well separated in frequency space, enabling fast and accurate discrimination of the qubit state. In practice, the outgoing microwave signal is extremely weak, so near-quantum-limited amplification is required. This is provided by superconducting parametric amplifiers such as Josephson parametric amplifiers (JPAs), Josephson parametric converters (JPCs), and traveling-wave parametric amplifiers (TWPAs), which operate at millikelvin temperatures (See Sections 3.36 and 3.37).

Shared resonators also entail backaction (Section 3.22): virtual photons created by one qubit polarize the resonator and feed that information back to the other qubit. Even measurement can induce backaction, as photons injected into a readout cavity Stark-shift nearby qubits and introduce dephasing, sometimes disturbing qubits that were not being measured at all.

In a coupled transmon–cavity system, the qubit and the resonator do not exchange excitations resonantly under normal conditions because they are far detuned. To deliberately make them exchange an excitation—i.e., to *swap* quantum information—you drive the system at a frequency that matches a sideband transition. The blue sideband corresponds to a process in which both the qubit and the cavity gain or lose an excitation together.

$$|g, n\rangle \longleftrightarrow |e, n+1\rangle,$$

where the qubit is excited *and* a photon is added to the cavity. The required drive frequency is approximately

$$\omega_{\text{drive}} \approx \omega_q + \omega_c$$

It is called "blue" because it lies *above* the bare qubit frequency. When this blue-sideband interaction is driven coherently for a precise duration, it produces Rabi oscillations between the two subsystems. At a specific interaction time—corresponding to half a Rabi period—the excitation initially in the qubit is transferred into the cavity (or vice versa). The operation is exactly a SWAP of quantum information between the qubit and the resonator. In contrast, a red-sideband SWAP (rSWAP) uses the complementary transition. Sideband transitions are used to move a photon into or out of a cavity, or to prepare specific Fock states.

3.31 Entangling Transmons Coupled via Resonators

Entangling transmons through a resonator can be done in both the strong coupling and dispersive regimes with different trade-offs. So choosing the right regime is an important architectural decision in cQED.

In the dispersive regime, the qubits and the resonator are detuned so that real photons are never exchanged. Each qubit slightly polarizes the cavity, and that polarization slightly shifts the energy of the other qubit. Even when it cannot exchange real photons with the cavity, its dipole moment distorts the cavity field slightly. When two transmons share a cavity in this regime, it produces an effective qubit–qubit interaction that can be turned into entangling gates. Crucially, the resonator remains nearly empty: it does not store or radiate real photons during the gate. This makes the interaction slower than in the resonant case, but cleaner and more stable, because the cavity no longer acts as a noisy channel to the outside world. It also allows large arrays of transmons to be entangled.

Imagine two qubits (Q_1 and Q_2) that are coupled to a shared resonator. The resonator acts as a "middleman," allowing the qubits to interact indirectly. When a qubit is coupled to the resonator, it can excite the resonator by "donating" a photon, transitioning itself from $|1\rangle$ to $|0\rangle$. Conversely, the resonator can "give" a photon back to the qubit, causing the qubit to transition from $|0\rangle$ to $|1\rangle$.

In the strong coupling regime, start with Q_1 in the $|0\rangle$ state and Q_2 in the $|0\rangle$ state. The resonator is initially empty (no photons, $n = 0$). Apply a single-qubit rotation to Q_1. This creates a superposition:

$$Q_1 = \alpha|0\rangle + \beta|1\rangle$$

Now the system is in the state:

$$\left(\alpha|0\rangle_1 + \beta|1\rangle_1\right) |0\rangle_2|0\rangle_{res} = \alpha|0\rangle_1|0\rangle_2|0\rangle_{\text{res}} + \beta|1\rangle_1|0\rangle_2|0\rangle_{\text{res}}.$$

This is still a product state. The coupling between Q_1 and the resonator allows Q_1 to transfer its excitation (if it is in the $|1\rangle$ state) to the resonator. After a short time, the state becomes

$$\alpha|0\rangle_1|0\rangle_2|0\rangle_{\text{res}} + \beta|0\rangle_1|0\rangle_2|1\rangle_{\text{res}}$$

Next, couple Q_2 to the resonator. This allows the resonator to transfer its photon (and the quantum information) to Q_2.

$$\alpha|0\rangle_1|0\rangle_2|0\rangle_{\text{res}} + \beta|0\rangle_1|1\rangle_2|0\rangle_{\text{res}} = |0\rangle_1\left(\alpha|0\rangle_2 + \beta|1\rangle_2\right)|0\rangle_2|0\rangle_{\text{res}}$$

At this point, all the excitation has been transferred from Q_1 to Q_2 via the resonator. This can be thought as a full swap operation between the two qubits mediated by the resonator.

It is also possible to apply fractional swap operations. For example, if one starts from a single excitation in the first cavity and no excitation on the other, i.e., $|1\rangle_1|0\rangle_2$, and performs a halfway swap (also known as $\sqrt{SWAP}$), the excitation gets shared between the two qubits creating an entangled state:

$$|\psi\rangle = \frac{1}{\sqrt{2}}(|01\rangle + |10\rangle)$$

which is also the Bell state [6].

In the strong coupling regime, Q_1 deposits a *real* photon into the resonator, and Q_2 later absorbs it. Entanglement comes from real excitation exchange.

The process allows energy and quantum information to be exchanged between a qubit and the resonator. This regime gives fast, strong interactions, but it also exposes the qubits to everything the resonator is exposed to: loss or thermal photons. If the cavity is not well isolated, coherence leaks away with the photons. Strong coupling is powerful but fragile, and it is typically used only for special tasks.

In the dispersive regime, prepare Q_1 in a superposition:

$$|\psi\rangle = (\alpha|0\rangle_1 + \beta|1\rangle_1)|0\rangle_2|0\rangle_{\text{res}}.$$

(Still a product state.) But now, because we are in the dispersive regime, this superposition shifts the resonator frequency by an amount $\pm\chi_1$, depending on Q_1's state.

Use a sideband drive to couple Q_1 to the resonator virtually. Apply a drive to Q_1 at the red sideband frequency:

$$\omega_d = \omega_{q1} - \omega_r$$

This activates the transition: $|1, 0\rangle \leftrightarrow |0, 1\rangle$, but only virtually—the resonator is never populated with a real photon. After a calibrated time, the state becomes

$$\alpha|0\rangle_1|0\rangle_2|0\rangle_{\text{res}} + \beta|0\rangle_1|0\rangle_2|0\rangle_{\text{res}}$$

but with a phase and amplitude imprint on the resonator's *virtual* excitation cloud. Even though the resonator never contains a real photon, the qubit–resonator interaction *still* creates a qubit-state-dependent distortion of the resonator's electromagnetic field. The qubit modifies the resonator's vacuum state in a qubit-dependent way, and this modified virtual state is what the second qubit interacts with, enabling entanglement without real photons. This is the dispersive analog of "putting the excitation into the resonator," but nothing real is transferred.

Next apply a complementary sideband drive to Q_2. Drive Q_2 at its red or blue sideband frequency. Because the resonator's virtual excitation depends on Q_1's state, Q_2's evolution becomes conditional on Q_1. After the appropriate interaction time, the joint state becomes

$$\alpha|0\rangle_1|0\rangle_2|0\rangle_{\text{res}} + \beta|1\rangle_1|1\rangle_2|0\rangle_{\text{res}}$$

This is an entangled state. The resonator ends in $|0\rangle$ because all excitations were virtual.

In the dispersive regime, Q_1 imprints information onto the resonator's virtual excitations, and Q_2 interacts with that virtual cloud. Entanglement comes from drive-activated transitions between JC rungs and state-dependent dispersive shifts, not real photons.

In the strong coupling regime, the qubits are tuned into resonance with the resonator one at a time, allowing real photons to shuttle back and forth between them. This produces fast gates: one qubit deposits an excitation into the resonator, the other retrieves it, and entanglement emerges from this direct exchange of energy. The simplicity and speed of these resonant iSWAP-style interactions makes them attractive in early cQED experiments. But there is a price. Bringing qubits into resonance requires active frequency tuning, which introduces flux noise, increases Purcell decay, and becomes harder to scale as more qubits are added. In a multiqubit device, avoiding unwanted resonances and crosstalk becomes an engineering challenge.

In the dispersive regime, entangling gates arise from these subtle dispersive effects—through drive-activated processes such as sideband transitions. Although these gates are typically slower and require more careful pulse engineering, they offer practical advantages. Because the qubits never need to be tuned into resonance, they can remain at fixed frequencies where coherence is optimized. The resonator stays unpopulated, reducing decoherence pathways, and the entire architecture becomes more scalable: many qubits can couple to the same resonator without frequency collisions or wiring complexity. For these reasons, dispersive regime entanglement has become more popular for modern superconducting qubit processors.

We focused above on how a resonator can mediate entanglement between a pair of transmons. The same underlying mechanism extends naturally to larger systems. Once a resonator can coherently share information with one

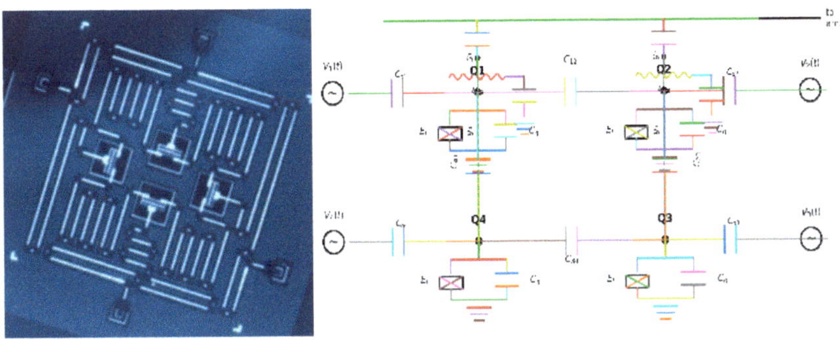

Figure 3.24 Superconducting transmon processor with resonator-mediated couplings (IBM-style architecture). The image detailed by the accompanying circuit diagram on the right shows a lithographically fabricated superconducting quantum processor in which multiple transmon qubits (the T-shaped capacitor pads) are arranged on a two-dimensional grid and interconnected by coplanar-waveguide microwave resonators. Each transmon consists of a Josephson junction shunted by large capacitive "antenna" pads that define its qubit frequency and electric dipole moment. The long, straight microwave lines running between neighboring transmons are bus resonators: high-Q CPW cavities that support quantized microwave modes and provide a coherent channel through which qubits interact. In this architecture, the qubits are detuned from the resonators and from one another so that they operate in the dispersive regime. Additional CPW lines at the chip edges serve as drive and readout ports, allowing individual qubits to be addressed and measured without disturbing their neighbors. This layout exemplifies the scalable "circuit QED lattice" used in IBM's superconducting quantum processors (adapted from Ref. [155]).

qubit, it can just as effectively serve as a communication channel among many (Figure 3.24), enabling interactions and multiqubit operations without requiring direct qubit-to-qubit wiring.

Thus, resonators may be used to act as mediators to couple distant qubits for multiqubit operations. Coupling multiple transmon qubits to a single resonator or a network of resonators allows for a more scalable QC architecture. The CPW acts as a communication channel between qubits, allowing them to be entangled or interact without requiring direct coupling between the qubits themselves. This reduces the complexity of wiring.

3.32 Engineering Qubit Coherence Through Resonator Coupling

Coupling a transmon qubit to a resonator places the qubit inside an engineered electromagnetic environment that can be tuned to protect it from unwanted decay channels. The resonator acts as a controlled interface between the qubit and the outside world: it enables high-fidelity readout and microwave control while filtering out environmental noise at frequencies far from its resonance. At the same time, this coupling opens a new pathway for energy loss, because the qubit can radiate into the resonator and then leak out through the resonator's ports—a process known as the *Purcell effect* (Section 3.35).

By adjusting the qubit–resonator detuning, the coupling strength, and the resonator's quality factor, this Purcell-induced decay can be strongly suppressed. A resonator that is sufficiently detuned and has a narrow linewidth makes it difficult for the qubit to emit energy into it, keeping the resonator nearly empty and minimizing energy leakage. In this way, the resonator becomes not just a readout tool but a tunable electromagnetic shield that can be optimized to preserve the qubit's relaxation time and overall coherence.

3.33 Resonators as Quantum Memories in cQED

The coupling between a transmon qubit and a high-quality resonator can be used as a *quantum memory* by leveraging the resonator's ability to store quantum information encoded in its electromagnetic modes. The transmon qubit's information (its quantum state) can be transferred to the resonator's electromagnetic field mode, where it can be stored for a longer duration due to a higher quality factor. A transmon acts as a control unit to write and read information into and from the resonator.

Whereas the transmon offers nonlinearity and control, its coherence time is fundamentally limited by its relatively low-quality factor due to dielectric loss, surface defects, and other factors (Chapter 6). A superconducting resonator, by contrast, can be engineered with extremely low loss—either as a 3D cavity or a high-Q planar resonator—allowing it to store quantum states for times far exceeding those of the transmon. By exploiting the coherent interaction between the two systems, the quantum state of the transmon can be mapped onto the resonator's electromagnetic mode, held there with minimal degradation, and later retrieved on demand. In this architecture, the transmon functions as the "processor" or control unit, while the resonator serves as a long-lived quantum memory.

When the transmon and resonator are appropriately coupled, the system supports coherent state-transfer operations—typically implemented through sideband transitions, parametric drives, or controlled swaps—that move excitations between the qubit and the resonator without loss of quantum information.

Beyond acting as passive memories, resonators can also serve as logical qubits when quantum information is encoded directly in their bosonic degrees of freedom (Chapter 4). In bosonic error-correcting codes—such as cat codes, binomial codes, and GKP codes—quantum information is represented by superpositions of resonator states. These encodings exploit the large Hilbert space of a harmonic oscillator and the long coherence times of high-Q cavities to achieve error protection that is difficult to realize with transmons alone. In such schemes, the transmon plays the role of a nonlinear ancilla that prepares, manipulates, and measures the encoded resonator states, while the resonator itself becomes the protected quantum memory.

3.34 Microwave Components

We now describe some of the necessary microwave components (filters, amplifiers, etc.) that allow the operation of the quantum system [161]. The final section of the chapter puts all these components together to show a complete setup for operating a quantum computer.

3.35 Purcell Filter

A Purcell filter enhances qubit performance by reducing spontaneous emission of the qubit while maintaining a strong signal coupling to the readout resonator for efficient qubit readout. It is named after the *Purcell effect* [162–164], which describes the spontaneous emission due to the presence of a resonant cavity. Qubits coupled to readout resonators can lose energy via spontaneous emission into the readout transmission line. This decay is exacerbated by the resonator's coupling to the external environment, especially when the coupling is strong to facilitate fast readout. While suppressing energy relaxation, the Purcell filter ensures that the resonator remains efficiently coupled to the measurement chain for readout purposes.

The Purcell effect was theoretically proposed by Edward Mills Purcell in 1946. Purcell noted that the interaction of an emitter with its electromagnetic environment could significantly alter its radiative properties. Specifically, he calculated that spontaneous emission could be enhanced if the emitter's resonance frequency matched that of a confined electromagnetic mode in a cavity. Similarly, a qubit coupled to a cavity or transmission line may experience enhanced spontaneous emission due to the Purcell effect. This enhancement can significantly reduce the qubit's lifetime (T_1). On the other hand, if the cavity suppresses modes at that frequency, emission is inhibited.

The Purcell filter [165–167] works by introducing a frequency-selective network between the qubit–resonator system and the transmission line. This network modifies the electromagnetic environment seen by the qubit, suppressing energy relaxation at the qubit frequency while allowing efficient coupling for readout at the resonator frequency. The filter ensures strong coupling at the resonator frequency, enabling fast and accurate qubit-state measurement.

The Purcell filter can be designed to introduce a notch or bandstop at the qubit frequency while maintaining high transmission at the resonator frequency. This is achieved using resonant structures, such as LC circuits or transmission line stubs, that create a sharp attenuation band at the qubit frequency. For example, a shorted $\lambda/4$ transmission line stub creates a bandstop at the qubit frequency. Or lumped-element inductors and capacitors form resonant circuits tuned to suppress qubit frequencies. The filter is introduced between the resonator and the transmission line.

Figure 3.25 shows the stepped impedance Purcell filter (SIPF), which consists of alternating sections of high and low impedance CPW transmission lines. The filter has a wide stopband and can provide protection for a large range of qubit frequencies. The qubit T_1 was improved by up to a factor of 9 compared to the predicted values for an unfiltered device [165, 166].

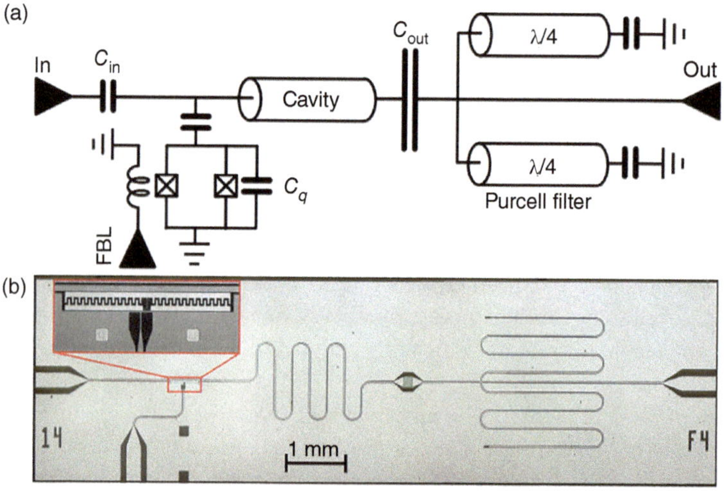

Figure 3.25 Design and realization of a Purcell-filtered cavity–transmon system. (a) Lumped-element circuit model illustrating the implementation of a Purcell filter for suppressing qubit radiative decay into the measurement line. The microwave input signal is coupled through the input capacitor C_{in} to a three-dimensional or planar cavity, which, in turn, is coupled to the output transmission line via the output capacitor C_{out}. A transmon qubit, characterized by its shunt capacitance C_q and Josephson junction, is dispersively coupled to the cavity. (b) Optical micrograph of the fabricated device showing the planar realization of the cavity, Purcell filter structures, and transmission lines. The meandered and $\lambda/4$ resonator sections implement the designed filter response. The inset highlights the transmon qubit region, including the capacitor pads and Josephson junctions, which are capacitively coupled to the cavity mode. The scale bar indicates 1 mm. *Source:* Matthew David Reed / https://rsl.yale.edu/sites/default/files/2024-08/2013-RSL-Thesis-Matthew-Reed.pdf / last accessed on Apr 25, 2026.

3.36 Amplifiers

Measurement signals leaving a quantum device are very weak and overwhelmed by thermal noise if the signals are not amplified. Therefore, it is critical to amplify the measurement signals within the cryogenic environment, before reaching room temperatures. Special amplifiers such as parametric amplifier, and high electron mobility transistor (HEMT) are typically employed for amplification within the fridge. A proper amplification is crucial for obtaining high measurement fidelity.

3.37 Josephson Parametric Amplifier

A JPA [169–171] is a superconducting device (Figure 3.26) used to amplify weak microwave signals with near-quantum-limited noise performance, making it

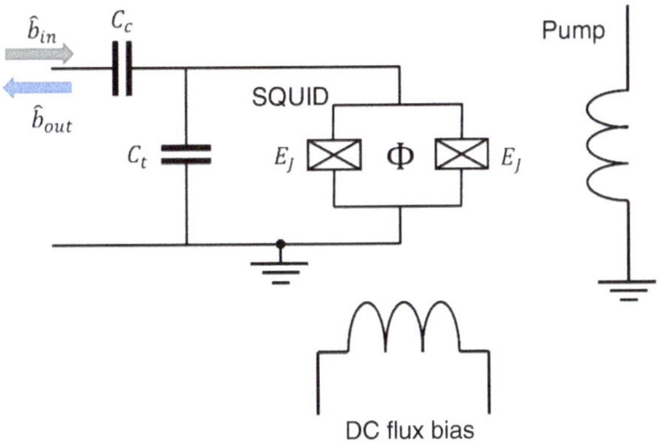

Figure 3.26 Josephson parametric amplifier (JPA) based on a flux-pumped SQUID resonator. A weak microwave signal enters and exits the device through the same transmission-line port, described by the incoming and outgoing field operators $\hat{b}_{in}$ and $\hat{b}_{out}$, and is coupled capacitively via the coupling capacitor C_c. Together with the shunt capacitance C_t, the nonlinear inductive element formed by a DC SQUID (Josephson energy E_J) defines a microwave resonator whose bare resonance frequency lies in the few-GHz range. The SQUID acts as a flux-tunable nonlinear inductor: a static (DC) magnetic flux bias Φ sets the operating point and resonance frequency, while a strong AC pump tone is applied through an inductively coupled pump line. The pump modulates the effective Josephson inductance at frequency ω_p, enabling parametric interaction between the resonator mode and the signal. Energy from the pump is coherently converted into signal and idler photons, resulting in microwave gain. The Josephson nonlinearity provides the mixing mechanism, while the pump supplies the amplification energy; depending on the pump frequency, the device operates in either degenerate (phase-sensitive) or nondegenerate (phase-preserving) amplification modes [172] (Creative Commons Attribution License CC-BY 4.0).

essential for high-fidelity qubit readout. It leverages the nonlinear inductance of JJs to achieve amplification through a parametric process, where energy is transferred from a strong pump tone to the signal of interest. JPAs are widely used where ultra-sensitive microwave signal detection is crucial.

An analogy with a "driven" pendulum is helpful to understand how a JPA works. In ordinary driven amplification, a force is applied directly in the direction of motion: push the pendulum at the right moment, and its swing grows because energy is injected along the trajectory. Parametric amplification works in a different way. Instead of pushing the pendulum sideways, imagine standing at the ceiling and periodically moving the suspension point up and down. This motion changes the effective length of the pendulum—one of the key parameters that determines its natural frequency. If the vertical modulation is timed correctly—specifically, at twice the pendulum's natural frequency—the amplitude of the swing grows dramatically. When the pendulum is moving outward, the suspension point should be raised (shortening the length), and when it is moving inward, the suspension point should be lowered (lengthening the pendulum). This coordinated modulation injects energy into the oscillation without applying a direct sideway force.

This is directly analogous to the role played by the pump in a JPA. The JJ acts like a tunable suspension point: its inductance depends on the superconducting phase across it, and therefore, the resonant frequency of the circuit can be modulated in time. The pump drive periodically modulates this inductance, effectively modulating the "stiffness" of the electromagnetic oscillator (in analogy, the oscillation mass on a spring). The weak microwave signal entering the JPA is analogous to the tiny initial swing of the pendulum. When the pump modulation is phase-matched to the signal, energy from the pump is coherently transferred into the signal mode, causing the signal amplitude to grow. A JPA does not amplify by brute force or by dissipating energy. Instead, it amplifies by reshaping the circuit's parameters so that the existing signal naturally grows, borrowing energy from the pump in a coherent and nearly lossless manner.

A JPA is a microwave resonator whose inductance is provided by a JJ, often a SQUID. With a SQUID, the effective Josephson energy becomes flux-tunable. The inductance depends nonlinearly on phase and can be modulated by a pump so that the resonance frequency can be varied in time. Pumping the circuit at the appropriate frequency causes energy from the pump to be coherently converted into signal and *idler photons*, amplifying the incoming microwave signal. The JJ supplies the nonlinearity; the pump supplies the power. Idler photons are the *partner photons* created alongside signal photons such that

$$\omega_p = \omega_s + \omega_i$$

where ω_p is the pump frequency, ω_s is the signal frequency, and ω_i is the idler frequency. Idler photons exit via the signal port, where they are separated by frequency using filters; they are usually absorbed in a cold load. Their vacuum fluctuations set the quantum noise floor of the amplifier.

Typical gain values range from 15 to 25 dB for boosting weak signals [173]. JPAs operate with negligible dissipation. However, the bandwidth of a standard JPA is narrow and generally is around 10 MHz. Even though this bandwidth is sufficient for single-qubit readouts, multiplexed scheme where multiple qubits are measured together requires broadband amplifiers. Recent advancements have enabled wideband JPAs that can show a bandwidth of several hundred MHz [174–177].

3.38 Attenuators

While the quantum device operates in cryogenic temperature, most control electronics operate at room temperature. It is thus unavoidable to bring in room-temperature signals, which contain a large amount of thermal noise. Attenuators, usually made of resistive (dissipative) components, are employed to attenuate the thermal noise. A dilution fridge is multistage, with each stage typically operating in temperatures 50 K, 4 K, 700 mK, 100 mK, and 10 mK. Attenuators are typically placed in one or more stages to damp the thermal noise successively [178]. Of course, the signals will also be attenuated. Thus, the input signals at room temperature are set to high amplitudes, and both signals and noise are attenuated as they propagate to the colder stages of the fridge.

3.39 Filters

Filters are usually made of inductive and capacitive (nondissipative) components, resulting in different transmission amplitudes for a different range of frequencies. In most experiments, we have a well-expected frequency range of signals for each control line. We can thus use the filters to filter out any transmission from unwanted frequencies, thus lowering the total thermal noise reaching the quantum device. A few common types of filters include low-pass, high-pass, and band-pass filters.

3.40 Isolator

A circulator is a passive nonreciprocal three-port device, in which a microwave or RF signal entering any port is transmitted to the next port in rotation. Circulators are essential for devices that are measured in reflection so that one can have separate input and output microwave lines. When one port of a three-port circulator is terminated in a matched load, it can be used as an isolator, since a signal can travel in only one direction between the remaining ports. The isolator is especially useful for the measurement output channel, where signals are traveling from the quantum device toward room-temperature electronics. The isolator blocks the thermal noise that can otherwise enter the quantum device through the measurement or output line that is not attenuated.

3.41 Microwave Electronics Components

3.41.1 Oscillator

An oscillator is used for outputting a set frequency and amplitude microwave signal. The switching times of these settings are typically rather long (much longer than qubit coherence time). Therefore, oscillators are used as a static source of microwave signal. Most generators support gated or switched output, meaning the on/off of signal output can be controlled by an external control signal. The control signal can be provided by an arbitrary waveform generator (AWG).

3.41.2 Arbitrary Waveform Generator

The AWG is a generator that allows the output of programmable voltage or current waveforms. There are constraints in amplitude, bandwidth, output resolution, and time precision of the waveforms. Some waveform generators have high enough clock rate that allows direct synthesis of microwave pulses at the qubit frequencies (4–8 GHz).

3.41.3 In-phase and Quadrature Mixer

An *I.Q. mixer* (in-phase and quadrature mixer) is a critical component in the microwave control electronics for controlling superconducting qubits. It enables precise generation and control of microwave pulses, which are required for driving qubit-state transitions and implementing gate operations. The mixer uses two input signals: the in-phase (I) and quadrature (Q) components. These components represent the real and imaginary parts of a signal in the complex plane. The mixer combines the inputs with a local oscillator (LO) signal to produce a modulated microwave signal:

$$S(t) = I(t)\cos(\omega_{\mathrm{LO}} t) - Q(t)\sin(\omega_{\mathrm{LO}} t)$$

where ω_{LO} is the frequency of the LO.

A stable microwave source (the LO) generates a CW signal at the desired carrier frequency (ω_{LO}). AWGs produce low-frequency (typically up to a few hundred MHz) control signals $I(t)$ and $Q(t)$, which encode the desired pulse envelope, amplitude, and phase. The I.Q. mixer combines the LO signal with the baseband I and Q signals to create a microwave pulse with precise amplitude and phase control. The output is a microwave pulse that allows arbitrary shaping of the microwave pulse,

$$S(t) = A(t)\cos(\omega_{\mathrm{LO}}t + \phi),$$

where the amplitude envelope is $A(t) = \sqrt{I(t)^2 + Q(t)^2}$.

And the phase is $\phi = \tan^{-1}\left(\frac{Q(t)}{I(t)}\right)$.

The I.Q. mixer enables customized envelopes, such as Gaussian or DRAG (derivative removal by adiabatic gate) pulses. These shapes reduce errors caused by leakage into higher-energy states and suppress unwanted frequency components. By controlling the $I(t)$ and $Q(t)$ signals, the mixer can generate microwave pulses with precise frequency modulation. This is used for resonant driving (matching the qubit frequency) or detuning (driving the qubit off-resonance) for gate operations. The phase of the microwave pulse can be modulated by adjusting the relative amplitudes of $I(t)$ and $Q(t)$. This is essential for implementing phase-sensitive gates.

Noise in the microwave control signals is a major source of decoherence. I.Q. mixers along with LOs allow for clean, low-noise signal generation.

3.41.4 Analog-to-Digital Converter

An analog-to-digital converter (ADC) is a system that converts an analog signal into a digital signal. It is used for qubit-state measurement, where we take the downconverted measurement signal (using the same LO) to ADC for digitization and use a computer to analyze the digital signals.

3.42 DC Sources

We use a DC voltage source to tune the frequency of tunable transmon qubits or other devices like the JPA. The stability of the DC voltage source is crucial for high qubit coherence since any voltage noise would result in qubit frequency noise. We also require the voltage source to be high precision and finely tunable, such that we can tune qubit frequencies with high precision.

3.43 Basic Experimental Setup

All the components together describe the basics of the experimental setup (Figure 3.27) needed for quantum computations.

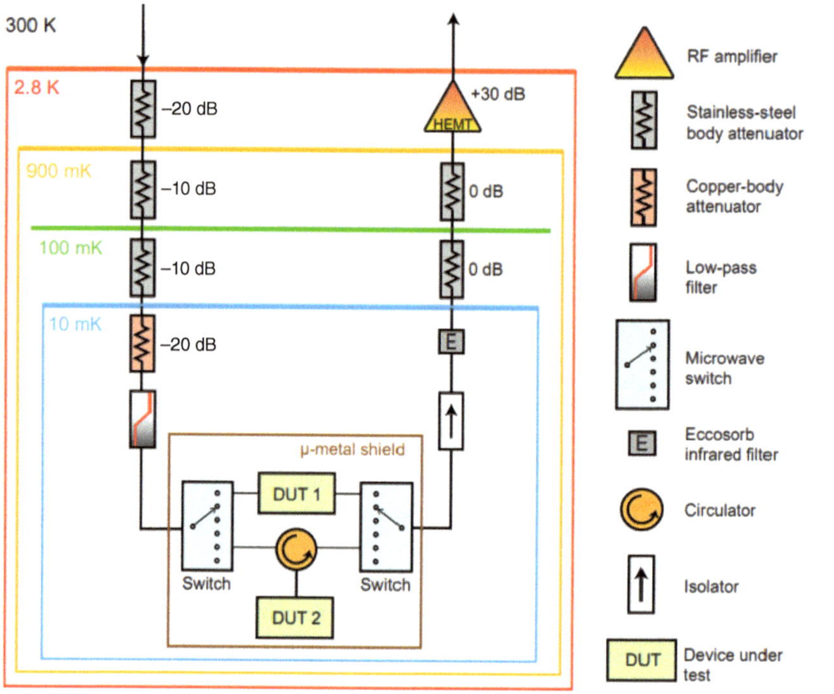

Figure 3.27 Schematic of a dilution-refrigerator–based microwave measurement setup used to control and read out superconducting transmon qubits coupled to microwave resonators. The system is organized in thermal stages spanning from room temperature (300 K) down to the base temperature (~10 mK), with progressively increasing attenuation and filtering to suppress thermal noise and spurious radiation. On the **input (drive) line**, microwave signals generated at room temperature are attenuated in stages as they descend through the cryostat: typically −20 dB at 2.8 K, −10 dB at 900 mK, −10 dB at 100 mK, and −20 dB at the 10 mK stage. These attenuators, implemented with stainless-steel and copper bodies, thermalize the line and reduce the effective noise temperature seen by the device under test (DUT), ensuring that the qubit is driven by near–quantum-limited fields. Additional low-pass filters and Eccosorb infrared absorbers remove high-frequency radiation that could otherwise generate quasiparticles or heat the device. At the base temperature stage (~10 mK), the DUT—typically one or more transmon qubits coupled to microwave cavities—is enclosed within a μ-metal magnetic shield to suppress stray magnetic fields that would degrade superconducting performance. Microwave switches allow routing of signals to multiple devices (DUT 1, DUT 2), enabling multiplexed measurements or device comparison within the same cryogenic environment. On the output (readout) line, signals emitted from the cavity (carrying information about the qubit state via dispersive shifts) pass first through a circulator and isolator, which protect the qubit from back-action noise originating from higher-temperature stages and amplifiers. The signal is then amplified by a high-electron-mobility transistor (HEMT) amplifier located at the 2.8 K stage. Further amplification occurs at room temperature before digitization.

3.44 Room-Temperature Wiring

The room-temperature control electronics start with some hardware capable of generating arbitrary waveforms. Typically, two RF channels are used to generate pulses for the qubit and for the readout. These pulses are combined before being sent inside the fridge. The output signal from the fridge is usually amplified at room temperatures through one or more low-noise amplifiers. The amplified signal is either demodulated or directly fed into a digitizer to record the readout signal. Sometimes, additional narrowband filters around the readout frequency are used to improve SNR. The control and digitization electronics are controlled through a generic computer.

3.45 Cryogenic Wiring

The quantum device is heat sunk via an OFHC copper post to the base stage of a DR (5–30 mK for best thermalization). The sample is also surrounded by a can contain in layers of μ – metal shielding and a layer of lead shielding, thermally anchored using an inner close fit copper shim sheet, attached to the copper can lid. The μ – metal shield and lead can (superconducting) protects the device from stray magnetic fields. Some designs also use thin coatings of eccosorb layer or foam for enhanced protection from the infrared (IR) radiation. Each device is connected to the rest of the setup through various ports: to apply qubit and readout drive tones, DC bias lines, and output ports for measuring the transmitted or reflected signal from the readout resonator. The input lines are attenuated at various stages of the fridge totaling 20–70 dB depending on the application.

The charge drive line also includes filters to block IR radiation and a low-pass filter. The transmitted or reflected signal from the readout resonator is passed through a set of cryogenic circulators (thermalized at the base stage) or isolators and amplified using a HEMT amplifier (thermalized at the 4 K stage) before leaving the fridge.

3.46 DR and Cryostat

Dry DRs operate without liquid helium baths. Their upper stages use a pulse-tube cryocooler, which precools the system to roughly 4 K using circulating helium-4 gas driven by a large, water-cooled compressor. Because pulse-tube cooling power falls steeply with decreasing temperature, only about 1–3 W of cooling is available at 4 K, and substantially less at lower stages. The first stage (≈40–50 K) provides tens of watts, which is essential for intercepting heat loads from wiring, shields, and mechanical supports.

The refrigerator and cryostat (Figure 3.28) provide vacuum isolation and a series of nested radiation shields that suppress thermal loading, vibrational coupling, and blackbody radiation from warmer stages. Importantly, the pulse-tube design places all moving mechanical components outside the cryostat, so no moving parts exist inside the cold volume. This eliminates vibration sources that could otherwise disturb superconducting qubits, which are extremely sensitive to mechanical and electromagnetic noise.

As heat is progressively removed through the 50 K, 4 K, still, and mixing-chamber stages, the system reaches the 10–20 mK regime where superconducting qubits operate. At these temperatures, the thermal excitation probability

$$p_{\text{th}} \sim e^{-hf/kT}$$

is exponentially suppressed, enabling long coherence times. The quantum processor is mounted at the mixing-chamber plate, enclosed within light-tight and magnetically shielded housings that further isolate the qubits from stray photons, magnetic fields, and other environmental disturbances.

Figure 3.29 illustrates the operating principle of a ^{3}He/^{4}He DR [180]. Below the superfluid transition at 2.17 K, the two helium isotopes behave very differently because of their quantum statistics. The ^{4}He atoms are bosons and condense into a superfluid with zero viscosity, while the ^{3}He atoms are fermions that remain in a normal liquid state. This distinction is essential: even as the temperature approaches absolute zero, a finite fraction of ^{3}He remains mobile in the ^{4}He-rich phase, enabling continuous refrigeration without solidification.

At temperatures below about 0.87 K, the mixture separates into two phases: a concentrated phase rich in ^{3}He and a dilute phase containing only about 6% ^{3}He in a background of superfluid ^{4}He. When a ^{3}He atom crosses from the concentrated phase into the dilute phase, it undergoes an effective evaporation-like process that absorbs heat from the surrounding environment. This enthalpy of mixing is the fundamental cooling mechanism of the DR.

To maintain continuous cooling, the system must continually remove ^{3}He from the dilute phase. This is achieved in the still, a distillation chamber held near 600–800 mK, where ^{3}He preferentially evaporates because its vapor pressure is much higher than that of ^{4}He. A turbomolecular pump extracts the evaporated ^{3}He, thereby lowering its concentration in the dilute phase and driving more ^{3}He across the phase boundary at the mixing chamber, which provides ongoing cooling.

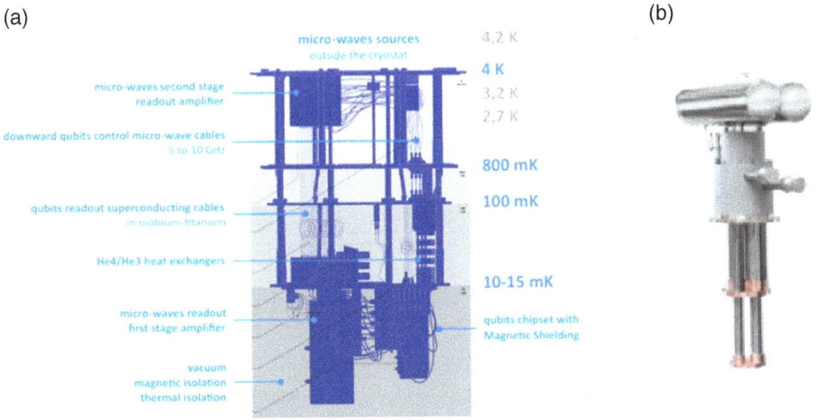

Figure 3.28 Cross-sectional schematic of a multistage DR and Pulse tube used to operate superconducting qubits at millikelvin temperatures, [18] / Ram C. Dhuley / Fermi Research Alliance, LLC. The cryostat consists of nested thermal shields anchored at successive temperature stages: 4.2 K, intermediate stages near 3.2 K and 2.7 K, an 800 mK stage, a 100 mK still plate, and the base temperature mixing chamber reaching 10–15 mK. The qubit chip is mounted at the mixing chamber stage inside magnetic shielding (e.g., Cryoperm and μ-metal shielding) to suppress flux noise and environmental magnetic fields. Microwave control and readout signals are delivered through carefully thermalized coaxial lines. Downward qubit-control microwave cables (5–10 GHz) are attenuated and heat-sunk at each temperature stage to reduce thermal photon population and noise. Superconducting readout cables connect the qubit chip to cryogenic amplifiers. The first-stage low-noise amplifier (typically a HEMT at 4 K) provides high-gain readout with minimal added noise. He^4/He^3 heat exchangers enable continuous circulation of the helium mixture that produces dilution cooling at the mixing chamber. (Right) Photograph of a two-stage pulse-tube cryocooler, a closed-cycle cryogenic refrigerator widely used as the precooling stage in modern cryogen-free dilution refrigerator (DR) systems. The pulse tube operates by cyclic compression and expansion of high-pressure helium gas, driven by a remote compressor, to generate cooling without moving mechanical parts at the cold end. This vibration-minimizing architecture makes pulse tubes particularly suitable for quantum computing applications where mechanical stability is critical. The upper assembly houses the regenerator and gas distribution components, while the lower vertical cold head provides two thermal stages: a first stage typically near ~40–60 K and a second stage near ~3–4 K. These stages precool radiation shields, heat exchangers, wiring, and the helium mixture in the dilution unit before it reaches the sub-kelvin stages. In DR systems, the pulse tube replaces liquid helium baths, enabling continuous operation without cryogen refilling and significantly reducing operational cost.

After leaving the still, the ^3He gas is purified in a liquid-nitrogen cold trap, compressed, and reliquefied as it flows back down through heat exchangers toward the mixing chamber. There it rejoins the concentrated phase, completing a closed-loop cycle that sustains millikelvin temperatures for extended periods.

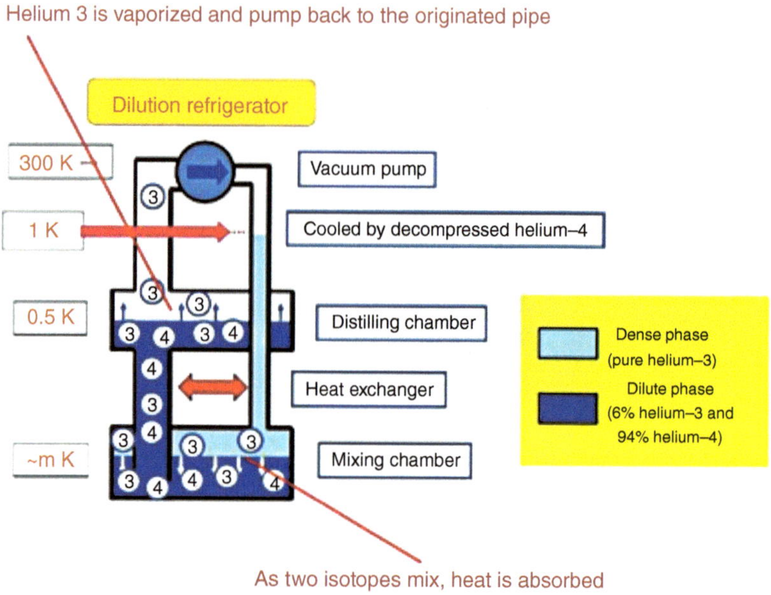

Figure 3.29 Schematic illustrating the thermodynamic cycle that enables continuous cooling to millikelvin temperatures in a dilution refrigerator [179] / ULVAC, Inc. The system exploits phase separation in a liquid mixture of helium-3 (^{3}He) and helium-4 (^{4}He) at temperatures below ~0.87 K. At low temperature, the mixture separates into two coexisting liquid phases: a concentrated (dense) phase consisting primarily of pure ^{3}He (light blue), and a dilute phase containing approximately 6% ^{3}He dissolved in ^{4}He (dark blue). The cooling power arises at the mixing chamber, where ^{3}He atoms cross the phase boundary from the concentrated phase into the dilute phase. This mixing process is endothermic: heat is absorbed as ^{3}He atoms dissolve in the ^{4}He-rich phase, providing the refrigeration effect that enables temperatures in the ~10 mK regime. The mixture then flows upward through heat exchangers that pre-cool the incoming concentrated ^{3}He using the outgoing dilute mixture, improving thermodynamic efficiency. Above the mixing chamber lies the still (distillation chamber), typically operated near 0.5–0.8 K. Because ^{3}He has a higher vapor pressure than ^{4}He at these temperatures, selective evaporation of ^{3}He occurs. A vacuum pump removes the evaporated ^{3}He vapor, which is then recondensed at higher-temperature stages (e.g., near 1 K and 4 K, often precooled by decompressed ^{4}He). The recondensed ^{3}He returns via capillaries to the mixing chamber, maintaining continuous circulation. The diagram also indicates thermal stages (300 K, 1 K, 0.5 K, and ~mK) and the direction of ^{3}He flow.

The still provides 10–20 mW. Its role is to evaporate ^{3}He and drive circulation through the dilution unit. The cooling power here is set by the vapor pressure of ^{3}He. At the cold plate/heat exchangers (~100 mK), cooling power is typically 200–400 μW. This stage supports attenuators, filters, and wiring thermalization for qubit experiments. The mixing chamber (~10–20 mK) provides 10–20 μW at 20 mK and 1–5 μW at 10 mK.

3.47 Summary and Conclusion: Key Points

This chapter traced the evolution of superconducting quantum circuits from their earliest charge-based realizations to the sophisticated cQED architectures that now form the backbone of QC. It began with a brief historical arc, showing how the CPB demonstrated the first electrical qubits but suffered from sensitivity to charge noise, motivating the invention of the transmon, which reduces charge sensitivity and increases coherence by operating in the regime $E_J \gg E_C$. In this regime, the qubit becomes an artificial atom whose degrees of freedom are collective phase and charge oscillations of a macroscopic superconducting condensate.

To make sense of this, the chapter reviewed the essential concepts from quantum physics to treat the quantum harmonic oscillator, first in its abstract form and then in its physical incarnation as an LC circuit, where voltage and current play the roles of position and momentum. The wavefunctions, ladder operators, and zero-point fluctuations introduced here provide the language for describing both microwave resonators and Josephson circuits.

The JJ introduces the crucial element of nonlinearity by replacing a simple quadratic inductive energy with the cosine potential $-E_J \cos \phi$. Physically, this junction acts as a lossless, nonlinear inductor whose phase-dependent energy couples the collective wavefunctions of two superconducting condensates, transforming a simple oscillator into a quantum system with discrete, anharmonic energy levels that can be used to define a qubit. With the addition of a nonlinear JJ, the LC oscillator levels become the transmon spectrum, from which a stable two-level qubit with ground and excited states $|g\rangle, |e\rangle$ that can be isolated to be treated as a two-level quantum system.

The chapter then showed how microwave electric fields interact with the transmon's electric dipole moment, producing Rabi oscillations, the basic mechanism for single-qubit control. These oscillations reveal both the coherence of the qubit and its coupling to its electromagnetic environment, motivating the study of coherence times and noise sources. Carefully shaped microwave pulses allow the implementation of quantum gates, turning abstract Hamiltonians into programmable operations.

From single qubits, the discussion moved to multiqubit systems. Direct capacitive coupling, tunable couplers, and the CR interaction provide different ways to convert weak, always-on interactions into controllable entangling gates. The ECR technique was introduced as a practical method to isolate the desired two-qubit interaction while canceling unwanted crosstalk.

At the heart of qubit–resonator physics lies the Jaynes–Cummings Hamiltonian, which describes how a two-level transmon interacts with a single-resonator mode. In the strong coupling regime, real photons are exchanged, producing dressed states, vacuum Rabi oscillations, and phenomena such as photon blockade, where quantum nonlinearity prevents multiple photons from entering the cavity. In the dispersive regime, energy exchange is frozen out, but virtual photons shift frequencies, enabling QND readout and resonator-mediated entanglement between distant qubits. This regime underlies nearly all scalable superconducting quantum processors.

To make quantum interactions observable, the chapter introduced the microwave hardware that surrounds the qubit. Purcell filters protect qubits from radiative decay. JPAs provide near-quantum-limited amplification of the outgoing microwave signals, while HEMT amplifiers, attenuators, filters, and isolators shape noise, suppress backaction, and preserve signal integrity as microwaves travel between millikelvin temperatures and room temperature.

Finally, the chapter described the microwave control electronics—oscillators, AWGs, I.Q. mixers, and digitizers—that synthesize phase-coherent pulses, route them to individual qubits, and reconstruct the returning signals. Together, these components form a complete quantum measurement chain, allowing superconducting circuits to be driven, entangled, and read out with precision. The pulse-tube cryocooler provides the thermal foundation upon which the subsequent ^{3}He/^{4}He dilution cycle operates, ultimately enabling millikelvin temperatures required for superconducting qubits. This dilution process forms the basis for achieving the ultralow temperatures required.

In sum, the chapter showed how a simple LC circuit, augmented by a JJ and embedded in a carefully engineered microwave environment, becomes a controllable artificial atom—and how networks of these atoms, coupled by resonators and guided by quantum optics, form the physical foundation of superconducting QC.

4

Cavity Quantum Processors: Qudits, Bosonic Codes, and Scaling

4.1 Introduction

This chapter shifts the focus from 2D CPW resonators to 3D superconducting cavities, discussing their transformative role in quantum computation. Unlike 2D resonators, where surface and dielectric losses degrade performance, 3D cavities confine electromagnetic fields primarily in vacuum, achieving lifetimes that exceed seconds in the best cavities. This makes them outstanding candidates for storing and manipulating quantum information with minimal decoherence.

The chapter introduces a variety of 3D cavity geometries under development and how 3D cavities couple with transmon qubits to form hybrid architectures that combine long-lived bosonic modes with fast controllability. Their large Hilbert spaces support qubits encoded in Fock states, higher-dimensional qudits, which compactly represent information and reduce gate counts, error detection encodings such as dual-rail, followed by more resilient encodings for error detection and correction such as binomial and cat codes and finally Gottesman–Kitaev–Preskill (GKP) codes, with the strongest protection approaching the regime of fault tolerance.

A cavity–transmon system provides an efficient platform for quantum storage as well as computation by combining the long coherence times of superconducting cavities with the fast, nonlinear control afforded by transmon ancillas. Algorithms and error-correcting operations can be implemented using fewer physical resources. The transmon functions as an ancilla, enabling control, conditional operations, and measurement, while the cavity mode hosts the quantum state, supports coherent superpositions, and enables interference between probability amplitudes.

The discussion here also highlights how 3D architectures reduce hardware overhead by enabling multimode operation within a single cavity and offer better scaling pathways than conventional 2D transmon arrays.

Thus, the chapter positions 3D superconducting cavities as a central technology for overcoming the challenges of coherence, error protection, error detection, and scalability, with the potential to form the backbone of next-generation quantum architectures.

4.2 Benefits of Superconducting Cavity-Based QC

We first turn to the development of 3D superconducting cavity-based digital QC systems [181]. 3D cavities involve various geometries like rectangular, cylindrical, coaxial, or cavities evolving from the TESLA geometry instead of 2D CPW resonators. The resonant modes of the 3D cavities have higher lifetimes than CPW resonators, even surpassing two seconds for the best Nb cavities originally designed for accelerator physics [182]. These 3D systems are coupled to JJ-based transmons for control of states and for readout. The high lifetimes render 3D cavities exceptional for storing and manipulating quantum information.

2D resonators show significant surface losses due to dielectric losses, TLS losses [183–185] from the oxide, and contaminant-induced losses stemming from the concentration of electromagnetic energy near the substrate surface (Chapter 5). In 3D cavities, the electric field is predominantly stored in the vacuum with lower surface field concentrations, leading to significant reduction in surface loss and longer coherence times [123].

Superconducting cavities show very high-Q factors [186] to enable storage times approaching seconds even in the few photon regime [182]. Such cavities are essential for many physics applications, especially particle accelerators, ultra-sensitive motion/displacement sensing [187], precise frequency stabilization [188], and testing fundamental physics [189]. 3D cavities have a large number of accessible energy eigenstates, known as *Fock levels*. A Fock state is a quantum state that has a definite number of particles (like photons in a cavity). If you have exactly n photons in a mode, the state is written as $|n\rangle$. $|0\rangle$ means there are zero photons (vacuum state). $|1\rangle$ means there is exactly one photon. $|2\rangle$ means there are two photons and so on. Fock states represent precise, discrete quantities of energy [190].

A qubit can be encoded using two or more Fock states of the cavity. One of the simplest ways to use cavity-based Fock states for quantum computation is to define a logical qubit subspace using the lowest two Fock states: $|0_L\rangle = |0\rangle, |1_L\rangle = |1\rangle$. This is known as the *Fock-state qubit encoding*. Higher photon number states can also be used, such as $|0_L\rangle = |n\rangle, |1_L\rangle = |m\rangle$. The choice of encoding depends on error resilience and ease of gate implementation.

Since the 3D cavity can support multiple frequency-separated modes, multiple qubits can be created simultaneously. For example, mode 1 can be used for qubit A with states $|0\rangle_1, |1\rangle_1$ and mode 2 as qubit B with states $|0\rangle_2, |1\rangle_2$. A single transmon can be used to control both modes, implementing entangling operations between them. This approach provides a path to realize large-scale QC.

3D resonators offer improved isolation between multiple qubits, reducing crosstalk and undesired qubit interactions. This isolation safeguards the integrity of quantum states, maintaining high coherence. The mode structure of a 3D resonator can be designed to achieve the optimal balance between coupling strength and environmental isolation, thereby enhancing their performance for quantum applications.

3D cavities' access to large Hilbert spaces opens avenues for higher-dimensional computational units known as *qudits* (Sections 2.25 and 4.7). A qudit is a d-dimensional quantum object—rather than just the two levels for qubits. Qudits offer advantages in simulations requiring fewer qudits with fewer gate operations and potentially better noise tolerance. For example, 100 qudits with $d = 32$ could accomplish a calculation that would require 500 qubits, since $2^{500} = 32^{100}$.

While one could argue that 2D CPW resonators with transmons can also be used for qudits, there are two major issues. 2D resonators have lower-quality factor and corresponding lower coherence unsuitable for creating for higher-dimensional states with multi-photon Fock states. 3D cavity systems offer longer-lived qudit states, improving error correction. A 2D coplanar resonator plus transmon can typically support only a few well-defined resonant modes, limiting qudit encoding. Practical control beyond four levels becomes challenging with less than 1 ms coherence for transmons. While qudits offer advantages in compact encoding and reduced connectivity needs, their challenges include higher control complexity, error susceptibility, and measurement difficulties.

Traditional 2D architectures with qubits use a linear chain or planar geometry with nearest-neighbor coupling to scale up so the Hilbert space grows exponentially as 2^N, where N is the number of qubits. Typically, these systems need one or two input lines per qubit, one readout line for one to seven (a practical limit) qubits, and additional input lines for tunable interqubit couplers. The number of room temperature and cryogenic electrical components increases linearly with the number of qubits.

By contrast, one d-dimensional *qudit* mode effectively encodes $\log_2 d$ *qubits*, requiring just one input and output line to control that many effective qubits. For example with $d = 16$, one qudit would be equivalent to four qubits. 3D cavity-based systems thus reduce cryogenic and room-temperature hardware overhead. Scaling up can be achieved by constructing multimode cavities, each supporting several high-coherence qudit modes, again without needing additional cryogenic lines or control hardware and with the benefit of obtaining a higher-dimensional Hilbert space per physical device.

Using cavity modes to store and manipulate quantum information is also favorable for quantum-error protection and correction, an essential aspect of scalable quantum computer architectures. The dominant loss mechanism is usually photon loss [146, 191] or cavity decay (also called amplitude damping). Photon loss in a cavity is a stochastic process, where each photon has an independent probability of being absorbed or scattered per unit time. Therefore, more photons means more opportunities for loss events. Consequently, the overall decay rate of the state increases linearly with photon occupation number, n. If the lifetime of a coherent state or the first Fock level (with one photon) is ${T_1}^{(1)}$, the lifetime of the nth Fock level is given by ${T_1}^{(n)} = {T_1}^{(1)} \times 1 / n$.

It is possible to design QEC codes that specifically correct for this most likely photon loss error. This is to be contrasted to more standard 2D QEC codes, such as the surface code, which aim at detecting and correcting both amplitude and phase errors. The higher accessible dimensional Hilbert space of a cavity can be exploited in bosonic encoding to provide the redundancy that is necessary for error protection and correction, allowing the use of fewer physical resources to protect quantum information than when using TLSs, transmon-based systems. Built from the foundation of Fock states, bosonic encodings, such as dual-rail, binomial codes, cat codes, and GKP codes, exploit superpositions of multiple Fock states for error protection and error correction.

Cat codes and related bosonic encodings are also fully compatible in principle with 2D transmon–CPW systems. However, their performance depends critically on photon lifetime, which are lower in planar geometries due to surface and dielectric loss. As materials improve, increasingly powerful bosonic encodings in 2D will become practical—but today, the most robust cat code and other bosonic encoding demonstrations still rely on high-Q storage modes typically found in 3D or quasi-3D architectures.

4.3 Examples for 3D Cavities and Transmons

A diverse array of options for cavity geometries has been explored. These encompass rectangular and cylindrical resonators, among other designs. The optimal placement of a qubit within a resonator is contingent upon the distinct characteristics of the resonator. Positioning the qubit at the electric field's crest ensures maximum coupling strength. Conversely, situating the qubit near where the electric field is nullified uncouples the qubit from the resonator mode, which can also offer advantages for certain applications.

In the pioneering 3D cavity QED effort [123], the transmon was introduced at the E-field maximum in a rectangular aluminum cavity operating in the lowest frequency TE101 mode at 8 GHz (Figure 4.1). T_1 of the transmon was in the

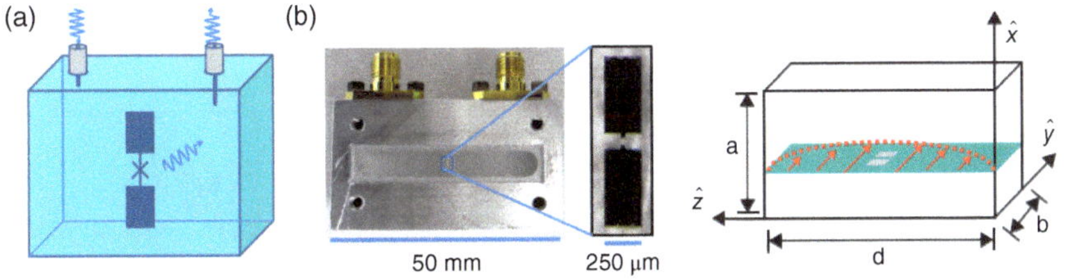

Figure 4.1 Schematic of a transmon qubit inside a 3D cavity operating in the TE101 mode. The qubit is coupled to the cavity through a broadband dipole antenna [123] / American Physical Society. (b) Photograph of a half of the 3D aluminum waveguide cavity. An aluminum transmon qubit with the dipole antenna is fabricated on a c-plane sapphires substrate and is mounted at the center of the cavity. (Inset) Optical microscope image of a single-junction transmon qubit. The dipole antenna is 1 mm long [123] / American Physical Society. (d) A cartoon of the E-field is shown in red arrows [161]. One important consideration while building 3D cavities is to minimize seam losses, which occurs at a region where two metal surfaces are joined. Several strategies were taken to avoid having seams in a design that include coaxial and flute geometries [192].

range of 25–50 μs corresponding to $Q_1 = 1 - 2 \times 10^6$. Dielectric and seam losses dominate[161]. While the initial experiments were performed to demonstrate the possibility of controlling and measuring a 3D-style transmon inside the cavity, it was soon realized that the same architecture would enable controlling the cavity mode, which can be made with much higher quality than a transmon.

The cylindrical options successfully explored were the solid Al cylinder [193] (Figure 4.2a) and the quarter wave coaxial resonator [146] (Figure 4.2 b). The coaxial cavity, also known as the post cavity, supports a fundamental TEM mode. The central post is short-circuited at one end (bottom) and open-circuited at the other (top). The fundamental frequency of the mode is determined by the length of the central post. It behaves like a $\lambda/4$ (quarter-wave) resonator. The system supports standing-wave patterns with

$$\ell = \frac{(2n-1)\lambda}{4}, \quad n = 1, 2, 3, \ldots,$$

with allowed TEM resonant frequencies: $f_n = (2n-1)f_0$, where f_0 is the fundamental (¼-wave). The next TEM harmonic $= 3f_0$(¾-wave).

The cavity is usually excited through an antenna inserted from the top lid. Strong coupling between an ancilla qubit and resonator is achieved by inserting the sapphire chip holding the transmon through a small hole in the side wall. Q values near 10^8 were achieved at about 8 GHz. T_1 for the Fock state $|1\rangle$ was measured to be above 1.2 ms and $T_2 = 0.72$ ms. Another recent work [194] demonstrated lifetime of 11.3 ms (internal $Q > 10^9$) at single-photon lifetimes after transmon integration inside a coaxial cavity resonator resonating at 6.5 GHz.

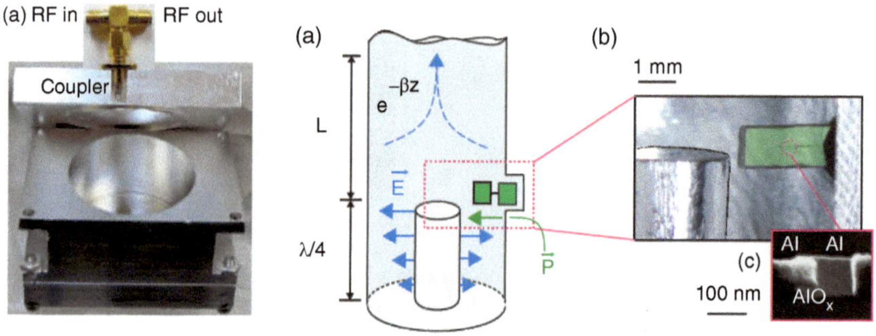

Figure 4.2 (a) Cylindrical resonator to achieve 10 ms photon lifetime [161] / Springer Nature. (b) A quarter-wave coaxial resonator is defined by shorting a coaxial transmission line's inner and outer conductors at one location on the transmission line (bottom) and open-circuiting the line a distance $\lambda/4$ away (upward). A superconducting transmon qubit (green) can be coupled to the $\lambda/4$ mode by aligning the electric dipole moment of the transmon to the electric field of the resonator E. The transmon qubit is inserted through a 1.5-mm hole. The qubit is also coupled to a second on-chip resonator used for control and readout. There is strong coupling between the qubit and the resonator; [161] / Springer Nature.

Another type of seamless cavities is known as the "flute" cavities (Figure 4.3). These are made by drilling holes in a solid block of metal so that there is a wide resonating structure in the middle. These multimode cavities can show lifetimes of several milliseconds.

TESLA-shaped elliptical cavities (Figures 4.4 and 4.5) [186], made from high-purity bulk Nb and fabricated through deep drawing (stamping) and electron-beam welding, are of particular interest for particle accelerators due to their exceptionally high intrinsic Q-factors. The single-cell TESLA cavity comprises an elliptical resonator flanked by two cylindrical sections known as beampipes. Reflection or transmission measurements can be conducted by introducing antennas through the beampipes. Recent work explored single-cell TESLA Nb cavities at low photon excitation regimes and temperatures of ~10 mK. Intrinsic Q-factors of the TM_{010} mode reached an astounding $Q = 2 \times 10^{11}$ at 1.2 K after proper surface treatments [195], with empty cavity (no coupled qubit) corresponding to photon lifetimes up to two seconds [182]. Upon cooling to 10 mK temperatures, the cavity exhibited a 10-fold decrease in Q-factor, attributed to TLS contribution (Chapter 5).

Quantum information can be stored in the fundamental mode (TM010) of 3- or 5-GHz TESLA cavities (Figures 4.4 and 4.5) [196, 197] to benefit from the very long lifetimes. The ancilla transmon is strategically positioned from one side, nearly reaching the center of the cavity to optimize electromagnetic interaction. The transmon is fabricated on a silicon or sapphire substrate, utilizing

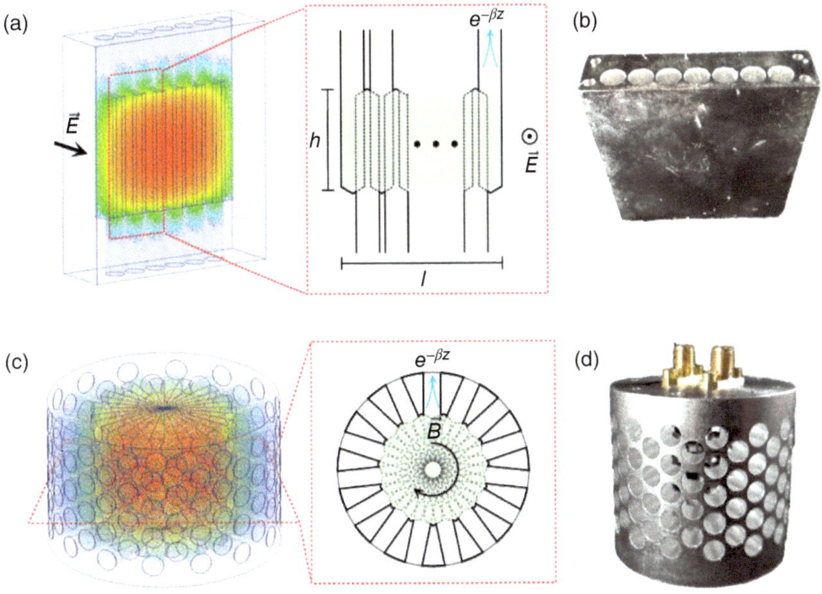

Figure 4.3 (a) Electromagnetic field for the operating mode (TE101) of the seamless rectangular cavity. The left panel shows a simulated electric field intensity map for the fundamental accelerating mode, illustrating the field confinement and uniformity. The central inset depicts the corresponding mechanical/electromagnetic model: a monolithic rectangular shell with an array of periodic flutes (green region) along the longitudinal length l used to control the field profile, frequency, and mechanical stiffness. The cavity height h and length l are indicated; the arrow $\vec{E}$ marks the direction of the electric field. (b) Photograph of a fabricated seamless rectangular cavity prototype showing the continuous, weld-free body and flute features, demonstrating the manufacturable realization of the design. The seamless approach eliminates equator welds, improves surface quality and mechanical integrity, and is intended to reduce RF losses. (c) Electromagnetic field for the operating mode (TM010) of the seamless cylindrical cavity. The left image shows the simulated electric field intensity pattern for the fundamental mode. The central inset presents the corresponding mechanical/electromagnetic model of the seamless structure: a monolithic cylindrical shell incorporating a periodic array of radial flutes or cells (green region). (d) Photograph of a fabricated seamless cylindrical cavity prototype illustrating the patterned internal cell structure. The seamless approach removes welds; [192] / American Physical Society.

superconducting materials such as aluminum, niobium, or tantalum to create the capacitor pads, while the JJ is commonly made from aluminum. The transmon's frequency ranges from 4 to 6 GHz. For the transmon readout, a higher harmonic (TM_{11}) of the cavity is employed. Several Fock states were established using appropriate gates (Section 4.4). The lifetime of the first Fock state $|1\rangle$ was determined by measuring the average population of $|1\rangle$ after varying waiting periods,

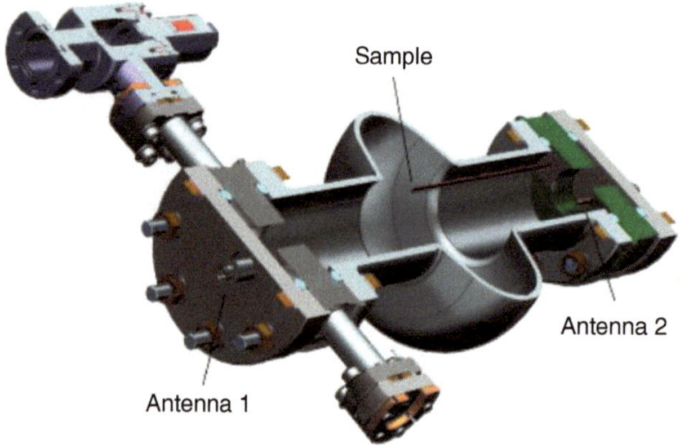

Figure 4.4 Cutaway view of a TESLA-style superconducting radio-frequency (SRF) cavity adapted for coupling to a transmon qubit. The figure shows a 3D elliptical SRF cavity (of the TESLA geometry) opened to reveal its interior volume and coupling ports. The central elliptical cell supports a high-Q microwave mode whose electric field is concentrated along the cavity axis, providing a coherent electromagnetic environment. A transmon or other superconducting quantum device is positioned inside the cavity (labeled "Sample"), where it couples strongly to the cavity's electric field. Two coaxial antenna ports ("antenna 1" and "antenna 2") penetrate the cavity walls and act as input and output couplers, allowing microwave signals to be injected into and extracted from the cavity. The flanges, vacuum seals, and mechanical support structures are part of the cryogenic RF assembly that maintains ultrahigh vacuum and superconducting conditions. In this architecture, the TESLA cavity serves simultaneously as a high-Q quantum memory and as a resonator bus that mediates coherent interactions with the embedded transmon, combining accelerator-grade SRF technology with cQED quantum hardware.

showing a T_1 of about 3.1 ms. By preparing a superposition $(|0\rangle + |1\rangle)/\sqrt{2}$, T_2 of 1.6 ms was determined.

Coupling a transmon to the maximum electric field of the fundamental mode in a TESLA cavity requires the chip to extend to the maximum electric field at the center to allow for sufficient coupling between the cavity and the transmon (Figure 4.5). This insertion approach exacerbates chip holder losses and also introduces vibrations. Another approach is to modify the TESLA geometry by bisecting the cavity along the equator (see Figure 4.6) to create a "mushroom"-shaped resonator [198]. This preserves the mode structure through reflection symmetry, while allowing the transmon to access the maximum of the electric field with minimal protrusion into the cavity [197].

The two parts of 4.3-GHz cavity are electron-beam welded to prevent seam loss. However, there is a seam between the cavity and the chip housing. Its loss contribution can be minimized by placing it at a depth of 6 mm in a 2-mm radius

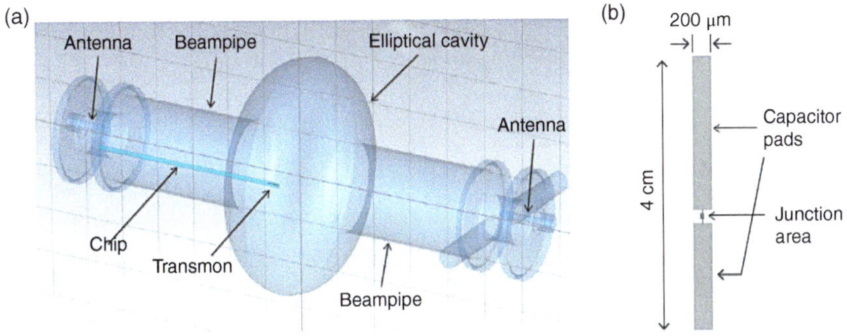

Figure 4.5 Transmon–elliptical cavity integration. (a) Cutaway schematic of an elliptical RF cavity (TESLA-type) with beampipes and weakly coupled antennas on each end. A planar transmon chip is mounted along the beam axis so that the transmon's dipole couples to the cavity mode; the end antennas provide input/output coupling for drive and readout. Labels indicate the elliptical cavity, beampipe, antennas, chip, and transmon. (b) Planar device sketch showing the transmon geometry: two elongated capacitor pads ($\approx$200 μm wide) separated at the center by the junction area that hosts the Josephson junction(s). The drawing indicates an overall dipole length of ~4 cm along the axis (schematic, not to scale) [197] / U.S. Department of Energy Office of Scientific and Technical Information / Public domain.

waveguide with cutoff frequency of 57 GHz. As a result, the magnetic field of the cavity mode is attenuated by three orders of magnitude, strongly suppressing currents through the seam. A compressed indium gasket is used during integration of the chip. The rest of the design is similar to the elliptical TESLA cavity geometry. The bare cavity showed a Q greater than 10^{10}. The transmon chip inserted had a lifetime of 110 micros. By preparing Fock states with zero and one photon populations, the single-photon lifetime of >25.6 ms was obtained, a world record at the time. The lifetime remained consistent across multiple cool-downs. By measuring the survival time of the superposition state $(|0\rangle + |1\rangle)/\sqrt{2}$, a T_2 of 34 ms was recorded.

4.4 Creating Fock States

A superconducting microwave cavity is far more than a passive resonator—it becomes a laboratory for counting and shaping individual quanta of the electromagnetic field. In this setting, the natural language is that of *Fock states*, the photon number eigenstates $|n\rangle$ that describe exactly how many photons occupy a given cavity mode. Unlike classical electromagnetic waves, which are characterized by continuously varying amplitudes and phases, Fock states are intrinsically discrete: a cavity that contains zero photons, one photon, two photons, and so on.

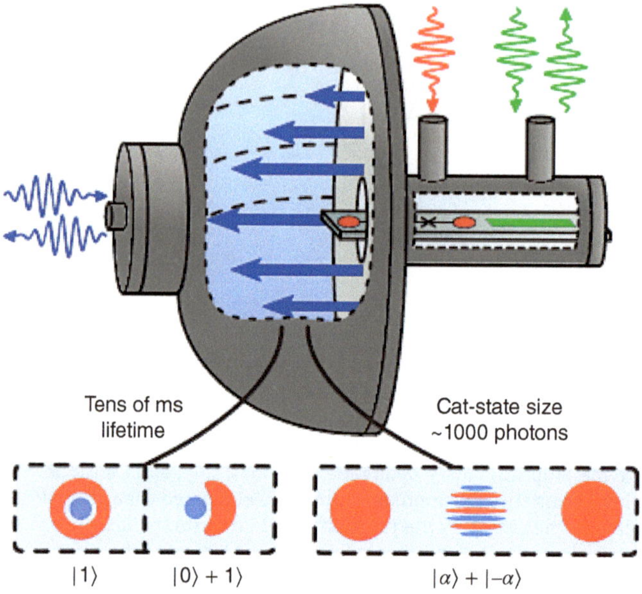

Figure 4.6 A superconducting cavity qubit. The high-purity niobium cavity (top) consists of a half-elliptical part and a flat part, which are electron-beam welded together. The fundamental TM_{010} mode is used, where the maximum electric field (straight blue arrows) is near the opening of the waveguide housing the transmon chip. This design allows for effective coupling to the transmon (red pads) with minimal chip protrusion into the cavity. The transmon is coupled to an on-chip stripline resonator (green rectangle) used for transmon and cavity measurements (green arrows). Classical ring-down measurements of the cavity (wavy blue arrows) are performed through an under-coupled cavity pin. The Wigner distributions (Section 4.26) of single-photon qubit states (bottom left) and a Schrödinger cat state (bottom right) illustrate the stored states demonstrated in this work [198] / American Physical Society / CC BY 4.0.

They're called Fock states in honor of the Russian physicist Vladimir Fock, who, in the 1920s, developed the formalism for describing quantum states of systems with variable particle number.

When we consider Fock states inside a cavity, the wave–particle duality of light becomes transparent. A Fock state has a definite photon number—and in that sense, it behaves like a particle with discrete, countable energy. If you look at its Wigner function (Section 4.26), you see smooth, wave-like rings and interference fringes. These concentric quantum interference fringes are the phase-space imprint of a standing-wave excitation of the electromagnetic field. These oscillations are the most literal manifestation of the state's wave character. The number of radial nodes corresponds to the photon number n. The radial oscillations contain negative regions that arise from interference between quantum amplitudes and are a signature of nonclassical wave behavior.

In this way, a Fock state embodies the dual nature of light: it is a discrete quantum of excitation living inside a continuous field. Measure it one way and you see a particle; probe it another way and you see a wave. If you perform a photon number measurement—"How many quanta are in this mode?"—a Fock state responds with an integer. But if you probe the cavity in a way that asks a wave-type question, you get a completely different answer. Measuring the field quadratures (Section 4.7)—the quantum analogs of the electric and magnetic field amplitudes (or the analogs of position and momentum)—reveals a continuous distribution.

In cavity QED, high-Q superconducting cavities trap these photon states for milliseconds or longer, allowing them to be prepared, manipulated, and measured with extraordinary precision [199]. Because many well-resolved Fock levels can coexist within a single cavity, these states also provide a natural platform for *qudits*—quantum systems with more than two levels—enabling quantum information to be encoded not just in $|0\rangle$ and $|1\rangle$, but across a larger Hilbert space of photon numbers, with potential advantages for error correction, gate efficiency, and hardware-efficient quantum memories.

In general, a Fock state with n photons is given by $|n\rangle = \frac{(a^\dagger)^n}{\sqrt{n!}} |0\rangle$.

The photon creation operator $\hat{a}^\dagger$ adds a photon:

$$\hat{a}^\dagger |n\rangle = \sqrt{n+1}\, |n+1\rangle$$

The photon annihilation operator $\hat{a}$ removes a photon:

$$\hat{a} |n\rangle = \sqrt{n}\, |n-1\rangle$$

The number operator representation for a qubit system is $\hat{n}|0\rangle = 0|0\rangle, \hat{n}|1\rangle = 1|1\rangle$, where $\hat{n} = \hat{a}^\dagger\hat{a}$ is the photon (excitation) number operator.

A Fock state has a fixed photon number and is, therefore, an eigenstate of the number operator. Because photon number and phase are conjugate variables, a state with zero number uncertainty must have maximal phase uncertainty. Consequently, a Fock state possesses no definite phase. This absence of phase manifests as rotational symmetry in phase space. The Wigner function of a Fock state (Figure 4.7) makes this visible: it is rotationally symmetric in phase space, meaning the state has no preferred phase angle.

Creating a specific Fock state $|n\rangle$ inside a superconducting cavity is nontrivial because microwave cavities naturally support coherent states (Section 4.6) rather than photon number states. However, transmon qubits act as auxiliary systems to generate and manipulate Fock states. A fast readout resonator measures the state of the transmon qubit.

There are several operations that enable creating quantum states in the cavity and enable arbitrary operations. Popular choices include the *displacement* gate and

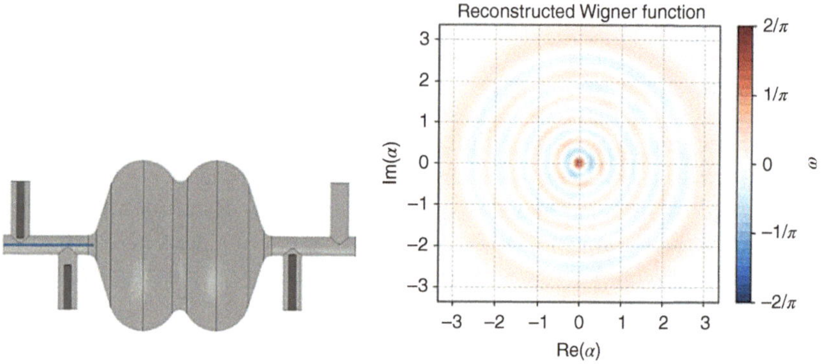

Figure 4.7 (a) Schematic of the double-cell elliptical SRF cavity with four ports for RF control. Three ports are used in this experiment. The chip (blue) contains an ancilla transmon, a Purcell filter, and a stripline resonator for measuring the transmon. (b) 2D Wigner tomograph (Section 4.26) for the number state $|n = 5\rangle$, obtained from displaced-parity measurements. The distribution is rotationally symmetric and exhibits the characteristic alternating red/blue rings (nonclassical fringes) with multiple radial nodes set by the Laguerre polynomial profile of a Fock state. Color scale spans $\pm 2/\pi$; negativity indicates nonclassicality. Axes show phase-space coordinates Re(α) and Im(α).

the *SNAP* gate [200, 201]. The first acts like shining a gentle light on the cavity to add or subtract photons in a specific way. The second shifts the phase of specific Fock states, allowing manipulation individually without disturbing other components. The SNAP gates, in conjunction with the displacement gate, constitute a universal gate set [201].

To create the $|1\rangle$ state, we start with the vacuum state $|0\rangle$, i.e., the cavity is empty. Apply a displacement gate to mix the vacuum state with higher Fock states: $|1\rangle$, $|2\rangle$, etc., creating a superposition. The SNAP gate adjusts the phase of specific Fock states. For example, we can apply a phase shift to just the $|0\rangle$ component, leaving the $|1\rangle$ and higher Fock components unchanged. By carefully choosing a reverse displacement, we cancel out unwanted contributions from the other states, isolating the desired or $|1\rangle$ state. Sometimes, this process is iterated (displacement and SNAP applied alternatively) to further refine the $|1\rangle$ state and improve its fidelity. A sequence of SNAP and displacements gates can also be used to prepare arbitrary qudit states and unitary gates. See further discussion under Section 4.8.

Significant progress has been made in experimentally creating multi-photon Fock states [202]. More than 20 Fock states were generated with a two-cell TESLA-type Nb cavity modified with enlarged aperture to increase mode separation, and coupled to transmons for control and readout through side tubes in the beampipe (Figure 4.7). The two modes had frequencies of 5.8 and 6.9 GHz. T_1 for Fock state $|1\rangle$ was measured to be 20.6 and 15.6 ms for the two bosonic modes. The lifetimes of the higher Fock states were found to decrease as expected

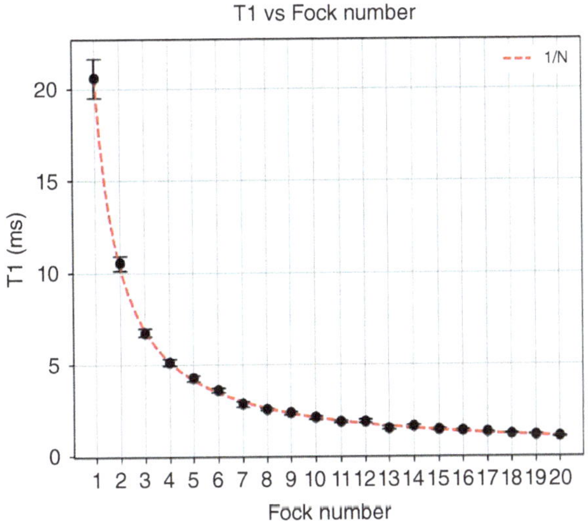

Figure 4.8 Relaxation times of Alice mode's Fock levels up to $|20\rangle$ following the expected $1/n$ behavior.

with increasing photon number occupation (Figure 4.8). They achieved 95% fidelity preparation of Fock states up to $N = 20$. An entangled state was also created using the Fock basis, namely, the Bell state $|01\rangle + |10\rangle$ with estimated coherence-limited fidelities up to 99.9%. These advancements mark a significant step in bosonic QC, laying the groundwork for exploring high-dimensional qudit encodings and gate operations.

4.5 Bosonic Encoding

At its most basic level, bosonic encoding [203–208] means that quantum information is stored in the Hilbert space of a 3D resonator rather than in a TLS. A cavity mode possesses an infinite ladder of energy eigenstates—the Fock states—and any quantum state of the mode can be expressed as a superposition of these number states. When we construct a qudit, we select a finite subset of the Hilbert space, for example, the lowest d Fock states, and information is encoded directly in photon number. Qudits represent the simplest possible bosonic encoding.

Quantum information stored in these systems still remains vulnerable to a small number of dominant physical error channels, most notably photon loss, dephasing, and phase-space diffusion (Section 4.9). In a bare Fock-state encoding, such errors act directly on the logical states, corrupting the stored information. This vulnerability motivates the introduction of more elaborate bosonic encodings. As additional structure is imposed on the Hilbert space, errors first become detectable, then correctable, and ultimately suppressible in a fault-tolerant manner. The various bosonic encodings discussed below—range from simple Fock-state qudits to dual-rail, binomial, cat, and GKP codes—can, therefore, be viewed as forming a hierarchy, distinguished by the degree to which they transform physical noise into identifiable and correctable logical errors. The encoded qubit is coupled to a transmon that plays the supplementary role in control and readout. Implementations of bosonic codes are making rapid progress in recent years by demonstrating QEC at the break-even point and also robust operations and fault-tolerant measurement of error syndromes, as we will discuss.

Traditional QIP schemes rely on coupling a large number of qubits. These protocols require significant resources for the many physical qubits. The drawback of this approach is that logical gate operations become more complicated because multiple physical systems need to be addressed.

Since bosonic codes encode quantum information into the states of a single mode, they reduce the need for multiple physical qubits, which is a major bottleneck in scaling up quantum computers. Implementing quantum gates and measurements with bosonic codes also requires fewer control lines and simpler operations than those needed for large-scale qubit arrays, reducing hardware complexity.

However, this elegance of bosonic codes comes with its own challenges. Bosonic codes rely on carefully engineered nonlinearity and high-fidelity control of a resonator's many-level Hilbert space, which can be technically demanding. Preparing and manipulating complex superpositions of Fock states or cat states requires precise pulse shaping and long coherence times. Even small imperfections can leak population into unwanted levels. Error correction, while more compact, often depends on repeated syndrome measurements or engineered dissipation, both of which introduce additional control complexity. Moreover, bosonic modes are particularly vulnerable to photon loss errors, and although the codes are designed to detect or correct such events, the performance of the entire scheme hinges on maintaining exceptionally low loss rates. In this sense, bosonic codes trade the spatial complexity of many qubits for the control complexity of a single, highly structured quantum system.

4.6 Coherent States

A driven superconducting resonator naturally produces coherent states that are the most classical-like states of a bosonic mode. A *coherent state* $|\alpha\rangle$ in a bosonic mode of a superconducting cavity is defined as an eigenstate of the annihilation operator:

$$\hat{a}\,|\alpha\rangle = \alpha\,|\alpha\rangle, \text{ where } \alpha \text{ is a complex amplitude.}$$

In the Fock basis, a coherent state takes the form:

$$|\alpha\rangle = e^{-|\alpha|^2/2}\sum_{n=0}^{\infty}\frac{\alpha^n}{\sqrt{n!}}\,|n\rangle$$

where α is a complex amplitude that determines the average photon number $|\alpha|^2$, showing that it is a superposition of number states rather than a state with a definite photon number. One can see the action of the annihilation operator as follows:

$$a\,|\alpha\rangle = e^{-|\alpha|^2/2}\sum_{n=0}^{\infty}\frac{\alpha^n}{\sqrt{n!}}a\,|n\rangle = e^{-|\alpha|^2/2}\sum_{n=0}^{\infty}\frac{\alpha^n}{\sqrt{n!}}\sqrt{n}\,|n-1\rangle.$$

Since $a\,|n\rangle = \sqrt{n}\,|n-1\rangle$.
Now shift the index $m = n-1$:

$$a\,|\alpha\rangle = e^{-|\alpha|^2/2}\sum_{m=0}^{\infty}\frac{\alpha^{m+1}}{\sqrt{m!}}\,|m\rangle = \alpha e^{-|\alpha|^2/2}\sum_{m=0}^{\infty}\frac{\alpha^m}{\sqrt{m!}}\,|m\rangle = \alpha\,|\alpha\rangle.$$

The resulting sum reconstructs the same coherent state multiplied by α. This is why $|\alpha\rangle$ is an eigenstate of a. The coherent state $|\alpha\rangle$ is an eigenstate of the annihilation operator a.

The photon number distribution follows a Poisson distribution:

$$P(n) = \frac{|\alpha|^{2n}e^{-|\alpha|^2}}{n!}$$

with mean photon number $\bar{n} = |\alpha|^2$and standard deviation $\sigma_n = \sqrt{\bar{n}} = |\alpha\,|$. The distribution is sharply peaked around: $n \approx |\alpha|^2$, which is the average photon number. As $|\alpha|^2 \to \infty$, relative quantum fluctuations vanish, phase becomes sharply defined, and field behaves deterministically. In that limit, quantum electrodynamics reduces to classical Maxwell theory.

As discussed earlier, a bosonic mode in a superconducting cavity is described by an infinite-dimensional Hilbert space $\mathcal{H}_\infty$, spanned by the Fock states $|n\rangle$. These states are eigenstates of the photon number operator $\hat{n} = a^\dagger a$, with eigenvalue n, meaning each $|n\rangle$ contains exactly n quanta of excitation in the electromagnetic field. Because the cavity mode is a harmonic oscillator, this ladder of number states extends without bound, giving access to a richer state space than the two-level structure of a qubit. Coherent states $|\alpha\rangle$, which are displaced vacuum states and eigenstates of the annihilation operator a, provide a complementary description: instead of having a definite photon number, they behave like classical electromagnetic waves with well-defined amplitude and phase. Together, Fock states and coherent states form two natural bases—one discrete, one continuous—for representing the same underlying cavity mode.

Bosonic codes take advantage of this large Hilbert space by encoding quantum information not in the discrete energy levels of a qubit, but in carefully engineered states of the oscillator itself. Logical states can be built from superpositions of Fock states, which exploit the discrete structure of the number basis, or from superpositions of coherent states, which leverage the continuous-variable nature of the field. Because all of this structure resides within a single physical mode, bosonic encodings dramatically reduce the number of physical qubits required to store and manipulate quantum information.

4.7 Qudits

In the Fock-state basis, a *qudit* is represented as a superposition of photon number states [209, 210]. The strength of a qudit versus a qubit is that a multiqubit state requiring many qubits can be mapped on to a *single*-qudit state. For example, a three-qubit state can be mapped into a single ($d = 3$) level qudit state. The scheme naturally provides all-to-all connected qubits, which otherwise require more hardware overhead (e.g., wiring) to realize in 2D superconducting qubit devices.

Expressed in terms of TLSs (qubits), a cavity with N levels can encode and process the same amount of information as $\log_2(N)$ TLSs (qubits). This brings two notable advantages of the qudit encoding over interconnected qubits. The number of control lines required is reduced from $\mathcal{O}(\log_2(N))$ to $\mathcal{O}(1)$, which makes the system more scalable. Entanglement is also simplified. To create the maximally entangled state in a collection of four TLSs, such as $|\psi\rangle = (1/\sqrt{2})(|0000\rangle + |1111\rangle)$ requires a series of three consecutive entangling operations to be performed. Creating the corresponding state with a qudit, as for example $|\psi\rangle = (1/\sqrt{2})(|0\rangle + |16\rangle)$, requires only a single operation.

While qudits offer advantages of compact encoding and reduced connectivity, their challenges include higher control complexity, error susceptibility, and measurement. Overcoming these issues requires advances in hardware, error correction, and quantum software frameworks. While qubits require a small set of standard gates (e.g., Pauli gates, Hadamard, CNOT), qudits need generalized multilevel gate operations. Qudit error correction is less mature than those of qubits and requires adapting existing protocols. Qubits typically use binary measurement (0 or 1), but qudits need precise discrimination between multiple states, requiring high-fidelity multilevel readout. Extracting the right state from a multilevel system thus could be more error-prone.

Although SRF cavities offer long coherence times, the number of usable levels in a qudit encoded in a single-cavity mode is still fundamentally limited. Even under optimistic assumptions—for example, a cavity lifetime of $T_1 = 1$ s and a minimum acceptable coherence of 200 μs per level—the effective Hilbert-space dimension is capped at roughly 5000 Fock states, corresponding to about 12 qubits' worth of computational space. Recall that for $n = 5000$, $T_1(n) = \frac{1\text{s}}{n} \geq 200$ μs. And the equivalent number of qubits is $\log_2(5000) \approx 12$.

In practice, however, the usable dimension is likely even smaller. As the photon number increases, the system gradually exits the dispersive regime, making selective control of individual Fock states increasingly challenging. The dispersive shift χ is not constant but varies nonlinearly with photon number n, reflecting the underlying anharmonicity of the transmon. At large n, the transmon experiences a stronger effective cavity field, which enhances higher-order interaction terms and modifies the effective coupling strength. Eventually, the interaction energy between the cavity and the transmon becomes comparable to their detuning, causing hybridization of the states and breakdown of the simple dispersive Hamiltonian. This defines a critical photon number beyond which the qudit levels are no longer well-resolved or controllable, imposing a practical ceiling on the maximum usable dimension of the cavity-encoded qudit. Although future transmons with coherence times approaching 100 ms or even 1 s would substantially extend the useful operating range and improve gate fidelities, the ultimate limit on the qudit dimension would still be set by dispersive breakdown and higher-order nonlinear effects rather than by coherence alone.

4.8 Qudit Control via Gates

Control of qudits is achieved via a sequence of analytic gates such as (un)conditional qubit rotation gates and (un)conditional cavity displacement gates. The system enables universal quantum control of each cavity mode using a combination of microwave pulses applied to the cavity to directly drive individual cavity modes to induce coherent displacement or rotations for preparing superpositions or displacing the modes in phase space.

As mentioned, the SNAP and displacement gates together form a universal gate set, allowing for the creation and transformation of any arbitrary qudit state. This means that, using only these two operations, one can prepare, transform and manipulate any arbitrary state within the d-dimensional Fock-space subspace of interest.

The displacement operator shifts the quantum state by a complex amplitude α

$$D(\alpha) = e^{\alpha a^\dagger - \alpha^* a}$$

Here, $a^\dagger$ and a are the photon creation and annihilation operators that raise or lower the photon number in the cavity. This operation coherently adds or removes photons, mapping Fock states to superpositions of other Fock states. It is implemented experimentally by applying a microwave drive to the cavity. Because the drive couples directly to the cavity field, the operation is unconditional—it acts the same way regardless of the state of any coupled qubit. Displacements are fast, high-fidelity operations and serve as the basic "continuous" control primitive for bosonic modes.

While displacement is an unconditional operation, the SNAP gate modifies phases conditioned on the Fock state. Each Fock level can obtain an independent and arbitrary phase:

$$S(\{\theta_n\}) = \sum_{n=0}^{d-1} e^{i\theta_n} |n\rangle\langle n|$$

where θ_n is an arbitrary phase applied to the Fock state $|n\rangle$. The SNAP gate, therefore, provides fine-grained, number-resolved control over the cavity. While displacement gates move the state around in phase space, SNAP gates "sculpt" the state by modifying the relative phases between Fock components. Together, these two operations—continuous translations and discrete phase adjustments—are sufficient to generate any unitary transformation on the truncated Fock space.

The *echoed conditional displacement (ECD) gate,* which is a conditional operation depending on the state of the qubit, forms a universal set along with unconditional single-qubit rotations ($\mathcal{R}$). Alternating ECD and $\mathcal{R}$ gates, any arbitrary unitary transformation in the Fock basis can be constructed, similar to how controlled operations and single-qubit rotations form a universal set in standard qubit architectures.

A standard conditional displacement gate $U_{\text{CD}}(\alpha) = |g\rangle\langle g| \otimes I + |e\rangle\langle e| \otimes D(\alpha)$ applies a displacement to the cavity state only when the qubit is in the excited state. This is achieved by applying a cavity drive whose frequency is shifted by χ to keep it on-resonance when the qubit is in $|e\rangle$. However, this procedure is very slow needing a time $\gg 2\pi/\chi$.

The ECD gate requires a much lower qubit–qubit coupling than the SNAP gate and can be realized faster than $2\pi/\chi$ limit. It has a more symmetrical expression given by

$$U_{\text{ECD}}(\alpha) = |g\rangle \left\langle g\right| \otimes D\left(-\frac{\alpha}{2}\right) + |e\left.\right\rangle \langle e| \otimes D\left(\frac{\alpha}{2}\right).$$

This operation applies a displacement of $\alpha/2$ when the qubit is in $|g\rangle$ and $-\alpha/2$ when the qubit is in $|e\rangle$.

One downside of the always-on qubit–cavity dispersive interactions is that it introduces unwanted dynamic phases, ECD cleverly uses echoing techniques (alternating conditional displacements with qubit flips) to cancel out unwanted interactions and achieve precise control over Fock states, leading to low errors while beating the $2\pi/\chi$ length limit.

4.9 Errors, Detection, and Correction

In bosonic QC, the three primary types of errors are photon loss errors, displacement errors, and phase errors.

Photon loss occurs when an energy decay process removes a photon from the system, if energy leaks out due to finite cavity lifetime, or if the transmon qubit coupled to the cavity introduces loss, or due to thermal excitations and coupling to other modes. When a photon is lost, the system moves from $|n\rangle$ to $|n-1\rangle$. This is modeled using the annihilation operator:

$$\hat{a}\,|n\rangle = \sqrt{n}\,|n-1\rangle$$

Photon loss typically leads to decoherence. Photon loss dominates in cavities. Certain bosonic encodings (cat, binomial, or GKP) are resistant to small photon loss since losing a single photon just flips the parity but does not completely destroy the encoded information. Loss of a single photon maps an even-photon state to an odd-photon state. Monitoring the parity of the state, therefore, allows for error detection. A QND measurement of photon parity is performed. Parity is determined by the photon number operator $\hat{n}$, and the parity operator as

$$\hat{\Pi} = e^{i\pi\hat{n}} = (-1)^{\hat{n}}$$

States with even-photon numbers have parity +1. States with odd-photon numbers have parity −1. To see this more clearly: If the state is in an unexpected parity, an active correction can be applied via an error correction pulse to restore the state.

The dispersively coupled ancilla qubit (transmon) (Section 3.30) can be used to measure the photon number parity without destroying the encoded state [211]. The interaction Hamiltonian of the transmon with the cavity $H_{\text{int}} = \chi a^{\dagger} a \sigma_z$ leads to a shift in the ancilla qubit frequency depending on the number of photons in the cavity. Here χ is the dispersive coupling strength, and σ_z is the Pauli-Z operator for the transmon. The parity measurement protocol involves a Ramsey-type experiment where a first $\pi/2$ pulse is applied to the qubit bringing it along the $+x$-axis, followed by wait period of π/χ. During this wait time, the qubit stays along the $+x$-axis if there is 0 photons in the cavity and rotates at a rate of χ and ends up along the $-x$-axis (π rotation about the z-axis) if there is one photon in the cavity. If there are any even number of photons, the qubit makes integer multiple of 2π rotations ending up along the $+x$-axis. In contrast, if there are odd number of photons, the qubit lands along the $-x$-axis. Then a second $\pi/2$ pulse is applied, which maps the even-photon number states to $|0\rangle$ and odd ones to $|1\rangle$.

If a photon loss is detected, there is a need to restore the lost photon while preserving quantum information. By applying a controlled photon raising operation $\hat{a}^{\dagger}$, we can move the state back into the logical encoding space. The system is restored by adjusting the relative phases and amplitudes, ensuring that the logical qubit remains intact. The system continues functioning as if no error occurred.

A *displacement error* occurs when a small, uncontrolled shift in phase space adds a small coherent displacement to the quantum state. A Fock state $|n\rangle$ has no definite phase and a fixed photon number. A small displacement mixes it with neighboring Fock states. So a displacement changes the photon number distribution. $|n\rangle$ becomes a superposition of other photon number states. The state is "pushed" toward neighboring number states, and photon number uncertainty increases.

Displacement error is modeled by the *displacement operator*:

$$D(\delta) = e^{\delta\hat{a}^{\dagger} - \delta^{*}\hat{a}}$$

where δ represents a small shift in position (X) and momentum (P) of the quantum state. If a small displacement error occurs in a coherent state $|\alpha\rangle$,

$$D(\delta)\;|\alpha\rangle = |\alpha + \delta\rangle$$

The physical causes are imperfect qubit or cavity control pulses. Measuring a system can create weak, uncontrolled displacements. Small thermal excitations displace the quantum state. Error detection is by dispersive readout. Feedback control applies the inverse displacement $D(-\delta)$.

A single number state does not encode information in a relative phase. There is no "second component" to interfere with. If information is encoded in superpositions like $|0\rangle + |1\rangle$, then phase matters. These encodings can suffer dephasing. A *phase error* affects the relative phase between superposition components. It is caused by stochastic shifts in the phase of the quantum state, changes in transmon frequency, dispersive interactions, charge noise, or temperature fluctuations. It is described by the phase operator $e^{i\theta\hat{n}}$ as

$$e^{i\theta}\hat{n}\;|n\rangle = e^{in\theta}\;|n\rangle$$

Parity measurements and echo pulses can detect phase drifts. After detection, a corrective phase shift $e^{-i\theta\hat{n}}$ can be applied.

4.10 Qudit Error Correction

Qudits can sometimes tolerate higher noise rates before logical error rates explode (because errors get "spread out" over more levels). Qudit surface codes can be defined on higher-dimensional Hilbert spaces and so can often pack more error-correcting power per physical unit. In qubits, there's only one type of shift error (flip $0 \leftrightarrow 1$). In qudits, there are multiple shift errors (move by $1, 2, \ldots, d-1$ steps). Error correction must detect and correct *any of these cyclic shifts.*

Although qudits pack more information into fewer devices, there can be higher error thresholds. Both gates and error correction are more complicated than with qubits. Crosstalk and leakage errors are harder to suppress because higher-dimensional Hilbert space offers more escape channels. Qubit error correction is simpler conceptually, but very resource-hungry (millions of physical qubits projected for fault-tolerant quantum computers). Qudit error correction potentially reduces hardware overhead by encoding more per element but increases operational overhead (complex gates, complex syndrome measurements).

4.11 Dual-Rail Encoding

Dual-rail encoding [212, 213] provides a conceptually transparent way to represent a logical qubit using cavity modes. Quantum information is stored in the distribution of a single excitation across two distinct modes. The modes can be linear like a cavity resonator or nonlinear like a transmon; however, we will continue our discussion with the former. Logical states are defined as

$$|0_L\rangle = |1,0\rangle, |1_L\rangle = |0,1\rangle,$$

where the two entries denote photon occupation in two separate resonators or "rails." Instead of storing the qubit in *one* mode with two energy levels, you store it in *two* modes with one photon shared between them. In this architecture (Figure 4.9), two high-Q microwave cavities—labeled Alice and Bob—serve as the two rails. $|1, 0\rangle$ means 1 photon in resonator 1 and 0 photons in resonator 2. $|0, 1\rangle$ means 0 photons in resonator 1 and 1 photon in resonator 2. The qubit is encoded in where the photon is, not in the internal state of a single mode.

Photon loss, the dominant error channel in superconducting cavities, causes transitions out of the code space—typically into $|00\rangle$ that lies outside the logical subspace. A loss event changes the total excitation parity or photon number, allowing error detection. By measuring total excitation number (or performing a nondestructive parity measurement), one can determine that the qubit is no longer in the one-photon manifold. The error is, therefore, *flagged*. In quantum error correction language, photon loss becomes an **erasure error** rather than an unknown Pauli error. An erasure error is easier to correct than a generic unknown error because its location and occurrence are known. By converting amplitude damping (photon loss) into a detectable erasure event, dual-rail encoding transforms the dominant physical noise channel into a tractable form.

However, dual-rail encoding does not automatically protect against all types of errors. For example, a phase error acting within the logical subspace—such as a relative phase between $|1, 0\rangle$ and $|0, 1\rangle$—does not take the state outside the one-photon manifold and is, therefore, not intrinsically detectable. Dual-rail is, therefore, best understood as a loss-detecting code (Figure 4.10), not a full Pauli-correcting code on its own. It converts amplitude damping into erasure, but it does not by itself correct arbitrary X, Y, or Z errors.

Each cavity is coupled to its own transmon ancilla, which enables both control and measurement, while additional couplers allow engineered interactions

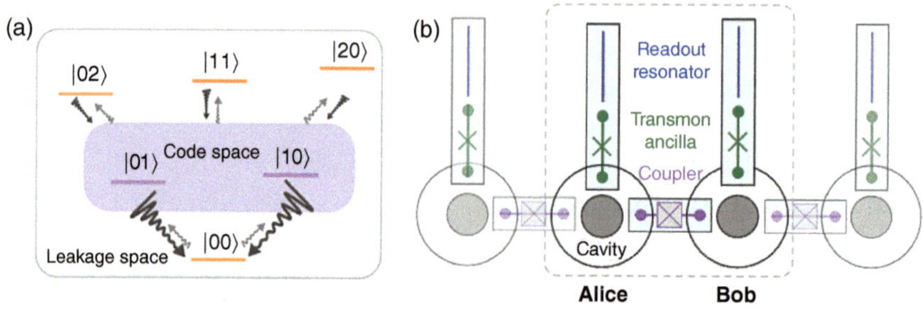

Figure 4.9 Dual-rail bosonic encoding and circuit QED implementation. (a) Schematic representation of the dual-rail code space defined within the two-mode Fock basis. The logical qubit is encoded in the single-excitation manifold, with logical states $|0_L\rangle = |01\rangle$ and $|1_L\rangle = |10\rangle$ spanning the protected code space (shaded region). States outside this manifold, including the vacuum $|00\rangle$, two-photon states $|11\rangle$, and higher-excitation states $|02\rangle$ and $|20\rangle$, constitute leakage spaces. Photon loss events map both logical states to the vacuum state $|00\rangle$, producing a detectable error without inducing a logical bit flip, while photon gain or excitation processes lead to transitions into higher-lying leakage states. (b) Physical realization of the dual-rail encoding in a circuit QED architecture. Two long-lived microwave cavity modes (labeled *Alice* and *Bob*) serve as the two rails of the logical qubit. Each cavity is dispersively coupled to a transmon ancilla, which enables state preparation, conditional control, and error-syndrome measurement. Tunable couplers between the cavities allow coherent photon-exchange (beam-splitter–type) interactions used to implement logical operations within the code space. A dedicated readout resonator provides high-fidelity measurement of the ancilla without directly disturbing the stored cavity states. In this architecture, quantum information is stored nonlocally across two bosonic modes, combining long cavity coherence times with fast transmon-based control; (Ref. [213] / Springer Nature).

between the modes. The encoding provides passive protection against certain errors and enables straightforward error detection. Photon loss events are mapped to transitions into known leakage states, which can be identified using ancilla-based syndrome measurements.

Dual-rail states are relatively easy to prepare, manipulate, and measure with high fidelity [214]. But when a photon is lost, the logical qubit is gone.

Binomial, cat, and GKP codes discussed in the following sections take the same idea further. They deliberately occupy many Fock states in a single resonator, engineering interference between photon number components so that dominant errors—especially photon loss—become more correctable or produce syndromes that preserve the logical information.

4.12 Binomial Codes

Binomial codes [215] are a type of bosonic quantum error-correcting code designed to encode a logical qubit into a superposition of Fock states. The

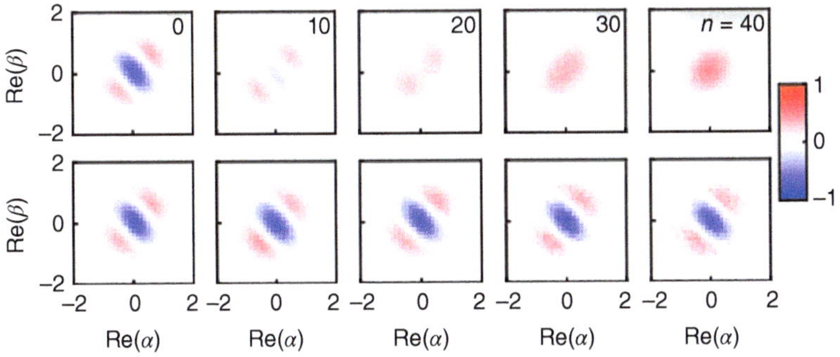

Figure 4.10 (Upper panels) Tomographic reconstructions of the dual-rail encoded state after increasing numbers of operations (or equivalently increasing evolution time), illustrating progressive decoherence and photon loss-induced leakage from the single-excitation manifold. The distributions broaden and gradually lose interference contrast, reflecting and sensitivity to photon loss. (Lower panels) After discarding trajectories where erasures (state $|00\rangle$) were detected, it is possible to recover the original information (Ref. [212] / American Physical Society). (See Section 4.26 for Wigner tomographs.)

code words are the patterns of Fock states selected. The patterns are chosen so that physical errors act in a controlled way. The idea is to separate the Fock components of the logical states by fixed gaps in photon number. These codes are specifically constructed to protect against photon loss errors and certain dephasing errors. They require only a finite number of photons, facilitating experimental feasibility.

A simple binomial code is the (2, 1) code, which encodes one logical qubit into a superposition of three Fock states. This code was introduced as the smallest member of a binomial codes, hence its name kitten or "baby-binomial" code.

Logical 0 state: $|0_L\rangle = \frac{1}{\sqrt{2}}(|0\rangle + |4\rangle)$ is a superposition of 0-photon and 4-photon states.

Logical 1 state: $|1_L\rangle = |2\rangle$ is encoded in the 2-photon state.

Photon loss ($\hat{a}\,|n\rangle = \sqrt{n}\,|n-1\rangle$) is the dominant error in bosonic modes modeled by the annihilation operator $\hat{a}$. This means that if a photon is lost, the system moves from $|n\rangle$ to $|n-1\rangle$.

The binomial code is designed so that a single-photon loss shifts the state into a different but correctable space, making recovery possible. If the oscillator undergoes small coherent displacements (noise that shifts the phase space slightly), the binomial code structure allows partial correction. Any photon loss moves the state into a known *error space*, allowing an error-syndrome measurement to detect and correct it. *Error space* refers to the set of physical states into which a code word is mapped when an error occurs. Different numbers of photon losses generate different error space. The spacing between Fock components defines the "distance" of the code in photon number space.

The *code size* in a binomial code quantifies how much photon loss the encoding can tolerate before logical information becomes ambiguous. It is determined by how widely spaced the Fock components of the code words are and how many such components are included. Increasing code size means increasing the extent or spacing of the photon number pattern. This allows the code to correct more loss events, but it also requires higher photon numbers and thus increases sensitivity to cavity decay. Code size in binomial encoding is tunable: it can be made larger by including higher Fock states, at the cost of greater physical resource usage.

More specifically, a photon loss gives

$$\hat{a}\,|0_L\rangle = \frac{1}{\sqrt{2}}(\hat{a}\,|0\rangle + \hat{a}\,|4\rangle) = \frac{1}{\sqrt{2}}(0 + 2\,|3\rangle) = \sqrt{2}\,|3\rangle$$

$$\hat{a}\,|1_L\rangle = \hat{a}\,|2\rangle = \sqrt{2}\,|1\rangle$$

The logical $|0_L\rangle$ moves into an error space spanned by $|3\rangle$.
The logical $|1_L\rangle$ moves into an error space spanned by $|1\rangle$.

The new states are outside the logical encoding space but are in a known *error space*, allowing us to detect and correct them. Since binomial codes separate logical states by parity, in order to detect the error (syndrome measurement), monitor photon number parity by measuring the ancilla qubit state. The loss of a photon changes the parity (odd/even) of the photon number.

If a photon loss is detected, we need to restore the lost photon while preserving quantum information. The simplest way is to apply a photon displacement operation using a controlled drive to apply a controlled unitary operation $\hat{a}^\dagger$ moving the state back into the logical encoding space. The system continues functioning as if no error occurred.

When a binomial code state is reconstructed through Wigner tomography (Section 4.26), the resulting phase-space distribution (Figure 4.11) reveals the structure engineered into the oscillator's Hilbert space. It tells us that that the intended Fock-state superpositions are coherently prepared. The pattern of negativity reflects the interference between selected photon number components that define the logical code words. The structure in the Wigner tomograph is experimental confirmation that the intended Fock-state superposition has been prepared coherently. The tomograph reveals the symmetry. For example, a code designed to protect against single-photon loss will show invariance under specific phase-space rotations.

Beyond passive protection, binomial encodings also introduce computational leverage. Because the logical states occupy only a discrete set of number states with well-defined frequency separations, many logical gates can be implemented

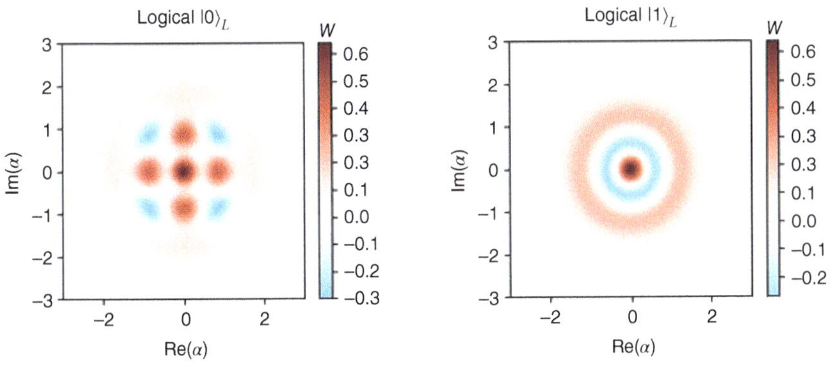

Figure 4.11 Wigner tomograph of binomial code states. Wigner quasiprobability distributions in phase space for logical qubit states encoded using binomial bosonic codes. The horizontal and vertical axes give the real and imaginary quadratures of the complex field amplitude α. Left: the logical state $|0\rangle_L$ exhibits a multilobed structure with pronounced interference fringes arranged in a rotationally symmetric pattern around the origin. The fine alternating red–blue features correspond to regions of positive and negative Wigner values arising from coherent superpositions of selected Fock states determined by the binomial code's photon number weighting. Right: the logical state $|1\rangle_L$ displays a more radially concentrated distribution with concentric interference rings, reflecting a different binomial superposition of Fock components that distinguishes it from $|0\rangle_L$. The presence of negative regions in both panels is a hallmark of nonclassicality and coherence of the encoded logical states. Color bars indicate the normalized Wigner function amplitude $W(\alpha)$. [216] / American Physical Society / CC BY 4.0.

through frequency-selective drives rather than long sequences of time-domain pulses. For example, a logical Z rotation can be realized by applying a dispersive interaction that shifts the phases of only the occupied Fock components—often achievable with a single-shaped cavity pulse instead of dozens of primitive transmon-level gates [217]. Experiments have demonstrated logical operations in binomial encodings where a single cavity-level operation replaces roughly 10–50 elementary transmon gates, highlighting the hardware efficiency of this approach [218].

4.13 Cat States and Cat Codes

Cat states (Schrödinger cat states) are superpositions of coherent-state [156, 157] wavepackets in phase space. In a superconducting cavity, these states arise naturally because a driven resonator produces coherent states as stable, classical-like field configurations. A cat state is formed by combining several such coherent states with well-defined relative phases, creating interference patterns that encode quantum information in global, symmetry-protected features of the electromagnetic field.

Cat states are especially useful for quantum error protection because they exploit the symmetries of multiphoton states. In particular, properties such as photon number parity or the relative phase between coherent components remain stable under many common noise processes. This intrinsic robustness means that cat states function as hardware-efficient logical qubits: part of their protection is built directly into the physics of the bosonic mode, even before any active error correction protocol is applied.

A key example is photon loss. Losing a photon changes the parity of the state but does not immediately destroy the encoded information. Because photon loss affects one logical degree of freedom much more strongly than the other, one can design gates that preserve this bias and avoid converting this dominant error into more harmful logical errors. As a result, computation can be performed in a way that naturally suppresses the most damaging noise, reducing the overhead required for fault tolerance.

Logical operations on cat qubits are implemented through smooth, analog transformations of the cavity field rather than long sequences of discrete qubit gates. Displacements correspond to applying a small classical microwave drive, nudging the state in phase space. Rotations arise from free evolution of the cavity field, producing a phase-space rotation that implements a logical Z-type operation. Controlled displacements implement logical X operations. *Kerr-induced evolutions*, by contrast, originate from the intrinsic nonlinearity of the JJs that couple to the cavity. This nonlinearity makes the cavity frequency depend slightly on photon number, so different Fock components accumulate phase at different rates.

When a cavity is coupled to a JJ, it introduces a tiny nonlinearity. This nonlinearity makes the cavity *slightly anharmonic*, meaning that the energy spacing between $|n\rangle$ and $|n+1\rangle$ is no longer perfectly constant, and the cavity frequency depends weakly on the photon number n. Mathematically, this shows up as a term in the Hamiltonian:

$$H_{\text{Kerr}} = -\frac{K}{2}a^{\dagger 2}a^2$$

Because of this term, each Fock component $|n\rangle$ in a superposition picks up a phase that depends on $n(n-1)$. So if the state is

$$|\psi\rangle = \sum_n c_n\,|n\rangle,$$

then under Kerr evolution it becomes

$$|\psi(t)\rangle = \sum_n c_n e^{-iKn(n-1)t/2}\,|n\rangle$$

Different photon number components rotate at different rates. This is the essence of Kerr-induced evolution.

As the state evolves under this Kerr Hamiltonian, the relative phases between number states shear the Wigner function, reshaping the interference structure that defines the encoded cat manifold. When the Kerr effect makes different Fock components accumulate different phases, the state's structure in phase space becomes distorted. Points at larger radius in phase space rotate faster, points near the origin rotate more slowly. The overall shape gets stretched and skewed, a distortion is called shearing. A circular blob becomes an ellipse, an ellipse becomes a banana shape. The Wigner function is twisted around the origin. Because the interference pattern *is* the logical information in a cat qubit, Kerr evolution becomes a powerful tool for implementing logical gates.

Bosonic codes are defined by two central concepts: code words and code size. Code words are the physical quantum states chosen to represent the logical basis states (e.g., logical $|0_L\rangle$ and logical $|1_L\rangle$). They are engineered so that dominant physical errors move the states in predictable and distinguishable ways. Logical information is, therefore, encoded not in a single configuration but in the structure and relationships among the code words. Code size measures how well separated the code words are in Hilbert space relative to the typical action of noise. If small errors can easily move one code word into another, the code is fragile. If errors must traverse a large "distance" before confusion occurs, the logical information is protected. Code size thus quantifies the tolerance of the encoding to noise.

Increasing the cat size means increasing the amplitude of the coherent states, which corresponds physically to storing more energy (more photons) in the cavity. As the average photon number grows, the coherent-state wavepackets move farther apart in phase space, while their intrinsic quantum width remains roughly constant. This increased separation reduces the overlap between components, making the logical states more distinguishable and improving error discrimination. Small cats (1–2 photons) are easy to prepare and tolerant of loss, but the coherent components overlap, limiting protection. For moderate cats (3–5 photons), the components are well separated, parity becomes robust, and logical states are clearly defined. Photon loss remains manageable, enabling repeated parity measurements and feedback. With Large cats (8–10+ photons), components are extremely well separated, and overlap is negligible. However, photon loss becomes more frequent because the loss rate scales with high photon number.

Larger cats, therefore, tolerate larger phase-space displacements before logical states become confused. Small noise causes only slight motion of the wavepackets, leaving them clearly associated with their original component. Only when displacements become comparable to the spacing between components does a logical error occur.

The *error space* consists of states reached when physical noise—primarily photon loss—displaces the cat state away from the ideal code manifold. Photon loss does not collapse the state to a single coherent component; instead, it shifts the state in phase space and flips its parity, altering its symmetry properties in a predictable way.

When stored in high-Q superconducting cavities, cat states exhibit coherence times exceeding those of transmons, making them long-lived quantum memories. More importantly, they serve as active computational units: logical gates, interference operations, and entanglement can be performed directly within the oscillator Hilbert space. Computation proceeds not by toggling a TLS between $|0\rangle$ and $|1\rangle$, but by coherently shaping probability amplitudes distributed across many photon number states or across continuous regions of phase space.

High-fidelity logical Pauli and Clifford gates have been demonstrated by tailoring phase-space transformations. Z operations and logical X operations often outperform the underlying physical qubits because the encoded states average over microscopic noise.

Cat states can also be entangled with each other or with ancillary qubits, enabling the generation of multipartite entanglement and facilitating communication between modules in a modular quantum processor. In such hybrid architectures, cat states act simultaneously as robust storage units and as flexible communication channels.

Cat-State Generation

The cavity is initially in the vacuum state $|0\rangle_c$ and the transmon is prepared in a superposition state: $\frac{1}{\sqrt{2}}(|g\rangle + |e\rangle)$, where $g\rangle$ and $|e\rangle$ are the ground and excited states of the transmon, respectively. Applying a conditional displacement $D(\alpha)|g\rangle\langle g| + D(-\alpha)|e\rangle\langle e|$ controlled by the transmon state leads to

$$\frac{1}{\sqrt{2}}(|g\rangle\ |\alpha\rangle_c + |e\rangle\ |-\alpha\rangle_c).$$

This means the cavity mode is in a superposition of displaced coherent states, entangled with the transmon. To create a pure cavity cat state, the transmon needs to be disentangled from the cavity. This is done using a π-pulse on the transmon, which transforms:

$$|g\rangle \rightarrow \frac{1}{\sqrt{2}}(|g\rangle + |e\rangle), |e\rangle \rightarrow \frac{1}{\sqrt{2}}(|g\rangle - |e\rangle).$$

Applying this transformation results in the combined state $\frac{1}{2}[|g\rangle(|\alpha\rangle + |-\alpha\rangle) + |e\rangle(|\alpha\rangle - |-\alpha\rangle)$. The transmon is then measured. If the outcome is the ground state, one gets the even cat $|C_+\rangle$ and $|C_-\rangle$ for $|e\rangle$. A conditional phase gate on the cavity based on the qubit measurement can be applied to deterministically prepare one of the cat states.

4.14 Two-Component Cat State

A two-component cat state is a superposition of two coherent states $|C^{\pm}(\alpha)\rangle = \frac{1}{\mathcal{N}^{\pm}}(|\alpha\rangle \pm |-\alpha\rangle)$, where $\mathcal{N}^{\pm} = \frac{1}{\sqrt{2(1\pm e^{-2|\alpha|^2})}}$ is a normalization factor.

Information is encoded in a basis spanned by cat states, i.e., the superposition of two coherent states $|\alpha\rangle$ and $|-\alpha\rangle$ of a resonant mode, where α is a complex amplitude.

The two-cat code (Figure 4.12) encodes logical qubits using two cat states, even and odd:

$$|0_L\rangle = |C^{+}(\alpha)\rangle = \frac{1}{\sqrt{2(1+e^{-2|\alpha|^2})}}(|\alpha\rangle + |-\alpha\rangle)$$

The above state is called the even-cat state because it is composed only of even-photon number components.

$$|1_L\rangle = |C^{-}(\alpha)\rangle = \frac{1}{\sqrt{2(1-e^{-2|\alpha|^2})}}(|\alpha\rangle - |-\alpha\rangle)$$

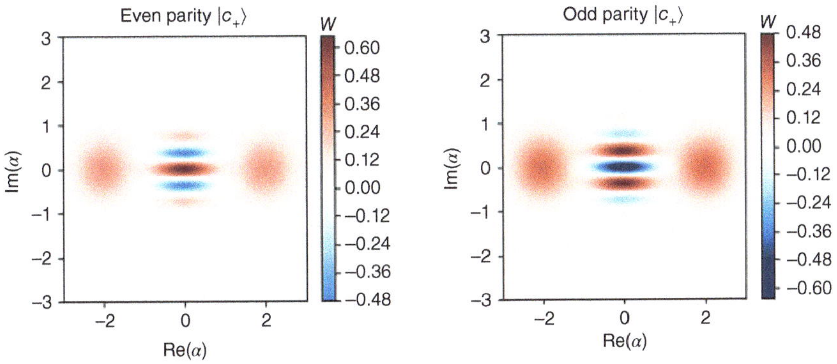

Figure 4.12 Wigner tomograph of logical two-leg cat states. Wigner phase-space quasiprobability distributions for two coherent-state superpositions (cat states) of opposite photon number parity. The horizontal and vertical axes correspond to the real and imaginary quadratures of the complex field amplitude α. Left: the even-parity cat state $|C_+\rangle$ exhibits two dominant Gaussian lobes centered symmetrically in phase space, with an interference fringe pattern of alternating positive (red) and negative (blue) Wigner values between them. The positive value at the origin reflects even-photon number parity. Right: the odd-parity cat state $|C_-\rangle$ shows a similar pair of coherent components but with a negative Wigner value at the origin, characteristic of odd parity, and a shifted interference pattern. The oscillatory fringes between the two lobes encode quantum coherence between the constituent coherent states, while the negative regions of the Wigner function are signatures of nonclassicality. Color bars indicate the normalized Wigner amplitude $W(\alpha)$. [220] / GUPEA.

The above odd-cat state contains only odd-photon number components. Since the states are superpositions of coherent states with opposite phases, they exhibit well-defined photon number parity, a crucial feature for error detection.

In the Fock basis representation, the even- and odd-cat states read

$$|C^{+}(\alpha)\rangle = \frac{2}{\mathcal{N}_{\alpha,+}} e^{-|\alpha|^2/2} \sum_{n=0}^{\infty} \frac{\alpha^{2n}}{\sqrt{(2n)!}} \, |2n\rangle$$

$$|C^{-}(\alpha)\rangle = \frac{2}{\mathcal{N}_{\alpha,-}} e^{-|\alpha|^2/2} \sum_{n=0}^{\infty} \frac{\alpha^{2n+1}}{\sqrt{(2n+1)!}} \, |2n+1\rangle$$

Photon loss transforms a cat state with even-photon number parity into an erroneous odd-parity state. This change in parity can be detected using *continuous parity monitoring* without disturbing the encoded information. Correcting photon loss is possible by applying a unitary operation conditioned on the detected error. Upon detecting a parity change, feedback control restores the logical state by displacing the system back to the correct superposition.

The strength of the two-component cat qubit lies in its simplicity and hardware efficiency. The states are relatively easy to generate, require only modest photon numbers, and can be naturally stabilized using two-photon drives. Once stabilized, the two-leg cat code changes the character of computation: the encoded qubit inherits a strong bias against bit-flip errors, and many logical operations can be implemented directly through bias-preserving drives rather than long gate sequences. Experiments have shown that this encoding achieves logical gate fidelities comparable to those of state-of-the-art physical qubits, while using only a single high-Q cavity mode instead of the 50–100 physical qubits typically required to build a surface code logical qubit with a similar lifetime [219]. This hardware-efficient protection is one of the key reasons cat qubits are viewed as a promising route toward scalable fault-tolerant architectures. Large-amplitude two-component cat states with average photon numbers exceeding 10 have been demonstrated experimentally in superconducting cavities, where increased coherent-state separation exponentially suppresses bit-flip errors without immediately converting photon loss into logical dephasing [219].

4.15 Four-Component Cat State

Two-component cat codes correct only single-photon loss errors. They are not resilient against higher-order errors like dephasing or multiple photon losses. A four-component cat state extends the encoding space using

$$|C_{\pm}^{(e)}\rangle = \frac{1}{\sqrt{2}}(|C_{+}\rangle \pm i\,|C_{-}\rangle)$$

$$|C_{\pm}^{(o)}\rangle = \frac{1}{\sqrt{2}}(|C_{+}\rangle \pm |C_{-}\rangle)$$

Or, expressed more simply as

$$|0_L\rangle = \frac{|\alpha\rangle + |i\alpha\rangle + |-\alpha\rangle + |-i\alpha\rangle}{2}$$

$$|1_L\rangle = \frac{|\alpha\rangle - |i\alpha\rangle + |-\alpha\rangle - |-i\alpha\rangle}{2}$$

This introduces an additional degree of freedom for error correction. Similar to the two-component cat state, the parity change is detected. Feedback is applied to restore coherence. The extra redundancy helps correct phase errors that arise from cavity mode fluctuations. Figure 4.13 shows the Wigner tomograph of the four-component cat state.

The key computational advantage of the four-leg cat state is that large classes of physical operations preserve the logical subspace. The code words are arranged with fourfold rotational symmetry in phase space, so any drive or Hamiltonian that respects this symmetry cannot mix the logical states or introduce certain logical errors. In practice, this means that even if a pulse is imperfect—slightly

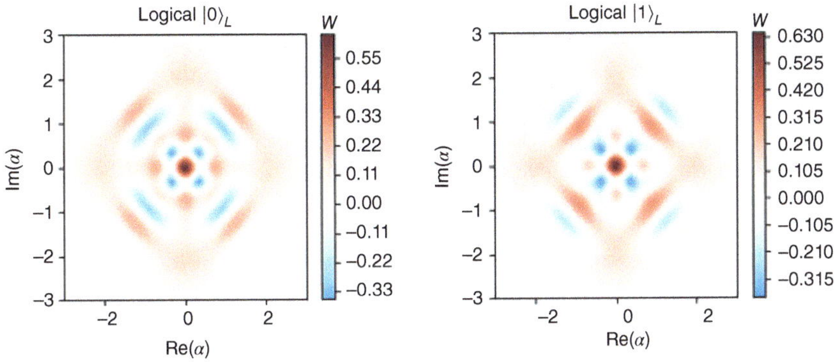

Figure 4.13 Wigner tomograph of logical four-leg cat states. Wigner quasiprobability distributions in phase space for logical qubit states encoded in four-component Schrödinger cat states. The horizontal and vertical axes correspond to the real and imaginary quadratures of the complex field amplitude α. Left: the logical state $|0\rangle_L$ is represented by a superposition of four coherent states arranged approximately at the corners of a square in phase space. The corresponding Wigner function displays four main positive lobes, surrounded by a lattice of interference fringes with alternating positive (red) and negative (blue) regions, evidencing strong quantum coherence among the four components. Right: the logical state $|1\rangle_L$ shows a similar four-lobe structure with a distinct internal interference pattern. The pronounced negative values of the Wigner function highlight the nonclassical nature of the states, while the fine checkerboard-like fringes near the origin encode the phase relationships that enable error-correctable logical encoding. Color bars indicate the normalized Wigner function amplitude $W(\alpha)$.

mis-timed, mis-shaped, or detuned—its action is constrained by symmetry to remain within the protected subspace. This sharply reduces the need for delicate pulse engineering. Quantitatively, simulations and early experiments show that several gate-induced error channels, especially those associated with phase-space distortions, are suppressed by factors of 5–10 compared with two-leg cats under the same control noise [221]. Computation becomes less about achieving high-precision control and more about ensuring that the applied operations respect the underlying symmetry—a shift that offers a powerful scaling advantage as systems grow more complex.

4.16 GKP Coding

GKP encoding [222] is a bosonic error-correcting code that encodes logical qubits into the continuous degrees of freedom of a quantum system, such as the position and momentum of a harmonic oscillator. This scheme is fundamentally different from binomial or dual-rail as those use Fock-level encodings. GKP encoding targets the dominant physical errors of bosonic systems—small, continuous displacements in position and momentum—at their source.

A cavity mode is an infinite ladder of energy levels and is naturally described by two continuous variables, analogous to the position q and momentum p of a mechanical oscillator. These variables form the *phase space* of the oscillator, a 2D plane in which any quantum state can be represented as a distribution. The electromagnetic analogs of q and p are the field *quadratures* X and P. For a classical electromagnetic mode with electric field

$$E(t) = E_0[X\cos(\omega t) + P\sin(\omega t)]$$

the coefficient X sets the amplitude of the cosine component (a "position-like" quadrature), while P sets the amplitude of the sine component (a "momentum-like" quadrature). Quantum mechanically, the operators $\hat{X}$ and $\hat{P}$ obey the same canonical commutation relation as $\hat{q}$ and $\hat{p}$:

$$[\hat{q}, \hat{p}] = i\hbar$$

$$[\hat{X}, \hat{P}] = \frac{i}{2}$$

Thus, the state of a cavity mode can be visualized as a point or distribution in the (X, P) plane. The annihilation and creation operators are then defined as *complex combinations* of these quadratures:

$$a = \frac{1}{\sqrt{2}}(X + iP), \quad a^\dagger = \frac{1}{\sqrt{2}}(X - iP).$$

$$X = \frac{a + a^\dagger}{\sqrt{2}}, \quad P = \frac{a - a^\dagger}{i\sqrt{2}}.$$

Cavity resonators do not suffer discrete bit-flip or phase-flip errors the way qubits do. Their dominant noise process is small, continuous displacements in the quadrature variables q and p. Environmental couplings—vacuum fluctuations, thermal excitations, weak interactions—nudge the state slightly in random directions. These shifts accumulate gradually, producing Gaussian diffusion in phase space, where small displacements are common and large ones are rare. Photon loss appears as a secondary effect when diffusion pushes the state toward lower-energy regions. The essential point is that oscillator noise is geometric drift, not discrete jumps.

The GKP code is engineered specifically to correct these small displacements. Instead of storing information in individual Fock levels or photon numbers, GKP states encode logical information in a periodic lattice in phase space. This structure allows the code to detect and correct continuous shifts by determining how far the state has wandered relative to the nearest lattice point.

To build such a lattice [223], one must first tailor the natural symmetry of oscillator noise. A typical oscillator state with equal uncertainty in both quadratures appears as a circular Gaussian blob in phase space—too delocalized to support a grid (Figure 4.14). Creating a GKP state requires *squeezing*, which redistributes uncertainty: noise in one quadrature (say q) is reduced below the vacuum level, while noise in the conjugate quadrature p increases (Figure 4.14). In phase space, the state becomes a narrow ellipse. These squeezed wavepackets serve as the "pixels" from which the GKP lattice is assembled (Figure 4.14).

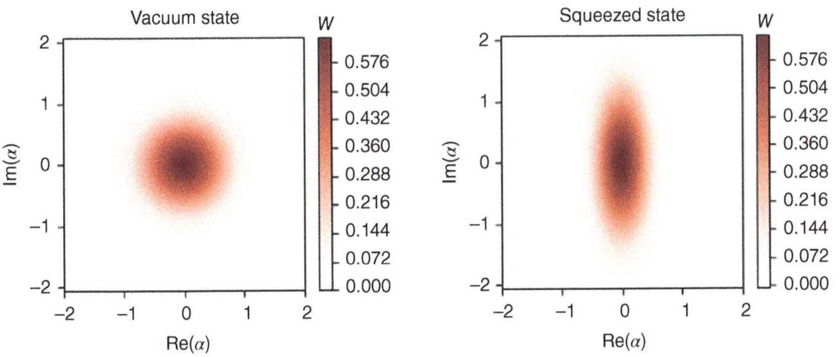

Figure 4.14 Wigner tomography of the vacuum state (left) and a single-mode squeezed state (right) plotted in phase space (Re(α), Im(α)). The vacuum state exhibits a symmetric Gaussian distribution reflecting equal quantum fluctuations in the two orthogonal field quadratures. In contrast, the squeezed state displays anisotropic uncertainty, with reduced variance along the x-quadrature [corresponding to Re(α)] and increased variance along the conjugate quadrature, consistent with the Heisenberg uncertainty principle. The color scale represents the Wigner quasiprobability density $W(\alpha)$.

In cQED, squeezing is achieved by using the nonlinearity provided by JJ-based elements. A common approach is parametric driving. By modulating the system at approximately twice the cavity frequency, the Hamiltonian acquires terms of the form:

$$\hat{a}^2 + \hat{a}^{\dagger 2},$$

which do not conserve photon number but instead create or annihilate photons in correlated pairs. These interactions selectively reduce fluctuations in one quadrature while amplifying fluctuations in the other. Physically, the system is driven in such a way that uncertainty is "pushed" from one direction in phase space into the orthogonal direction.

Let's discuss this in more detail.

If the cavity frequency is modulated as $\omega_c(t) = \omega_c + \delta \cos(2\omega_c t)$, then in the interaction picture the fast oscillations cancel and the only surviving resonant terms are

$$a^2 e^{-2i\omega_c t} + a^{\dagger 2} e^{+2i\omega_c t}$$

Under the RWA, these become time-independent, giving exactly

$$H_{\text{eff}} \propto a^2 + a^{\dagger 2}$$

Since a lowers photon number: $a^2 \,|n\rangle = a(\sqrt{n}\,|n-1\rangle) = \sqrt{n(n-1)}\,|n-2\rangle$. Two photons are removed. This takes $|n\rangle \rightarrow |n-2\rangle$.
Since $a^\dagger$ raises photon number:

$$a^\dagger \,|n\rangle = \sqrt{n+1}\,|n+1\rangle$$

$$a^{\dagger 2} \,|n\rangle = \sqrt{(n+1)(n+2)}\,|n+2\rangle$$

This takes $|n\rangle \rightarrow |n+2\rangle$. Two photons are added giving the Hamiltonian above. Now use the quadratures X and P to rewrite the Hamiltonian

$$X = \frac{a + a^\dagger}{\sqrt{2}}, \quad P = \frac{a - a^\dagger}{i\sqrt{2}}.$$

$$a^2 + a^{\dagger 2} = (X^2 - P^2)$$

$$H_{\text{param}} \propto X^2 - P^2.$$

The Hamiltonian increases curvature (restoring force) along one quadrature and decreases curvature along the orthogonal quadrature. The result is fluctuations in one quadrature shrink (squeezed) and fluctuations in the other expand (anti-squeezed). Thus, the circular Gaussian becomes an ellipse.

In transmon–cavity architectures, the Josephson nonlinearity of the transmon can be harnessed to engineer effective parametric interactions on the cavity mode.

By modulating the system at approximately twice the cavity frequency, one realizes an effective Hamiltonian $\propto a^2 + a^{\dagger 2}$, which creates and annihilates photons in pairs and thereby squeezes one quadrature of the cavity field while antisqueezing the conjugate quadrature. In parallel, the dispersive coupling $\chi a^{\dagger} a \sigma_z$ provides a state-dependent frequency shift of the cavity, enabling conditional phase accumulation. By preparing the transmon in superposition and applying sequences of cavity displacements, SNAP-type number-selective phase operations, and controlled evolutions, one imprints periodic phase patterns on the oscillator wavefunction. When these non-Gaussian phase patterns are combined with Gaussian operations such as squeezing, repeated interactions generate the comb-like structure in phase space that characterizes GKP states. In this way, the engineered squeezing Hamiltonian and the dispersive transmon-mediated controls form a unified architecture for generating and manipulating grid-encoded logical states in a superconducting cavity.

GKP codes encode a logical qubit into a periodic lattice of sharply peaked wavefunctions in the oscillator's phase space (Figure 4.15). The logical $|0_L\rangle$

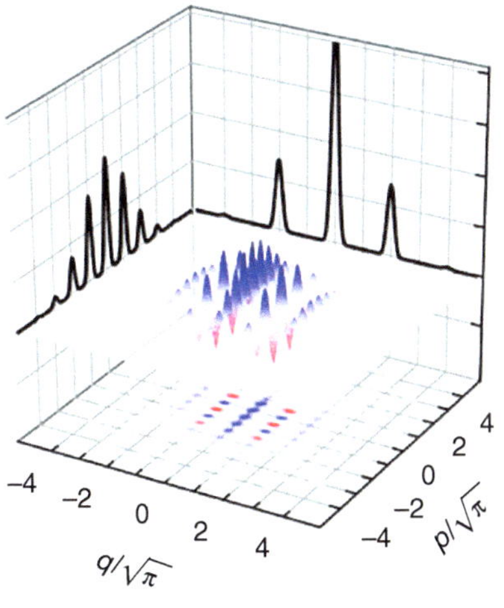

Figure 4.15 Phase-space lattice structure of a Gottesman–Kitaev–Preskill (GKP) encoded state. 3D representation of the Wigner quasiprobability distribution of a single-mode GKP logical state. The horizontal axes correspond to the dimensionless oscillator quadratures $q/\sqrt{\pi}$ and $p/\sqrt{\pi}$, where $q \equiv X$ and $p \equiv P$ denote the canonical position- and momentum-like field quadratures, satisfying $[X, P] = i$. The vertical axis shows the value of the Wigner function. The encoded state forms a periodic square lattice in phase space, composed of sharply localized peaks ("combs") whose spacing defines the logical grid. Alternating positive and negative regions reflect quantum interference between lattice components and constitute a signature of strong nonclassicality. Finite peak widths arise from realistic finite squeezing and determine the tolerance to small displacement errors [223] / IOP Publishing / CC BY 4.0.

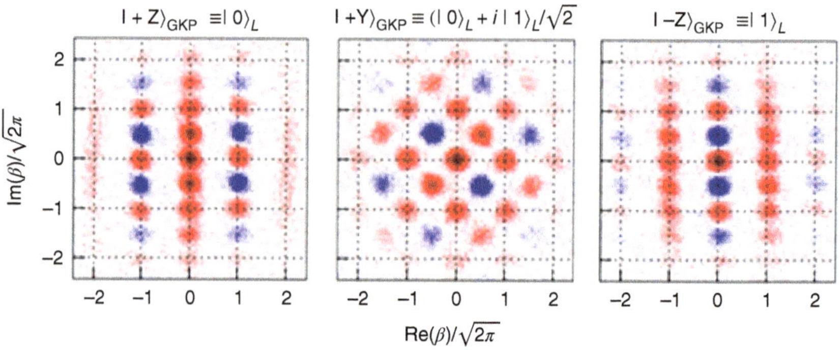

Figure 4.16 Wigner tomographs of experimentally realized logical states in GKP encoding. The three panels correspond to the logical states $|+Z\rangle_{GKP} \equiv |0\rangle_L$ (left), $|+Y\rangle_{GKP} \equiv (|0\rangle_L + i\,|1\rangle_L)/\sqrt{2}$(center), and $|-Z\rangle_{GKP} \equiv |1\rangle_L$ (right). The phase-space axes are normalized as $Re(\beta)/\sqrt{2\pi}$ and $Im(\beta)/\sqrt{2\pi}$, reflecting the periodic grid structure of the GKP code in dimensionless quadrature units. The lattice of alternating positive (red) and negative (blue) peaks arises from interference between displaced squeezed states forming the logical grid state. Logical $|0\rangle_L$ and $|1\rangle_L$ are distinguished by a half-grid displacement along the position quadrature, while the $|+Y\rangle_{GKP}$ state exhibits the characteristic phase-shifted interference pattern corresponding to a $\pi/2$ rotation in the logical Bloch sphere. The finite width of the peaks reflects experimental squeezing and residual noise, while the overall periodic structure encodes protection against small displacement errors in phase space [224] / Springer Nature.

consists of peaks at positions $q = 2k\sqrt{\pi}$, while $|1_L\rangle$ is shifted by half a lattice spacing (Figure 4.16). This half-period offset places the two code words on interleaved grids—much like two checkerboard patterns shifted relative to one another—ensuring that they remain distinguishable even under small displacements. In any physical realization, infinite squeezing is impossible, so each ideal delta-function spike becomes a narrow Gaussian. Nevertheless, the underlying periodic structure survives, and logical information is stored in which lattice cell the state is closest to. As long as noise displaces the state by less than half the lattice spacing, the logical value is preserved.

Because the code words form a geometric lattice, error correction becomes a decoding problem in phase space. A modular quadrature measurement determines the oscillator's position modulo the lattice spacing, revealing how far the state has drifted within a single-unit cell without collapsing the encoded logical information. Once this displacement syndrome is known, a compensating shift is applied to return the wavefunction to the nearest lattice point. In hardware such as a transmon–cavity system, the transmon mediates this modular measurement: it becomes sensitive only to the displacement modulo the grid spacing, and a calibrated microwave pulse then applies the corrective displacement. Repeating this cycle of syndrome extraction and correction stabilizes the encoded state against continuous-variable noise.

This geometric alignment with the dominant noise processes is what distinguishes GKP codes from other bosonic encodings. Cat and binomial codes excel at protecting against discrete events such as photon loss, but they are not naturally matched to the Gaussian diffusion that characterizes oscillator noise. GKP encoding, by contrast, is built around the geometry of that diffusion: small displacements simply move the state within a unit cell, where they can be detected and reversed.

The lattice structure also endows GKP qubits with powerful computational properties. Clifford operations on a harmonic oscillator correspond to *symplectic transformations*—linear transformations of phase space that preserve the canonical commutation relations and, crucially, the geometry of the GKP lattice. Rotations, squeezes, and shears of the (q, p) plane map the square lattice onto itself without changing its spacing or orientation. Because these transformations act uniformly across the entire lattice, they implement logical gates transversally: a single physical operation on the oscillator induces the correct logical action without requiring multiqubit interactions or elaborate fault-tolerant gadgets. This stands in contrast to most quantum codes, where fault-tolerant logical gates typically require complex entangling circuits.

Geometric symmetries allow GKP codes to recast quantum computation as the manipulation of a lattice in phase space. Logical operations correspond to global transformations—translations, rotations, and squeezes—that move or reshape this lattice in controlled ways. Many logical Clifford gates, therefore, reduce to Gaussian operations, which are among the fastest and highest-fidelity controls available in bosonic hardware. For example, a logical Hadamard gate is simply a $\pi/2$ rotation of the entire lattice, implemented through a straightforward Hamiltonian evolution rather than a long sequence of qubit-level pulses.

Finally, GKP codes integrate naturally into larger fault-tolerant architectures. Gaussian operations supply the full logical Clifford group, while a small number of non-Gaussian elements—typically implemented through ancilla-assisted measurements or engineered nonlinearities—complete the universal gate set. When combined with repeated syndrome extraction, GKP qubits can be concatenated with qubit-based codes, such as the surface code. In such hybrid schemes, the GKP layer suppresses small continuous-variable displacement errors at the oscillator level, while the outer code corrects the remaining discrete Pauli-type errors. This two-tier structure allows continuous-variable protection to feed directly into digital fault tolerance, improving thresholds and reducing overall error rates.

Recent experiments have demonstrated that GKP-encoded qubits can achieve logical lifetimes roughly twice as long as the underlying physical cavity mode, and, more importantly, that their logical gate implementations require fewer layers and lower circuit depth than comparable transmon-based constructions

[225, 226]. This combination of geometric structure, Gaussian control, and experimentally validated error suppression is what makes GKP encoding a compelling route toward scalable fault-tolerant quantum.

In summary, Bosonic codes offer a powerful alternative to conventional qubit-based QC by exploiting the large Hilbert space of a harmonic oscillator. Instead of storing information in individual energy levels or single excitations, these encodings distribute logical states across structured patterns in phase space or Fock space. In binomial codes, the pattern consists of carefully engineered superpositions of Fock states; in cat codes, it appears as interference between well-separated coherent states; and in GKP codes, it takes the form of a periodic lattice spanning the continuous quadratures. Although these realizations differ in appearance, they share a unifying principle: logical information resides in the relationships among states, not in any single state itself. Errors may distort or shift these patterns, but as long as the global structure remains identifiable, the logical information can be detected and restored. In this way, bosonic encodings replace fragile state occupation with resilient pattern formation as the fundamental carrier of quantum information.

Different bosonic codes emphasize different symmetries to combat specific noise channels. Rotation-symmetric codes—such as four-component cat codes (four-legged cats) and certain binomial codes—are naturally robust against both photon loss and dephasing because their logical states are invariant under discrete-phase rotations. Translation-symmetric codes like the GKP code, by contrast, are powerful against small displacement errors and can be made highly resilient to photon loss, though they are more sensitive to dephasing. In superconducting cavities, however, pure dephasing is typically weaker than in qubits, so dephasing becomes relevant mainly through ancilla-induced noise rather than through the cavity itself. These complementary strengths allow bosonic codes to be tailored to the dominant noise processes of a given hardware platform.

The control toolbox available in cavity-based architectures makes these encodings not only robust but also highly programmable. Displacements, SNAP gates, and Kerr-induced evolutions provide a universal set of operations for manipulating oscillator states with high precision. These tools enable the preparation, stabilization, and logical manipulation of cat, binomial, and GKP states directly within a single-resonator mode. As a result, bosonic codes can achieve hardware-efficient fault tolerance: a single high-coherence cavity can host an encoded qubit whose protection would otherwise require many physical qubits in a traditional circuit-based architecture.

Experimentally, these ideas have progressed from theory to practice. Bosonic quantum error correction has already surpassed the break-even point [191], meaning that the lifetime of a logical qubit encoded in a cavity exceeds that of any individual physical component of the system. This result has also been extended [227] to GKP qudits with up to $d = 4$. Moreover, universal gate sets and fault-tolerant logical operations have been demonstrated on bosonic encodings, marking a significant step toward scalable, error-corrected quantum processors. These advances highlight the practical viability of bosonic codes and their potential to serve as foundational building blocks for future quantum technologies.

Viewed from a broader perspective, bosonic encodings shift the conceptual picture of quantum computation. As one moves from dual-rail qubits to cat codes and ultimately to GKP lattices, the system begins to resemble a designed quantum material, where information is embedded in geometric or symmetry-protected structures rather than in isolated TLSs. This architectural shift—toward storing and processing information in patterns, symmetries, and lattices—opens new pathways for robust, hardware-efficient, and scalable QC.

4.17 Architectures for Scaling Quantum Systems

4.17.1 Multiple Qubits

A central challenge in moving from few-qubit demonstrations to practical quantum processors is scaling the hardware in a way that preserves coherence, maintains high-fidelity control and readout, and remains compatible with the resource constraints of a cryogenic environment. Conventional cQED systems typically require at least two coaxial lines per qubit—one for control and one for readout—and often a third line to bias the first-stage parametric amplifier [175]. As the number of qubits grows, this approach becomes untenable: the number of cables grows linearly with qubit count, the heat load on the dilution refrigerator increases, and the complexity of routing and integration becomes prohibitive. At the same time, as systems become larger, variability in device parameters, increased crosstalk between control channels, and reduced readout fidelity all become more pronounced. Clearly, this is not a practical solution beyond a few hundred qubits. These scaling pressures motivate new hardware architectures that can address multiple qubits more efficiently while preserving the performance advantages that superconducting qubits and 3D cavities offer.

The architecture shown in Figure 4.17 represents a possible route in this direction. In this design, four transmon qubits are each embedded in their own single-mode 3D aluminum cavity, providing long coherence times and isolation between qubits. Rather than wiring each cavity individually, the cavities are aperture-coupled to a common rectangular waveguide feedline. By appropriately sizing and positioning the apertures milled into the cavity walls, the external coupling rates can be tuned independently for each cavity, allowing careful control of both drive strength and readout bandwidth. All four cavities are then addressed

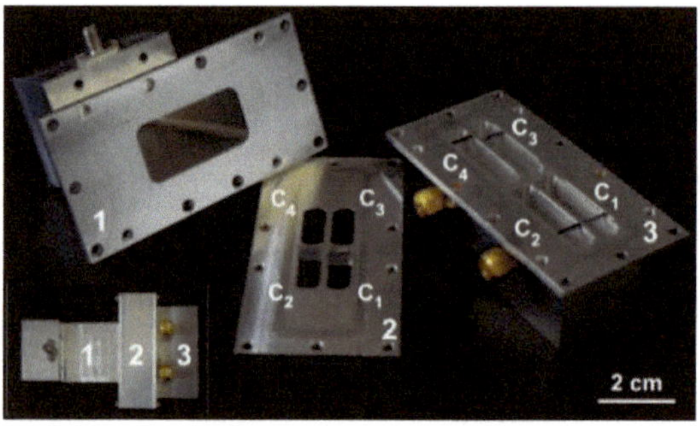

Figure 4.17 Photograph of a scalable 3D architecture in which four transmon qubits (C_1–C_4) each reside in their own single-mode aluminum cavity, and all four cavities are aperture coupled to a common rectangular waveguide feedline. The rectangular slots milled in the cavity walls set the external coupling to the feedline; by sizing/positioning the apertures, the coupling rate for each cavity can be tuned independently. A single coaxial input/output on the feedline thus addresses all four cavities (for drive and readout) using frequency multiplexing, reducing cryogenic I/O while maintaining cavity-to-cavity isolation. Each cavity is machined in two halves (center and left), split along the electric field antinode plane so that a planar transmon chip can be mounted precisely at the E-field antinode of the fundamental mode. The two halves bolt together to form a seam away from current maxima, minimizing seam loss; the window cutouts provide access for the transmon substrate and alignment features. This module exemplifies one route to scaling 3D cQED: replicate many small, high-Q cavities, couple them through a shared waveguide bus, and control/read them over a single line. Scale bar: 2 cm [229] / AIP Publishing LLC.

through a single coaxial input/output port on the shared feedline using frequency multiplexing: each cavity is designed with a distinct resonance frequency spaced by roughly 70–100 MHz. A broadband Josephson parametric amplifier (Section 3.37) at the output enables simultaneous, frequency-multiplexed readout, achieving single-shot fidelities above 98% without requiring dedicated readout lines per qubit.

Equally important is the mechanical design of the cavities. Each cavity is machined in two halves and split along the electric field antinode plane of the fundamental mode. This allows the planar transmon chip to be placed exactly at the electric field maximum, optimizing coupling while keeping the mechanical seam away from current antinodes, where it would contribute loss. The resulting module illustrates how 3D cQED systems—traditionally designed around single qubits—can be replicated, tiled, and coupled through a shared bus to create larger integrated processors.

At the same time, this approach highlights the challenges that must be addressed on the road to truly large-scale quantum systems. As more cavities and qubits are coupled to a common waveguide, spectral crowding becomes increasingly difficult to avoid. Shared-line control introduces new pathways for microwave crosstalk and backaction between qubits. Amplifiers must operate over wider bandwidths and larger dynamic ranges without distortion. Finally, fabrication tolerances and variability in JJs become more critical as the number of devices increases.

Another challenge is to achieve interqubit connectivity in order to minimize gate count and implement algorithms efficiently with reduced error. Quantum data needs to be transported across several connected qubits in a grid, leading to additional operations and consequently more errors. Processors today mostly provide nearest-neighbor connectivity by direct capacitive coupling, tunable couplers, or bus resonators. All-to-all connectivity would be the ideal with uniform interqubit coupling strength from pair to pair, and low crosstalk from proximity of qubits.

A scalable architecture for a network of superconducting qubits with significantly enhanced interqubit connectivity is the use of a ring resonator as a bus resonator to mediate the interaction between several qubits. The qubits are distributed throughout the circumference of the ring resonator and thus spatially separated. This provides good microwave isolation resulting in individual qubit addressability and negligible qubit crosstalk. Figure 4.18 shows an example with

Figure 4.18 Ring resonator ("whispering-gallery-mode") bus for high-connectivity superconducting qubits. Photograph of a machined aluminum ring resonator used as a shared bus to mediate interactions among many spatially separated qubits. Twelve transmon sites are milled as equally spaced qubit slots (every 30° around the circumference); an example slot and chip are shown in the inset (scale bars: 10 mm main, 5 mm inset). Quarter-wave ($\lambda/4$) readout resonators are implemented as radial transmission-line sections extending outward from the ring. The ring supports a uniform standing-wave amplitude around its perimeter, making the qubit–bus coupling independent of position and enabling clean frequency allocation. Because the qubits are distributed around the ring, they remain well isolated from one another for individual addressability with negligible crosstalk. In the example shown, 12 qubits are arranged so that each qubit couples to nine others through the bus, with only two distinct nonzero coupling strengths, providing a scalable route to enhance interqubit connectivity in 3D cQED [228] / American Physical Society.

12 qubits where each qubit is connected to nine other qubits in a ring resonator or whispering-gallery-mode resonator [228]. It has a uniform standing-wave amplitude independent of the position along the resonator. Hence, the qubit–resonator coupling is independent of its position in the ring resonator. The qubit slots (magnified in the inset) are placed 30° apart and the readout resonators are $\lambda/4$ sections of transmission line extending radially outward. In this scheme, each qubit is connected to nine other qubits with only two values of finite coupling strength. Extensions are possible to 36 qubits with 10° angular spacing where every qubit is connected to 27 other qubits in the network with only two different values of coupling. Other extension schemes are possible.

4.18 Scaling 3D Cavity-Based Quantum Processors

One way to grow from today's few-cavity, few-transmon demonstrations to a large-scale processor is to treat each high-quality 3D cavity /transmon pair as a small "quantum processing unit" (QPU) and then "wire" many of these units together. Figure 4.19 sketches two complementary scaling concepts that are being pursued at the SQMS (Fermilab Superconducting Quantum Material and Systems Center). In the left panel, each colored fundamental block on top of the long central cavity (the hub) represents a two-cell (or few-cell) 3D QPU module[230]: a small piece of hardware containing a handful of very high-Q modes plus dedicated ancilla transmons for control, readout, and error correction. Within each module, the ancilla can prepare, manipulate, and read out Fock, cat, binomial, or qudit states encoded in the local modes. All of this "intramodule" activity is largely self-contained.

The modules are then docked onto the central hub cavity that supports a set of extended modes extending across its length. At specific positions along the hub, intermodule couplers (purple elements) can be turned on or off to hybridize one mode of a local module with one of the hub modes. By tuning the coupler strengths in frequency and time, it is possible to perform SWAP-like operations between a local mode and a hub mode, or to mediate entangling gates between distant modules via the shared hub, as well as to configure which modules are "connected" at any given moment. Local operations stay within a module; long-range operations go through the hub.

Scaling up means replicating QPU modules and plugging them into the hub that provides tunable long-range connectivity. Since each module is a compact, shielded 3D cavity system, excellent lifetimes and low crosstalk demonstrated in small devices can be preserved. If a few transmons are bad, they can be swapped out without scrapping the QPU. However, the downsides are multiple flanges, couplers, and mechanical interfaces requiring stringent engineering conditions. To

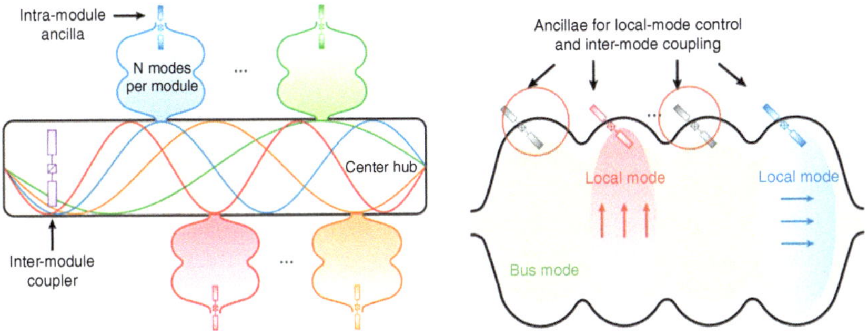

Figure 4.19 Two complementary architectures for scaling three-dimensional (3D) superconducting quantum processing units (QPUs). (Left) Modular, hub-and-spoke architecture in which multiple 3D cavity modules, each supporting N long-lived electromagnetic modes, are connected through a central multimode bus or "hub." Within each module, an ancilla transmon provides local control, readout, and error correction operations on the cavity modes, enabling bosonic encoding and qudit-level storage. Intermodule couplers link the individual modules to the shared hub, allowing quantum information to be exchanged between distant cavities via selected hub modes while keeping most modes localized and protected. This approach emphasizes modularity, long coherence times, and scalable connectivity by separating local memory from global communication. (Right) Distributed local-mode architecture in which a continuous 3D structure supports both extended bus modes and spatially localized cavity modes within the same physical enclosure. Ancilla transmons placed at strategic locations couple to nearby local modes for control and readout, while also mediating intermode and intermodule interactions through shared bus modes. In this scheme, the geometry of the cavity itself defines a hierarchy of modes: localized regions act as protected quantum memories, while delocalized modes provide coherent transport and entanglement pathways across the processor. Together, these two approaches illustrate distinct but complementary strategies for scaling 3D QPUs—one emphasizing modular interconnection via a central hub, the other embedding locality and connectivity directly into a unified multimode cavity structure.

grow the Hilbert space, more physical modules and couplers are needed. As more modules are attached, frequency crowding may pose limits. Routing all the control, readout, and drive lines into the many separate volumes can still be challenging,

Instead of many small cavities attached to a hub, a more radical approach would be to build one monolithic 3D cavity engineered to support a rich spectrum of modes. Within this volume are defined local modes (red and blue field areas) that live in specific "pockets" of the geometry and function as storage or processing modes for logical qubits/qudits. A bus mode (green background) would extend across the whole structure, providing a global communication channel. Ancilla transmons placed at selected points would be used both to control individual local modes and to tune couplings between local modes and the bus.

In this concept, the scaling lever is the mode structure, not the number of separate enclosures. By adjusting the cavity shape, and coupling cells, it may be possible to pack many well-resolved qudit modes into a single 3D volume. Multiple ancillas give local control "handles" and tunable interconnectivity.

The advantage here is that a single 3D volume can support a large number of high-Q modes, each usable as a qubit or qudit, dramatically increasing the Hilbert space per piece of hardware. There would be fewer seams and flanges. However, sculpting a single cavity to have many high-Q, well-separated modes with desired field profiles is a nontrivial electromagnetic and mechanical design problem. Small fabrication errors can shift many modes at once. As the number of modes grows, the calibration space explodes. A defect, contamination, or machining error in the monolithic cavity affects the entire processor; it is not possible to "swap out" a bad submodule. Many control and readout lines for multiple ancillas will still be necessary.

4.19 Multimode Cavities

A viable scaling strategy is an important part of any quantum architecture in order to access an even larger Hilbert space as required to run the more relevant quantum algorithms. One approach to scaling is to use multi-cell cavities as multimode quantum systems. Niobium multi-cell cavities are routinely used in particle accelerators. These couple nominally equal single cells, resulting in closely spaced fundamental modes, all with very high-quality factor. Figure 4.20 shows a nine-cell 1.3-GHz cavity design extensively used in worldwide accelerators. For QC 4× higher frequency cavities would be proportionally smaller. A single-transmon qubit is dispersively coupled to the set of fundamental modes, each corresponding to a specific field distribution over the various cells of the cavity and acting as an independent harmonic oscillator. The geometric design of the cavity can be optimized to achieve properties better suited to encoding quantum information. Increased spacing between the fundamental pass-band mode frequencies of the nine-cell cavity can be achieved by enlarging the aperture to modify the cell-to-cell coupling; similarly, equal coupling for all the pass-band modes can be achieved through tapering [231]. As a very optimistic example for the future, the very long coherence (two seconds) in SRF cavities offers transformational advances, ultimately allowing creation, control, and manipulation up to 2^{14} (about 16,000) microwave photon-energy levels in each cavity mode (limited by residual nonlinearities and photon loss), which in turn corresponds to 14 qubits for a cavity mode. The highest Fock state would have a lifetime of 125 μs, which may be acceptable for gate times of 100 ns, assuming also a 0.1% error budget per gate.

A major advantage of the multicell cavity architecture is the reduced number of input–output lines for a very large controllable Hilbert space.

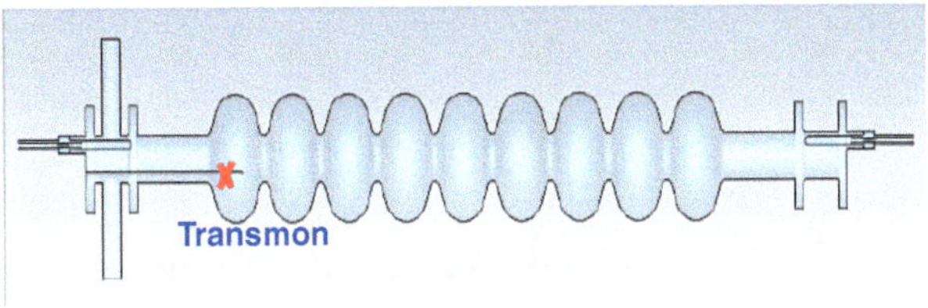

Figure 4.20 Nine-cell superconducting cavity used as a multimode quantum register. Schematic of a nine-cell elliptical niobium cavity (TESLA-type geometry widely used in accelerators) illustrating a route to scaling via multimode storage. The chain of nominally identical cells forms the fundamental pass-band of closely spaced normal modes; each mode has a distinct axial field distribution across the nine cells and behaves as an independent harmonic oscillator with high intrinsic Q. A single-transmon qubit (blue label; red "x" indicates a representative coupling location), introduced through the beam tube or a side port with a small capacitive antenna, is dispersively coupled to the set of fundamental modes. Detuning and antenna geometry set the per-mode dispersive shifts χ, enabling number-selective control, Fock-state preparation, and intermode gates while keeping Purcell loss low. By optimizing cavity geometry (cell-to-cell coupling, end-group/beam-tube dimensions, and external ports), the mode spacing, participation, and coupling can be tailored for quantum information tasks—turning a standard accelerator cavity into a multimode quantum memory/bus addressed by a single nonlinear element.

The flute technique [192] creates a cavity through the overlap of holes drilled from different directions of a monolithic piece of Al superconductor, resulting in the generation of a cavity volume with no seams. This is illustrated in Figure 4.21 for a rectangular and cylindrical cavity. Cavity modes are controlled using a superconducting transmon circuit that couples to all the modes. With an average measured T_1 cavity ~2 ms and T_1 qubit ~80–100 μs, it was possible to control the Hilbert spaces of ~10 modes before errors from multiplexing posed a limit. The superconducting transmon circuit is simultaneously coupled to all the cavity modes by placing it at one end of the multimode flute cavity.

4.20 Quantum Memory and Quantum Processor

A cavity/transmon system provides an efficient quantum memory [232] and computational unit by leveraging the long coherence times of superconducting cavities and the fast control capabilities of transmons. Quantum states are offloaded into the cavity's large Hilbert space, where they can persist far longer than the coherence time of the nonlinear element that created them. The transmon mediates all essential operations.

Quantum information is stored in the Fock-state basis as shown in Figure 4.22 [233, 234]. When the transmon's frequency is tuned (by parametric modulation to

Figure 4.21 (a) Seamless multimode "flute" cavity in a monolithic superconductor (Ref. [192] / American Physical Society). Photograph of a single-piece aluminum block machined by the flute technique: arrays of partially overlapping cylindrical bores are drilled from the top and bottom faces so that their overlap defines a closed cavity volume with no joints or seams. Eliminating seams suppresses seam loss and supports high-Q operation. The resulting 3D cavity hosts a manifold of electromagnetic modes; a superconducting transmon circuit (mounted in a side pocket or recess, not shown) is capacitively coupled to all modes for control and readout. Geometry (bore diameter, spacing, and pattern) provides a simple route to scaling and mode engineering in multimode cQED. (b) Cylindrical "flute" cavity—seamless multimode resonator in a monolithic superconductor. The three SMA feedthroughs on the top cap provide drive/readout (and optional bias) to a superconducting transmon circuit mounted beneath the lid (not shown), which capacitively couples to the full set of cavity modes. (Adapted from [192].)

a specific memory mode (one of the resonant modes of a superconducting cavity), they resonantly interact, allowing energy (quantum states) to swap. An important feature of this architecture is selective addressing of any memory mode without disturbing others achieved by dynamically tuning the transmon frequency so that only the desired memory resonates. A SWAP gate transfers the state into the resonator $|1\rangle$ Fock state. The memory mode retains the quantum state. Each memory mode is one electromagnetic eigenmode of a high-Q superconducting cavity (or planar resonator) that stores quantum information in the form of photonic Fock states $|0\rangle, |1\rangle, |2\rangle, \ldots$.

The transmon is free to interact with other modes or perform other operations. To retrieve quantum information, the process is reversed: the transmon swaps the stored state back out of memory and into its $|e\rangle$ state for further processing or measurement. This hybrid processor–memory architecture enables scalable quantum storage by storing many qubits' worth of information across multiple modes. It retrieves or modifies any mode selectively. One central processor manages multiple storage channels.

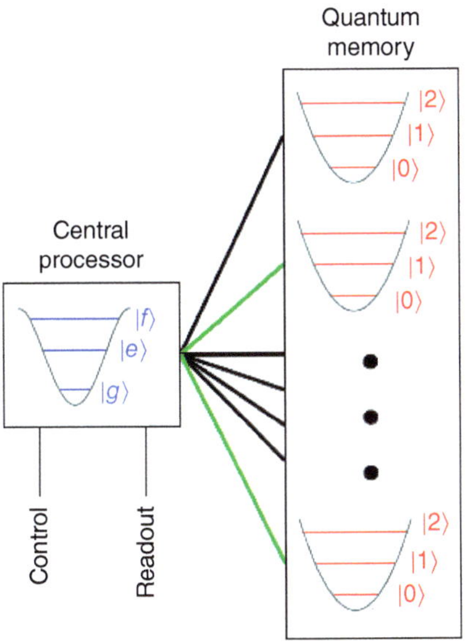

Figure 4.22 The transmon qubit (left box in the diagram) is the core quantum processor acting as a control hub with energy levels $|g\rangle, |e\rangle, |f\rangle$. It writes quantum information into the memory modes, retrieves it on demand, and performs limited local processing. Control lines (bottom-left) deliver microwave pulses to perform quantum gates on the transmon. Readout lines measure the transmon state using dispersive shifts in an auxiliary resonator. On the right, the diagram shows multiple quantum memory elements, i.e., superconducting resonators with long-lived cavity modes. Each memory mode has its own harmonic ladder of Fock states: $|0\rangle, |1\rangle, |2\rangle, \ldots$ labeled in red. The memory modes function like quantum hard drives. The black and green lines connecting the transmon to the memories represent tunable couplings that allow controlled exchange of excitations. The transmon couples dispersively to each memory mode. SWAP operations enable writing and retrieving quantum states between the processor and the memory modes, allowing random access storage and retrieval of quantum information (Ref. [233] / Springer Nature / CC BY 4.0).

The transmon and cavity are operated in the dispersive regime. State transfer between the transmon and the cavity is achieved using a state-swapping interaction:

$$H_{\text{swap}} = g(a^{\dagger}\sigma_{-} + a\sigma_{+})$$

Here g is the coupling strength, $a^{\dagger}$ and a are the cavity creation/annihilation operators, respectively, and σ_{+}, σ_{-} are the transmon raising/lowering operators. A resonant pulse tunes the transmon and cavity into resonance. A full oscillation transfers the quantum state from the transmon to the cavity:

$$|e\rangle_{\text{transmon}}|0\rangle_{\text{cavity}} \rightarrow |g\rangle_{\text{transmon}}|1\rangle_{\text{cavity}}$$

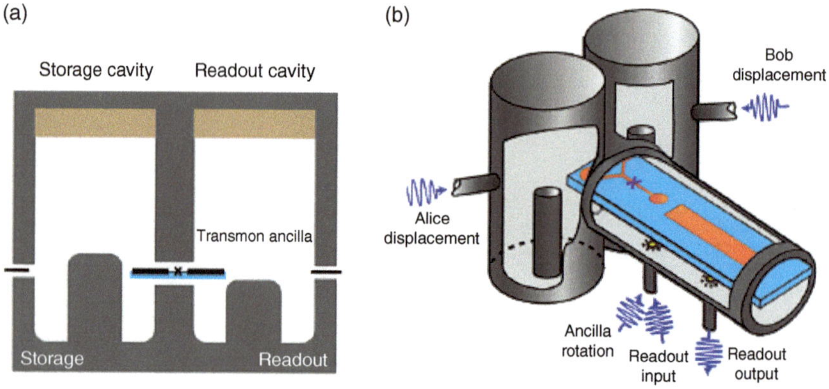

Figure 4.23 3D cavities for bosonic-encoded qubits. (a) Two coaxial microwave resonators (gray) bridged by a single-transmon superconducting qubit (black) fabricated on a sapphire chip (Ref. [235] / Springer Nature). (b) 3D view of the device consisting of two coaxial cavities (Alice and Bob), a Y-shaped transmon with a single Josephson junction (marked by ×), and a stripline readout resonator (Ref. [236] / American Association for the Advancement of Science).

This enables state storage in the long-lived cavity mode. Cavity lifetimes are much longer than transmon qubits lifetimes, reducing error rates. Entangling operation can be performed between any two pair of modes, enabling all-to-all connectivity in this architecture.

Figure 4.23 shows examples where a superconducting cavity coupled to a transmon qubit provides an attractive architecture for quantum memory and computation.

4.21 Cavity as a Computational Hilbert Space

It is important to recognize Fock states in the storage cavity are not just for passive storage but also as an active computational Hilbert space in their own right. Because cavity Fock states have coherence times far longer than those of transmons, computations carried out within the memory can be more robust against decoherence. Operating in a higher-dimensional Hilbert space allows certain algorithms and error-correcting operations to be implemented with fewer physical resources.

In this architecture, the transmon (ancilla) and the cavity play complementary roles in the processor. Through dispersive coupling, different photon number states shift the transmon frequency by different amounts, allowing the ancilla to selectively address, phase-shift, or entangle specific Fock levels. This selectivity is the basis for SNAP and other gates, conditional displacements, that constitute a universal gate set for bosonic qudits. Thus, the transmon is responsible for control, conditional logic, and measurement. The cavity mode stores and supports interference. Quantum information is processed not by flipping discrete two-level states, but by allowing probability amplitudes stored in the cavity to overlap, acquire relative phases, and interfere with one another.

In a conventional transmon-only system, each physical element is restricted to a two-level subspace, and computation proceeds by applying single- and two-qubit gates that flip or entangle these discrete levels. Scaling such a system requires adding more physical qubits, more couplers, and more control lines, all of which increase hardware complexity and introduce additional decoherence channels. Error correction in qubit-only architectures is also resource-intensive: protecting a single logical qubit typically requires dozens or even hundreds of physical qubits arranged in a surface code or similar stabilizer code.

By contrast, a single cavity can host a high-dimensional computational space or even a fully error-corrected logical qubit using bosonic codes. This substantially reduces the number of physical components required for a given logical capacity. Operations such as displacements, SNAP gates, and conditional phase shifts allow one to sculpt and interfere amplitudes across many Fock levels at once.

As discussed in Sections 2.23 and 4.9, the error landscape differs significantly. Qubit-only systems must contend with bit-flip and phase-flip errors on each physical qubit. Bosonic encodings can tailor their logical states to the dominant error channel of the cavity—typically photon loss—and use the transmon to implement error-correcting operations aligned with that channel.

As a computational example, consider the simple Grover search over eight items. For qubits, this requires three qubits and a sequence of oracle and diffusion operations. In a cavity-based approach, the same Hilbert space can be encoded directly in Fock states of one single-cavity mode, which serves as a qudit with logical basis states $0\rangle$, $|1\rangle$, $|2\rangle$, $|3\rangle$, $4\rangle$, $|5\rangle$, $|6\rangle$, $|7\rangle$. The transmon is used as the fast, nonlinear control hub that can imprint phases on selected number states, implement reflections, and finally read out which Fock level the cavity occupies.

The algorithm begins by preparing the cavity with a coherent equal-amplitude superposition across eight Fock states

$$|\psi_0\rangle = \frac{1}{\sqrt{8}}(|0\rangle + |1\rangle + \cdots + |7\rangle),$$

Physically, this is carried out by using the transmon's dispersive coupling and shaped microwave control to essentially "paint" amplitudes and phases onto the desired subset of number states using number-selective operations. Next comes the oracle, which marks the target item by flipping its phase. If the marked entry corresponds to, say $|m\rangle$, the oracle is the operation:

$$U_{\text{oracle}}\,|m\rangle = -\,|m\rangle, \quad U_{\text{oracle}}\,|n \neq m\rangle = |n\rangle.$$

In the dispersive regime, the transmon's frequency depends on the cavity photon number n. The system is addressed in a number-selective way: by choosing an appropriate microwave tone (or sequence) to arrange for only the $n = m$ component to acquire a πphase shift, leaving all other components unchanged.

Then comes the diffusion step, the "inversion about the mean." In the cavity–qudit picture, it is again a reflection operation—but now the reflection occurs in an 8D number-state space. The algorithm's speedup emerges from coherent interference among photon number amplitudes rather than from many operations on three separate physical qubits.

4.22 Entangling Fock States

When Fock states in a storage cavity are elevated from passive memory to an active computational Hilbert space, the ability to entangle those Fock states becomes essential. Computation in this setting is no longer about manipulating isolated photon number levels, but about shaping and coordinating superpositions across many of them. Entanglement is what allows different Fock components to share phase relationships, interfere, and participate collectively in logical operations. Without entanglement, the cavity would behave as a collection of independent energy levels; with it, the oscillator becomes a genuinely quantum processor whose computational power arises from correlations spread across its entire Fock-space structure. Entanglement transforms a long-lived cavity mode from a high-quality storage device into a fully programmable quantum register capable of supporting gates, algorithms, and error-correcting encodings.

A standard protocol for generating entanglement between the two lowest Fock states of memory mode A and memory mode B could be as follows: prepare the transmon in the superposition:

$$|\psi_T\rangle = \frac{1}{\sqrt{2}}(|g\rangle + |e\rangle).$$

Use a C-SWAP or a controlled-phase (CZ) gate. If the transmon is in $|e\rangle$ swap an excitation into memory mode A. If in $|g\rangle$ do nothing. After this, we have an entangled state between the transmon and memory A.

$$|\Psi\rangle = \frac{1}{\sqrt{2}}(|g\rangle|0_A\rangle + |e\rangle|1_A\rangle).$$

Now perform a transmon-to-memory B SWAP, mapping the transmon's $|e\rangle$ amplitude into memory B's $|1_B\rangle$. Final state after this transfer is

$$|\Psi_{AB}\rangle = \frac{1}{\sqrt{2}}(|1_A\rangle|0_B\rangle + |0_A\rangle|1_B\rangle).$$

This is a maximally entangled Bell state between the lowest two Fock states of memory modes A and B.

4.23 Preparing Qudits

When the cavity is treated as an active, high-dimensional computational register, the ability to *populate* specific Fock levels on demand becomes important. Preparing a qudit within a single resonator requires controlled access to many photon number states. Here the transmon plays a crucial operational role. Acting as a tunable "photon loader," the transmon can inject excitations into the cavity one at a time, with each transfer conditioned on its own internal state and the dispersive shift it imparts. By sequencing these controlled exchanges, one can selectively build up population in chosen Fock levels and sculpt arbitrary qudit states inside the memory mode. In this way, the cavity's Hilbert space is harnessed for genuine multilevel QIP.

We present a simple protocol to prepare a *qudit* in the memory resonator by using more Fock levels of the same cavity mode with the transmon acting as a "photon loader" [202].

Let's say we want to prepare the qutrit basis state $|2\rangle$. We start by initializing the system in $|g\rangle_{\text{transmon}} \otimes |0\rangle_{\text{memory}}$. First, we apply two π pulses (bottom blue and red arrows in Figure 4.24) to excite the transmon to $|e\rangle$ and then to the $|f\rangle$ level so that the total state becomes $|f\rangle\,|0\rangle$. Then a sideband pulse (green arrow) is applied to bring the system to $|g\rangle\,|1\rangle$ so that effectively a single-photon Fock state is created in the cavity. A direct transmission from $|e\rangle\,|0\rangle$ to $|g\rangle\,|1\rangle$ is not allowed due to selection rules. Starting from $|g\rangle|1\rangle$, a similar sequence of two π pulses and a sideband pulse are applied to go to $|g\rangle\,|2\rangle$.

An arbitrary single-qutrit state $|\psi\rangle = c_0\,|0\rangle + c_1\,|1\rangle + c_2\,|2\rangle$ can also be prepared using a similar scheme where instead of π_{ge} pulses some partial rotations are applied on the transmon.

4.24 Density Matrix and Trace

When we begin treating cavity states as quantum resources—superpositions, mixtures, and entangled combinations of Fock states—we need a language that can describe not just amplitudes, but the full statistical and coherence structure of the state. Here the *density matrix* becomes indispensable. It captures everything the oscillator can hold: populations along the diagonal, coherences in the off-diagonals, and the effects of noise and measurement through operations expressed as traces. Once the state is represented in this operator form, we can ask deeper questions: How pure is it? Which coherences survive? How do interactions reshape its structure? These questions naturally lead to *phase-space representations*, where the same density matrix can be visualized as a quasiprobability distribution. Among these, the *Wigner function* stands out because it

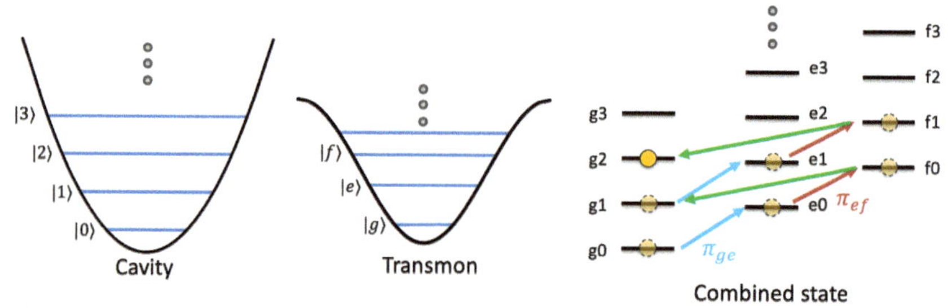

Figure 4.24 Sideband-assisted preparation of cavity Fock-state qudits using a multilevel transmon ancilla. Left: Energy-level structure of a harmonic cavity, showing evenly spaced Fock states $|0\rangle, |1\rangle, |2\rangle, |3\rangle, \ldots$. Middle: Weakly anharmonic transmon spectrum with nonequally spaced levels $|g\rangle, |e\rangle, |f\rangle, \ldots$, which enables selective transitions among adjacent manifolds. Right: Dressed (combined) cavity–transmon eigenstates $|gn\rangle, |en\rangle, |fn\rangle$ are shown, where n denotes the photon number in the cavity. Controlled sequences of *carrier* π-pulses on the $|g\rangle \leftrightarrow |e\rangle$ and $|e\rangle \leftrightarrow |f\rangle$ transitions (blue and red arrows) together with *red/blue-sideband pulses* (green arrows) exchange excitations between the transmon and cavity. By alternating these operations, population is coherently "stepped" up the Fock ladder, deterministically preparing higher photon number states $|n\rangle$of the cavity. A combination of transmon π and $|f, n\rangle \rightarrow |g, n+1\rangle$ sideband pulses is used to climb the Fock ladder. (Adapted from Ref. [202].)

provides a complete, experimentally accessible picture of the quantum state. Understanding density matrices and the trace, therefore, sets the stage for Wigner tomography—the practical method by which we reconstruct and interpret the full quantum state of the cavity in phase space.

Accordingly, this section is introductory material for discussing Wigner tomography in Section 4.26 and examples in Section 4.27.

The density matrix is a description of a quantum state, covering both pure superpositions and statistical mixtures. Its diagonal entries tell us the probabilities of different outcomes, while its off-diagonal entries reveal the presence of quantum coherence and interference. The density matrix tells us how likely the system is to be in each possible state and whether those states are coherently superposed (quantum) or just randomly mixed (classical).

The trace is the simple mathematical operation that guarantees total probability is one and allows us to compute measurable averages. Together, the density matrix and trace provide the practical toolkit for understanding—and diagnosing—the behavior of quantum systems.

A pure quantum state is described by a density matrix of the form: $\rho = |\psi\rangle\langle\psi|$.

For a pure state, the squared density matrix satisfies: $\mathrm{Tr}(\rho^2) = 1$. This means the system is in a definite quantum state with no classical uncertainty.

To better illustrate construction of the density matrix for a simple case, consider a qubit in an equal superposition of the computational basis states

$$|\psi\rangle = \frac{1}{\sqrt{2}}(|0\rangle + |1\rangle)$$

The bra version of this state is $\langle\psi| = \frac{1}{\sqrt{2}}(\langle 0| + \langle 1|)$.

We construct the density matrix: $\rho = |\psi\rangle\langle\psi|$, which involves taking the outer product.

$$\rho = \left(\frac{1}{\sqrt{2}}(|0\rangle + |1\rangle)\right) \cdot \left(\frac{1}{\sqrt{2}}(\langle 0| + \langle 1|)\right)$$
$$= \frac{1}{2}(|0\rangle\langle 0| + |0\rangle\langle 1| + |1\rangle\langle 0| + |1\rangle\langle 1|)$$

Now, we represent the basis states in matrix form:

$$|0\rangle = \begin{bmatrix}1\\0\end{bmatrix}, \quad |1\rangle = \begin{bmatrix}0\\1\end{bmatrix}$$

$$|0\rangle\langle 0| = \begin{bmatrix}1\\0\end{bmatrix}[1 \ \ 0] = \begin{bmatrix}1 & 0\\0 & 0\end{bmatrix}$$

$$|1\rangle\langle 1| = \begin{bmatrix}0\\1\end{bmatrix}[0 \ \ 1] = \begin{bmatrix}0 & 0\\0 & 1\end{bmatrix}$$

$$|0\rangle\langle 1| = \begin{bmatrix}1\\0\end{bmatrix}[0 \ \ 1] = \begin{bmatrix}0 & 1\\0 & 0\end{bmatrix}$$

$$|1\rangle\langle 0| = \begin{bmatrix}0\\1\end{bmatrix}[1 \ \ 0] = \begin{bmatrix}0 & 0\\1 & 0\end{bmatrix}$$

Summing these terms:

$$\rho = \frac{1}{2}\left(\begin{bmatrix}1 & 0\\0 & 0\end{bmatrix} + \begin{bmatrix}0 & 1\\0 & 0\end{bmatrix} + \begin{bmatrix}0 & 0\\1 & 0\end{bmatrix} + \begin{bmatrix}0 & 0\\0 & 1\end{bmatrix}\right)$$
$$= \frac{1}{2}\begin{bmatrix}1 & 1\\1 & 1\end{bmatrix}$$

This structure arises because the superposition gives equal probability to both basis states. The off-diagonal elements (coherences) are 1/2, indicating quantum superposition. The diagonal elements represent probabilities, which sum to 1, as required for a density matrix.

Now, we compute ρ^2 as

$$\rho^2 = \left(\frac{1}{2}\begin{bmatrix}1 & 1\\1 & 1\end{bmatrix}\right) \cdot \left(\frac{1}{2}\begin{bmatrix}1 & 1\\1 & 1\end{bmatrix}\right) = \frac{1}{4}\begin{bmatrix}1+1 & 1+1\\1+1 & 1+1\end{bmatrix} = \frac{1}{2}\begin{bmatrix}1 & 1\\1 & 1\end{bmatrix}$$

The **trace** of a square matrix A, denoted as Tr(A), is the sum of its diagonal elements: In quantum mechanics, the **trace** of a density matrix $\rho$is always **1** for a properly normalized quantum state:

Taking the trace:

$$\mathrm{Tr}(\rho^2) = \frac{1}{2} + \frac{1}{2} = 1$$

This indicates that the state is pure. In contrast, a nonpure or mixed state has $\mathrm{Tr}(\rho^2) < 1$ as we will see below. Any pure state $|\psi\rangle$ can be described as $|\psi\rangle = \alpha|0\rangle + \beta|1\rangle$, with$|\alpha|^2 + |\beta|^2 = 1$.

Its *density matrix* representation is

$$\rho = |\psi\rangle\langle\psi| = \begin{bmatrix} |\alpha|^2 & \alpha\beta^* \\ \alpha^*\beta & |\beta|^2 \end{bmatrix}$$

Here, the off-diagonal terms ($\alpha\beta^*$ and $\alpha^*\beta$) encode quantum coherence, meaning that the system retains phase information.

Example 4.1 ***Trace of Mixed States and Decoherence***

Mixed states arise when we lack complete knowledge about a quantum system or when it interacts with an uncontrolled environment. If a quantum system is prepared in one of several possible states, but we don't know which one with certainty, we must describe it using a mixed state. For example, a mixed state where chances of being in $|0\rangle$ and $|1\rangle$ are equal with no phase coherence present is expressed as

$$\rho = \frac{1}{2}\,|0\rangle\Big\langle 0\,| + \frac{1}{2}\,|1\Big\rangle\langle 1\,|$$

$$\rho = \frac{1}{2}\begin{bmatrix} 1 & 0 \\ 0 & 1 \end{bmatrix}$$

A qubit initially in $|\psi\rangle = \frac{1}{\sqrt{2}}(|0\rangle + |1\rangle)$ starts interacting with a noisy environment. The environment "measures" the qubit, slowly collapsing it into either $0\rangle$ or $|1\rangle$. Over time, the density matrix evolves as

$$\rho = \begin{bmatrix} \frac{1}{2} & e^{-\gamma t}\cdot\frac{1}{2} \\ e^{-\gamma t}\cdot\frac{1}{2} & \frac{1}{2} \end{bmatrix}$$

where $e^{-\gamma t}$ represents decoherence. As $t \to \infty$, the off-diagonal elements vanish, leading to a fully mixed state as shown above. This occurs in QC, where decoherence is a major challenge.

A mixed state is represented by a *density matrix*: $\rho = p_1\,|0\rangle\langle 0|\, + p_2\,|1\rangle\langle 1|$, where p_1, p_2 represent classical probabilities ($p_1 + p_2 = 1$). For example, if a system is randomly prepared in either $|0\rangle$ or $|1\rangle$ with equal probability:

$$\rho = \frac{1}{2}\begin{bmatrix} 1 & 0 \\ 0 & 1 \end{bmatrix}$$

Here, the off-diagonal elements are zero, meaning that there is no coherence.

Squaring it: $\rho^2 = \left(\frac{1}{2}\begin{bmatrix}1 & 0\\0 & 1\end{bmatrix}\right)\cdot\left(\frac{1}{2}\begin{bmatrix}1 & 0\\0 & 1\end{bmatrix}\right) = \frac{1}{4}\begin{bmatrix}1 & 0\\0 & 1\end{bmatrix}$.

Taking the trace: $\mathrm{Tr}(\rho^2) = \frac{1}{4} + \frac{1}{4} = \frac{1}{2} < 1$.

Since $\mathrm{Tr}(\rho^2) < 1$, this confirms that the system is in a mixed state.

A generic mixed state is represented by a density matrix: $\rho = \sum_j p_j \,|\,\psi\rangle\langle\psi\,|$, where p_j represent classical probabilities and $\sum_j p_j = 1$. For example, if a system is randomly prepared in either $|0\rangle$ with probability p_0 or in $|1\rangle$ with probability $p_1 = 1 - p_0$:

$$\rho = \begin{bmatrix}p_0 & 0\\0 & 1-p_0\end{bmatrix}$$

Here, the off-diagonal elements are zero, meaning that there is no coherence.

Squaring it:

$$\rho^2 = \left(\begin{bmatrix}p_0 & 0\\0 & 1-p_0\end{bmatrix}\right)\cdot\left(\begin{bmatrix}p_0 & 0\\0 & 1-p_0\end{bmatrix}\right) = \begin{bmatrix}p_0^2 & 0\\0 & (1-p_0)^2\end{bmatrix}$$

Taking the trace: $(\rho^2) = p_0^2 + (1-p_0)^2 = 1 - 2p_0 + 2p_0^2 < 1$ for $p_0 > 1$.

Figure 4.25 shows the density-matrix representations of the four Bell states. A Bell state has only two basis components, but its density matrix contains four nonzero elements—two diagonal populations and two off-diagonal coherence terms. For $|\Phi^+\rangle = \frac{1}{\sqrt{2}}(|00\rangle + |11\rangle)$, the density matrix is

$$\rho_{\Phi^+} = \frac{1}{2}\begin{pmatrix}1 & 0 & 0 & 1\\0 & 0 & 0 & 0\\0 & 0 & 0 & 0\\1 & 0 & 0 & 1\end{pmatrix}$$

Diagonal elements show equal probability (½ each) of measuring 00 or 11. Off-diagonal elements are the coherence terms—the signature of quantum superposition between 00 and 11. Their presence makes the states entangled.

For $|\Phi^-\rangle = \frac{1}{\sqrt{2}}(|00\rangle - |11\rangle)$, the density matrix looks identical, except the off-diagonal terms change sign:

$$\rho_{\Phi^-} = \frac{1}{2}\begin{pmatrix}1 & 0 & 0 & -1\\0 & 0 & 0 & 0\\0 & 0 & 0 & 0\\-1 & 0 & 0 & 1\end{pmatrix}$$

The populations are unchanged.

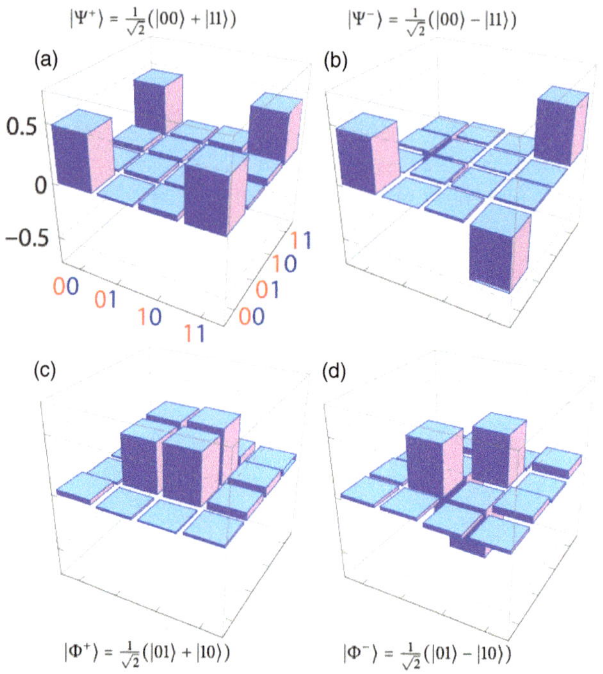

Figure 4.25 Density-matrix representations of the four Bell states, highlighting populations and coherences. Three-dimensional bar plots show the real part of the two-qubit density matrix $\rho = |\Psi\rangle\langle\Psi|$ in the computational basis $\{|00\rangle, |01\rangle, |10\rangle, |11\rangle\}$; the imaginary parts vanish for these real Bell states and are, therefore, not shown. Each bar represents a single-matrix element $\rho_{ij} = \langle i|\rho|j\rangle$. Diagonal elements ($i = j$) correspond to state populations, while off-diagonal elements ($i \neq j$) represent quantum coherences between basis states. (a) $|\Psi^+\rangle = (|00\rangle + |11\rangle)/\sqrt{2}$. Two of the four prominent bars correspond to equal populations in $|00\rangle$ and $|11\rangle$ ($\rho_{00,00} = \rho_{11,11} = 1/2$), while the other two arise from the off-diagonal coherences $\rho_{00,11} = \rho_{11,00} = 1/2$, which encode the phase-coherent superposition linking these states. (b) $|\Psi^-\rangle = (|00\rangle - |11\rangle)/\sqrt{2}$, with identical populations but opposite sign coherences, reflected as inverted off-diagonal bars. (c) $|\Phi^+\rangle = (|01\rangle + |10\rangle)/\sqrt{2}$, showing populations and coherences confined to the $|01\rangle$–$|10\rangle$ subspace. (d) $|\Phi^-\rangle = (|01\rangle - |10\rangle)/\sqrt{2}$, again with equal populations but negative coherences. In all cases, the presence of nonzero off-diagonal matrix elements connecting distant basis states provides a clear visual signature of maximal two-qubit entanglement; [17] / Jerry Moy Chow.

For $|\Psi^+\rangle = \frac{1}{\sqrt{2}}(|01\rangle + |10\rangle)$, the weight shifts to the middle two basis states:

$$\rho_{\Psi^+} = \frac{1}{2}\begin{pmatrix} 0 & 0 & 0 & 0 \\ 0 & 1 & 1 & 0 \\ 0 & 1 & 1 & 0 \\ 0 & 0 & 0 & 0 \end{pmatrix}$$

This state encodes perfect anticorrelation: when one qubit is 0, the other is 1. For $|\Psi^-\rangle = \frac{1}{\sqrt{2}}(|01\rangle - |10\rangle)$, only the sign of the coherence terms changes:

$$\rho_{\Psi^-} = \frac{1}{2}\begin{pmatrix} 0 & 0 & 0 & 0 \\ 0 & 1 & -1 & 0 \\ 0 & -1 & 1 & 0 \\ 0 & 0 & 0 & 0 \end{pmatrix}$$

If you take the trace for one qubit, each individual qubit is maximally mixed.

$$\rho_A = \mathrm{Tr}_B(\rho_{\mathrm{Bell}}) = \frac{1}{2}I$$

4.25 Wigner Tomography

Wigner tomography [237, 238] is a technique to visualize and reconstruct the quantum state of a system, particularly in continuous-variable quantum systems like superconducting cavity states. It provides a quasiprobability distribution of a quantum state in phase space, similar to how classical probability distributions describe a system. The Wigner function [239] is a quasiprobability distribution in phase space (position q and momentum p or equivalently, in quadratures X and P).

The Wigner function is not an ordinary probability distribution. It is a quasiprobability distribution. It behaves like a probability in many ways—except that it is allowed to go negative. In classical physics, probabilities are always ≥ 0. These represent ignorance about underlying reality. Negativity is the mathematical signature that classical probability theory cannot explain the state. Negativity arises from quantum interference. The negative regions are the imprint of coherent superposition between classically distinct states. When states decohere, the interference disappears—and so does the negativity.

In the Wigner tomogram picture, bright positive regions show where the state is likely to be found, while alternating red–blue regions of positive and negative value reveal the underlying quantum interference. The negative regions have no classical analogue; they are signatures that the system is truly quantum and in a coherent superposition rather than a simple probabilistic mixture.

Just like an MRI scan builds a 3D image of a body by taking 2D slices, Wigner tomography reconstructs the quantum state by taking many measurements from different angles and combining them into a full phase-space picture. The Wigner function reveals the quantum properties of the state, such as negativity, oscillations.

A Wigner tomogram is determined by systematically displacing a quantum state in phase space, measuring photon number parity at each point, and combining those measurements to reconstruct a full phase-space picture of the state. Microwave cavity measurements extract parity information. The system's state is modified by applying known quantum operations. This allows different projections of the Wigner function to be sampled. The full Wigner function is reconstructed from the measured quadrature data.

The *Wigner function* $W(\alpha)$ is defined as

$$W(\alpha) = \frac{2}{\pi}\text{Tr}[D(\alpha)\rho D^{\dagger}(\alpha)P]$$

ρ is the quantum state of the resonator, $D(\alpha)$ is a displacement operator applied to the cavity field, and P is the parity operator, which measures whether the photon number is even or odd.

The desired quantum state is initialized by applying displacement pulses to create coherent states, or using conditional qubit operations to generate Fock states or other interactions to create cat states.

The displacement operator $D(\alpha)$ is applied to the cavity field using a carefully designed microwave pulse. This pulse coherently shifts the state in phase space

$$D(\alpha) = e^{\alpha a^{\dagger} - \alpha^{*} a}$$

where α is a complex number representing a point in phase space. This step allows probing the quantum state at different phase-space locations.

The operator displaces the position and/or momentum of the harmonic oscillator depending on the value of the complex number α. Real values of α correspond to pure position displacements, while imaginary values of α correspond to pure momentum ones.

The parity is measured using the transmon qubit as a probe. The qubit is dispersively coupled to the cavity, causing a phase shift proportional to the photon number. A controlled-phase gate between the cavity and the transmon encodes the parity information into the qubit. The parity operator is

$$P = e^{i\pi a^{\dagger} a}$$

A final qubit measurement reveals the parity. This measurement is repeated many times to extract the expectation value of the parity operator.

Displacement applications and parity measurements described above are repeated for a grid of displacement values α spanning phase space to construct a full phase-space representation of the state.

Another form for The Wigner function $W(X,P)$ is

$$W(X,P) = \frac{1}{\pi\hbar}\int_{-\infty}^{\infty} dy e^{\frac{2iPy}{\hbar}} \langle X - y|\rho|X + y\rangle$$

$\hat{\rho} = |n\rangle\langle n|$ is the density matrix of the quantum state.

X and P are the quadrature variables (position and momentum in a quantum harmonic oscillator system).

4.26 Wigner Tomography Examples

The Wigner function of a coherent state (Figure 4.26) is a Gaussian-shaped distribution centered at $(\mathrm{Re}(\alpha), \mathrm{Im}(\alpha))$ in phase space. The coherent state is localized in phase space

$$W(X, P) = \frac{2}{\pi} e^{-2(X-X_0)^2-2(P-P_0)^2}$$

Recall that a coherent state is the quantum analog of a classical wave. It has a well-defined amplitude and phase. The coherent state looks like a smooth bell curve. A coherent state has a smooth Gaussian Wigner function without oscillations, without negative regions. Hence, coherent states are often called "classical-like."

A cat state is a quantum superposition of two coherent states $|\alpha\rangle$ and $|-\alpha\rangle$. The Wigner function shape (Figure 4.26) has two Gaussian lobes at $\pm\alpha$ (like two coherent states). Interference fringes in the middle, where the function goes negative → indicates quantum superposition. The negative regions between the two peaks confirm quantum interference.

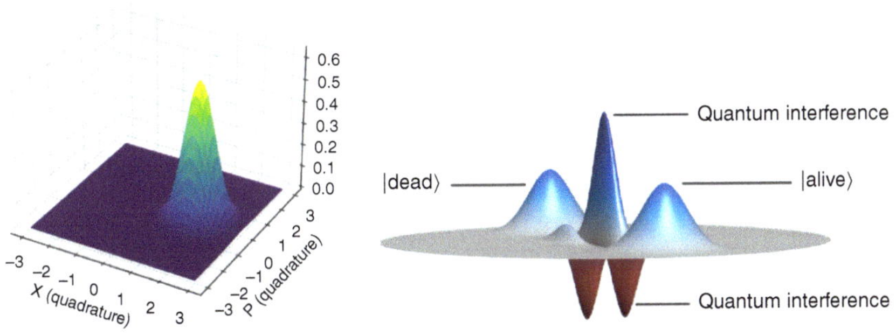

Figure 4.26 Left: Wigner distribution for a *coherent state*, the closest quantum analogue of a classical oscillating field. The distribution forms a single, approximately Gaussian peak in phase space, centered at a nonzero value of the field quadratures (X, P). The function is everywhere positive and localized, reflecting well-defined amplitude and phase with minimal uncertainty allowed by quantum mechanics. Right: Wigner function for a *Schrödinger cat state*, a quantum superposition of two coherent states with opposite phase-space displacements. Two main lobes appear in phase space, representing the two classical components of the superposition, while the central region exhibits a pattern of alternating positive and negative interference fringes. The bell shapes represent the "alive" and "dead" possible states for the "cat." The oscillations between them indicate the quantum coherence between these states (i.e., the classic "both alive and dead" statement). The negative regions are signatures of nonclassicality and coherence between the two components, distinguishing the cat state from any classical mixture of coherent states [240] / American Physical Society.

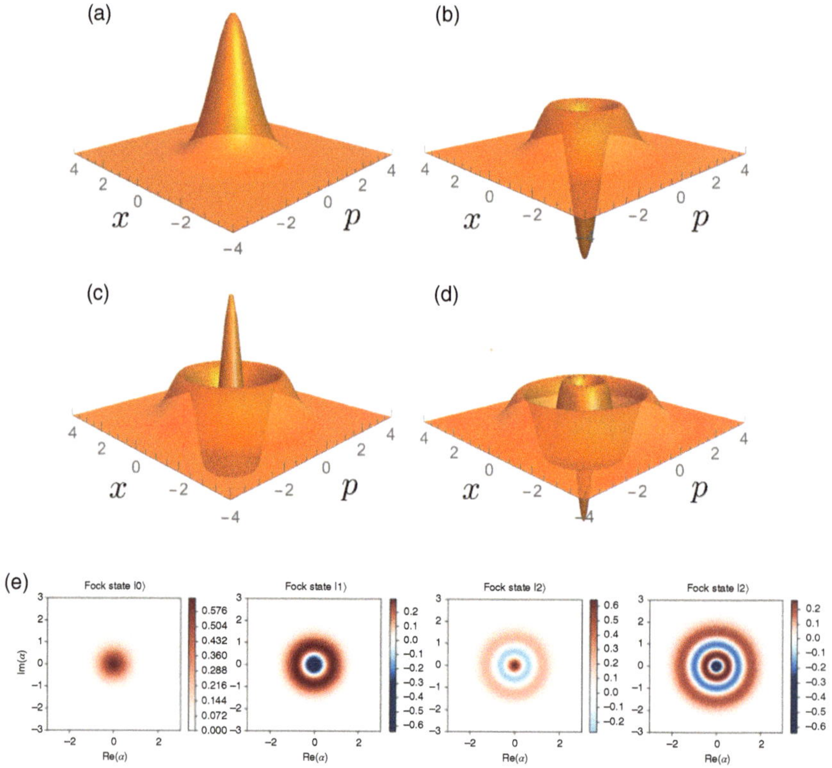

Figure 4.27 Three-dimensional Wigner functions of the first four Fock states. Panels (a)–(d) show the Wigner quasiprobability distributions in phase space for photon number (Fock) states $|0\rangle$, $|1\rangle$, $|2\rangle$, and $|3\rangle$, respectively, plotted as functions of the field quadratures x and p. In (a), the vacuum state $|0\rangle$ exhibits a single positive Gaussian-like peak centered at the origin, corresponding to the minimum uncertainty ground state of the harmonic oscillator. In (b), the $|1\rangle$ state displays a central negative dip surrounded by a positive ring, illustrating the appearance of Wigner negativity as a hallmark of nonclassicality. Higher Fock states in (c) and (d) show increasingly complex radial oscillations: alternating positive and negative rings that grow in number with photon number. These oscillations reflect the Laguerre-polynomial structure of the Wigner function for Fock states and encode the increasing energy and nodal structure of higher photon number states (Ref. [241] / Andreas Ketterer). (e) Two-dimensional Wigner function projections for the first four Fock states. (Adapted from Ref. [242] / Oxford University Press.)

For a Fock state $|n\rangle$, the Wigner function (Figure 4.27) can be explicitly written as

$$W_n(X,P) = \frac{2}{\pi\hbar}(-1)^n e^{-2(X^2+P^2)/\hbar} L_n(4(X^2+P^2)/\hbar)$$

$L_n(x)$ is the Laguerre polynomial of order n. The Laguerre polynomials are solutions to Laguerre's differential equation and are given by the Rodrigues formula:

$$L_n(x) = \frac{e^x}{n!}\frac{d^n}{dx^n}(x^n e^{-x})$$

Order 1: $L_1(x) = 1 - x$

Order 2: $L_2(x) = 1 - 2x + \frac{x^2}{2}$

Order 3: $L_3(x) = 1 - 3x + \frac{3x^2}{2} - \frac{x^3}{6}$

Order 4: $L_4(x) = 1 - 4x + 3x^2 - \frac{2x^3}{3} + \frac{x^4}{24}$

The Wigner function for $|0\rangle$ (vacuum state) is a simple Gaussian centered at the origin:

$$W_0(X, P) = \frac{2}{\pi\hbar}e^{-2(X^2+P^2)/\hbar}$$

This is a classical-like probability distribution with no negative regions and oscillations.

The Wigner function for

$|1\rangle$ (single-photon state) has a doughnut shape with a central dip (negative region), indicating quantum nonclassicality.

$|2\rangle$ (two-photon Fock state): A two-ring structure with oscillations. More pronounced negative regions, highlighting increased quantum interference.

Higher Fock states $|n\rangle$ exhibit alternating concentric rings of positive and negative values.

The number of oscillations increases with n. The negativity grows, making them more useful in quantum applications.

The Wigner tomograph is often shown as 2D projection instead of a 3D rendering for simplicity.

4.27 Conclusion

The development of 3D cavity architectures has opened a powerful pathway for advancing QIP. A central advantage of 3D cavities is their ability to support coherence times far beyond those achievable in planar CPW resonators. By suppressing surface participation and associated losses, these structures achieve high-quality factors, translating into photon lifetimes ranging from milliseconds up to seconds. Such long-lived photons form the foundation for reliable storage of quantum information.

Within these cavities, several complementary encoding strategies are available. The most direct approach uses Fock-state qubits in which information is stored simply in photon number states and manipulated with the ancilla transmon using displacement operations and SNAP gates. Qudit encodings exploit the full multilevel Hilbert space of a cavity mode, enabling compact algorithms that require fewer physical subsystems than standard qubit approaches. Sophisticated bosonic encodings—such as dual-rail, binomial, and cat codes—rely on superpositions of multiphoton states to passively protect against photon loss and correct errors (except for dual-rail).

Protection from noise remains essential, as photon loss continues to be the dominant error channel in these systems. The availability of transmon ancilla qubits allows parity-based error detection and active error correction. Notably, both cat and binomial codes have now reached the break-even point, where the lifetime of the encoded logical qubit exceeds that of the underlying physical cavity mode—an important milestone.

3D architectures also offer compelling scaling advantages. Because a single cavity can host multiple logical qubits or qudits, wiring complexity and cryogenic overhead are reduced. New structures such as ring resonators supply natural connectivity as well as enlarged Hilbert spaces. Hybrid processor–memory systems, in which transmons provide fast logic while cavities serve as long-lived memories connected via SWAP operations, illustrate practical routes toward scalable architectures.

State preparation and characterization tools have also matured. Protocols now exist for generating Fock states, coherent states, cat states, and for diagnosing them using Wigner tomography. These methods not only verify state preparation but also provide a visual insight into decoherence processes.

Wigner tomographs play a central diagnostic role in the bosonic-encoding paradigm by providing a direct, phase-space visualization of how quantum information is stored and manipulated within cavity modes. Unlike simple population measurements, Wigner tomography reconstructs the full quasiprobability distribution of a bosonic state, revealing both its classical structure and its quantum features, such as interference fringes and regions of negativity. In dual-rail encodings, Wigner tomographs primarily serve to confirm mode occupancy and coherence between rails. For binomial and cat codes, Wigner functions expose the defining structure of the code words. For more complex GKP states, Wigner tomography provides a window into the underlying lattice or grid structure, revealing periodic features that encode logical information and indicate the degree of finite squeezing or diffusion from noise.

A cavity–transmon system provides a natural division of labor between control and storage, combining fast nonlinear processing with long-lived quantum memory. Finally, the use of cavities for computation extends beyond error protection alone. Multilevel logic has been demonstrated experimentally, including qudit-based implementations of algorithms.

This chapter makes clear that 3D superconducting cavities are not just an incremental improvement over 2D architectures, but a paradigm shift. By uniting ultralong coherence times, bosonic error-correcting codes, and scalable architectures, they enable quantum processors that are more resilient, hardware-efficient, and algorithmically versatile. This positions cavity-based QC as a promising route toward large-scale, fault-tolerant quantum computation.

Summarizing the comparative capabilities, 3D cavity plus transmon architectures and 2D qubit arrays represent two distinct visions for scaling: one based on many simple qubits arranged in a planar grid, the other based on fewer, more powerful bosonic modules with long coherence and rich internal structure. Both approaches continue to advance, and the eventual landscape of QC may incorporate elements of each.

In a planar 2D transmon array, each physical element is restricted to a two-level subspace, and computation is built from sequences of single- and two-qubit gates. Scaling such systems requires adding more qubits, more couplers, and more control lines, which increases hardware complexity and introduces additional decoherence channels. Error correction in qubit-only systems is resource-intensive: protecting a single logical qubit typically requires dozens or hundreds of physical qubits arranged in a surface code-like structure.

A cavity-based processor offers a different scaling philosophy. A single cavity can host a high-dimensional computational space or even a full logical qubit encoded in a bosonic code, dramatically reducing the number of physical components required. The cavity provides long coherence and a large Hilbert space; the transmon supplies conditional control. This division of labor allows one physical module to perform the work of many qubits. Certain algorithms, state-preparation routines, and error-correcting operations become more compact because they exploit the structure of the harmonic oscillator.

At present, 2D superconducting circuits lead in demonstrated processor-scale performance. They support multiqubit algorithms on tens to hundreds of qubits, early demonstrations of logical qubits, and a mature control stack optimized for large-scale integration. By contrast, 3D cavity-based systems have demonstrated high coherence, bosonic control, and high-fidelity operations on single cavities or small hybrid modules, but have not yet reached the scale of multi-logical qubit processors.

Yet the long-term potential of 3D cavity architectures is compelling. Their exceptional coherence, combined with the ability to encode logical qubits in a single bosonic mode, suggests a path toward processors built from fewer, higher-quality logical units rather than large arrays of modest coherence qubits. The challenge lies in engineering scalable 3D integration, managing mode crowding, and developing modular architectures that can network multiple cavity–transmon units. If these engineering hurdles are overcome, cavity-based processors could offer a route to fault-tolerant QC with dramatically reduced physical overhead.

5

Materials for High Coherence

5.1 Introduction

The raw performance of superconducting qubits is ultimately limited by the quality of the materials from which they are fabricated. Achieving long coherence times [46]—necessary for reliable quantum computation—requires a deep understanding of how materials contribute to energy loss and decoherence. This chapter opens by establishing that transmon qubits behave as nonlinear resonators, where energy relaxation (T_1) and dephasing (T_2) are constrained by multiple loss channels: dielectric losses, conductor losses, radiative losses, and quasiparticle excitations. The framework of *quality factors (Q) and participation ratios* provides the language to quantify how each material region or interface contributes to overall dissipation.

The chapter highlights how interfaces and oxides—such as niobium pentoxide or aluminum oxide—harbor microscopic TLSs that interact with qubit fields and significantly reduce coherence. Simulation and experimental tools allow researchers to map these loss mechanisms, refine fabrication processes, and explore alternative materials. Beyond 2D transmon architectures, the discussion expands to tantalum-based devices, where crystalline oxides and fewer suboxides lead to longer lifetimes, as well as to 3D niobium cavities, where TLS-driven loss in surface oxides is linked to the mysterious "low-field Q-slope" (LFQS) frequently discussed in accelerator cavity applications.

5.2 Materials for 2D

5.3 Quality Factors

In a transmon qubit, energy relaxation (characterized by T_1) and dephasing (T_2) are limited by various loss mechanisms. These include dielectric losses, conductor losses, radiative losses, and quasiparticle losses. The coherence time T_1 is directly linked to the total quality factor of the resonator:

$$T_1 = \frac{Q_{\text{total}}}{\omega_0}$$

where ω_0 is the resonance frequency in radian units. The quality factor, Q, of a resonator is defined as

$$Q = | \omega_0 \frac{\text{average energy stored}}{\text{dissipated power}} |$$

Q defines the bandwidth of the resonance from $= \frac{1}{Q}$ but also quantifies the different loss mechanisms. The quality factor can be written as

$$Q = \frac{\omega_0}{\Delta\omega_0} = \frac{f_0}{\Delta f_0}$$

Here f_0 is the linear resonance frequency. Q can also be defined in terms of impedance, Z, or admittance, Y, as

$$Q = \frac{\text{Im}[Z]}{\text{Re}[Z]} = \frac{\text{Im}[Y]}{\text{Re}[Y]}$$

For example, to determine the quality factor of a parallel LCR circuit, we evaluate the power dissipated by the resistor, $P_d = V^2/2R$. The average stored energy in the electric field by the capacitor, $E_c = \frac{1}{4}V^2C$, and the average stored energy in the magnetic field by the inductor will be $E_i = \frac{1}{4}\frac{V^2}{\omega_0^2 L}$. Accordingly,

$$Q_{\text{LCR}} = \omega_0 \frac{\frac{1}{4}V^2C + \frac{1}{4}V^2/\left(\omega_0^2 L\right)}{\frac{V^2}{2R}}$$

$$Q_{\text{LCR}} = \frac{1}{2}\omega_0 RC + \frac{1}{2}\omega_0 \frac{R}{\omega_0^2 L}$$

$$Q_{\text{LCR}} = \frac{1}{2}\omega_0 RC + \frac{1}{2}\omega_0 RC$$

$$Q_{\text{LCR}} = \omega_0 RC = \omega_0 \tau$$

Losses in a transmon can be separated into contributions from various sources in its materials stack (Figure 5.1), each with its own Q:

$$\frac{1}{Q_{\text{total}}} = \sum_i \frac{1}{Q_i}$$

This multilayer stack highlights the combination of superconducting films, dielectric interfaces, and microfabrication steps that together define the coherence properties and functionality of the transmon qubit. Here each Q_i corresponds to a different loss mechanism. These individual Qs can be estimated from material properties and from simulations, or fit from experimental data.

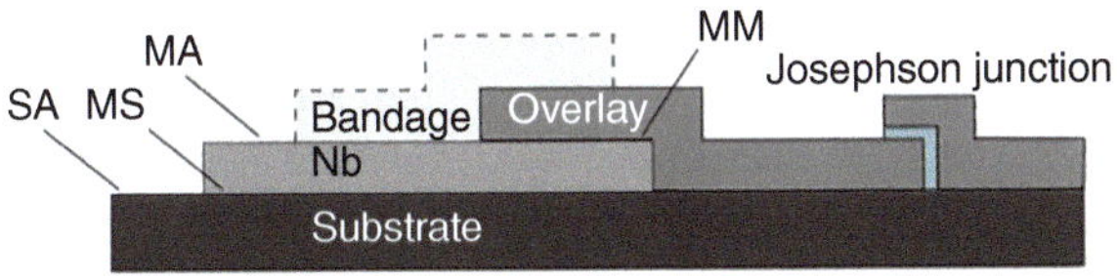

Figure 5.1 Materials stack of a transmon qubit device. Schematic cross-section showing the layered fabrication stack used to realize a planar transmon. A substrate (typically high-resistivity Si or sapphire) forms the mechanical and dielectric base. A patterned niobium (Nb) thin film defines the superconducting circuitry, including the capacitor pads and ground plane. The Josephson junction is formed by two overlapping superconducting electrodes separated by a thin oxide tunnel barrier (blue), providing the nonlinear inductance that defines the qubit. The drawing also illustrates multiple metal layers and processing steps used to ensure low-resistance, well-sealed electrical contacts: MS/SA denote initial surface preparation or seed/adhesion layers, MA indicates the first metal layer forming the main wiring, and MM denotes the subsequent metal layer used to complete crossovers or contact extensions. A bandage layer and overlay metallization are applied to reinforce junction leads and reduce contact resistance, improving device robustness and microwave performance; adapted from Ref. [243].

5.4 Participation Ratios

An independent way to describe dissipation in an RF circuit is by participation ratios, which quantify how much of the qubit's electric or magnetic field energy is stored in a given region or in a material with loss tangent $\tan\delta_i$ occupying that region. The total Q can be expressed as

$$\frac{1}{Q_{\text{total}}} = \sum_i p_i \cdot \tan\delta_i$$

A higher participation ratio for a specific material or region within the resonator implies a greater contribution to the overall energy stored in the resonator. In the case of resonators with multiple materials or interfaces, it is important to minimize the participation ratio of lossy materials.

Knowing the participation ratios can guide materials engineering for achieving cleaner interfaces and making optimal substrate choices [244]. Ratios help to optimize device geometry, for example, larger pads → lower electric field per unit volume.

The participation ratio p_i for a dielectric material or surface i is defined as

$$p_i = \frac{\int_{\text{material}i} \varepsilon_i |E(\vec{r})|^2 dV}{\int_{\text{all space}} \varepsilon(\vec{r}) |E(\vec{r})|^2 dV}$$

The total dielectric loss is then given as

$$\frac{1}{Q_{\text{total}}} = \sum_i p_i \cdot \frac{1}{Q_i}$$

Q_i is the quality factor of region i, and p_i is the participation ratio of region i.

Surface dielectric losses arise from thin oxide layers and interfaces, which host TLSs. Though the volumes are tiny, the electric fields are often strongest here to yield high participation ratios. These are determined by simulating electric fields at material boundaries using full 3D electromagnetic simulation with tools, such as ANSYS, HFSS, and COMSOL, and by using known or measured $\tan\delta_i$ (loss tangent) for each material/interface.

5.5 Substrate Loss (Bulk Dielectric)

Consider participation ratios of a parallel capacitor filled with a material with a complex dielectric constant with complex dielectric constant, $\varepsilon = \varepsilon_r + i\varepsilon_i$ ($i = \sqrt{-1}$). The admittance for a capacitor is $Y_c = i\omega C$, where C is the capacitance and ω is the angular frequency. The capacitance for this parallel plate is $C_{pp} = \varepsilon\frac{A}{d}$, where ε is the complex permittivity, A is the area of the plates, and d is the distance between the plates. The admittance for this lossy capacitor is $Y_{cpp} = j\omega\varepsilon\frac{A}{d}$. Writing out the complex permittivity,

$$Y_{cpp} = i\omega\varepsilon_r\frac{A}{d} + \omega\varepsilon_i\frac{A}{d}$$

$$Y_{cpp} = i\omega C_r + G_{\text{shunt}}$$

The first term stores energy in the E-fields and acts as an idealized capacitor. The second term is a dissipative term. To determine the quality factor of this capacitor, we can take the ratio between the imaginary and real parts of equation:

$$Q_{cpp} = \frac{\text{Im}[Y_{cpp}]}{\text{Re}[Y_{cpp}]}$$

$$Q_{cpp} = \frac{\omega C_r}{G_{\text{shunt}}}$$

$$Q_{cpp} = \frac{\omega\varepsilon_r\frac{A}{d}}{\omega\varepsilon_i\frac{A}{d}}$$

$$Q_{cpp} = \frac{\varepsilon_r}{\varepsilon_i} = \frac{1}{\tan\delta}$$

Thus, the quality factor of this circuit is related to the loss tan of the dielectric.

In planar transmons, a significant fraction of the electric field energy lives in the dielectric environment—especially near edges where fields concentrate. Hence, a substrate with an intrinsically higher loss tangent (or a dirty/defective surface layer) can dominate. Silicon and sapphire are both capable substrates. High-resistivity silicon is extremely good $\tan\delta \sim (1-5)\times 10^{-7}$, but it comes with the complication of its native oxide (SiO2), plus water/hydrocarbon adsorbates, plus processing residues—exactly the kinds of disordered layers that host TLS loss

(Section 5.9) and can increase the losses to tan $\delta \gtrsim 10^{-5}$or 10^{-6}. Processes such as buffered oxide etch (BOE) or HF-based treatments can lower the effective loss tangent. When the surface is aggressively cleaned/etched and the downstream process prevents regrowth and adsorption, silicon can move closer to sapphire in effective microwave loss [139].

High-quality sapphire (as-received HEMEX sapphire wafers $\tan\delta \sim 3\times10^{-7}$) is an excellent microwave dielectric when properly prepared, and because its surfaces can be made clean and stable. Annealing is not merely a cleaning step, but a transformation of the substrate's microwave properties. SQMS studies measure the substrate loss tangents directly with an ultra-high-Q niobium SRF resonator, at millikelvin temperatures and GHz fields. The studies are sensitive enough to resolve dielectric loss in the 10^{-8}–10^{-7} regime. They find unannealed HEMEX sapphire has a bulk loss tangent of about 3.15×10^{-7} (implying a substrate-limited T_1 of only ~0.24 ms), while HEMEX sapphire after annealing drops to $\tan\delta \approx 3.98\times10^{-8}$—roughly an 8x reduction—corresponding to a substrate-limited T_1estimate near 1.9 ms. The sapphire preparation anneal is at 1100 °C for 3 hours (with RCA cleaning and solvent cleaning workflows contrasted against "as-received HEMEX" and "solvent clean only").

The anneal is understood to reduce the density (or activity) of the microscopic defect populations—near-surface disorder, impurity/defect complexes, and adsorbate-related dipoles—that otherwise contribute to dielectric loss in the quantum regime.

5.6 Metal Losses

Conductor losses arise due to surface resistance of the superconducting film and can be simulated via surface current density and known surface resistance. The metal participation ratio is analogous to the dielectric participation ratio, but instead of energy stored in electric fields, it's based on magnetic (surface current) energy:

$$p_m = \frac{\int_{\text{metal surface}} R_s |\vec{H}_{\parallel}|^2 \, dA}{\omega W}$$

R_s denotes the surface resistance of the superconductor (frequency and temperature dependent).

$\vec{H}_{\parallel}$ is the magnetic field tangential to the metal surface.

ω is the angular frequency.

W is the total stored energy in the qubit mode.

This gives the fraction of power dissipated due to conductor loss and contributes to the inverse quality factor:

$$\frac{1}{Q_m} = p_m \cdot \frac{R_s}{Z_0}$$

where Z_0 is a characteristic impedance.

There are variations for Nb deposition techniques. Rigetti is using HIPIMS to deposit Nb films for qubits. Other deposition methods such as MBE, PLD, and CVD are explored. There are also variations from final chemical treatment (etch) of the (Nb and Ta) metal films.

5.7 JJ and Associated Losses

The junction and its immediate electromagnetic environment—particularly, the junction leads—can make measurable contributions to the transmon's energy relaxation time T_1. The thin amorphous tunnel barrier, typically aluminum oxide, hosts a high density of microscopic TLSs. These TLSs couple directly to the strong electric field across the junction and can absorb energy from the qubit. Hence, junction quality [245, 246], oxidation conditions, and barrier uniformity remain important. In transmons, the shunt capacitor is designed so that only a small fraction of the total electric energy resides across the junction. This reduces the junction participation ratio and protects coherence. Residual junction loss becomes visible only once other dielectric channels have been pushed low. In the best devices, junction loss is, therefore, not dominant, but it can still set a floor for T_1. As substrate, surface, and packaging losses are reduced, junction-related dissipation increasingly emerges as a limiting factor.

Closely related are losses associated from the junction leads. The narrow superconducting wires that connect the junction to the larger capacitor pads concentrate electric fields and host surface oxides similar to those on the main capacitor electrodes. Because these leads are geometrically small but experience large current densities and strong field gradients, their dielectric participation can be disproportionately large. Surface TLS residing on the side walls and native oxides of the leads can couple strongly to the qubit, contributing both to steady dissipation and to temporal fluctuations in lifetime.

Imperfections in the oxide can produce localized conduction channels with energies below the superconducting gap. These channels behave as dissipative elements and introduce a small but finite resistance in parallel with the ideal Josephson inductance. Although difficult to measure directly, such subgap states contribute to an effective quality factor of the junction that can become relevant once other losses are suppressed.

The junction and its leads are also implicated in noise-induced lifetime variability.

5.8 Example Transmon Loss Budget

In the case of 2D QPU transmons, niobium is deposited as a thin film on an underlying dielectric substrate of silicon or sapphire. There are various interfaces and surfaces that can limit coherence times. These include the interface between the metal thin film and the underlying substrate (metal/substrate interface), the surface oxide [247] that forms on the unexposed substrate (substrate/air interface), and the niobium pentoxide that forms at the surface of the niobium metal (niobium/air interface). Losses from Nb_2O_5 are influenced by the presence of TLSs [248, 249], which are atomic or molecular configurations that can exist in two distinct states and contribute to energy loss mechanisms. Niobium pentoxide is the most detrimental part of qubit's coherence. The next section discusses TLS losses in more depth.

Lithography processes used to define the superconducting circuit can also introduce interstitial impurities in the niobium metal. Aluminum is traditionally used for the two superconducting electrodes in the JJs. These electrodes are separated by an ultra-thin insulating barrier through which electrons tunnel. The most common method for fabricating this tunneling barrier involves a diffusion-limited oxidation process of aluminum, which leads to the formation of an aluminum/aluminum oxide/aluminum junction. Spatial variations in the oxide barrier width can lead to inhomogeneity in the tunneling currents and induce variability in operating frequency of the qubit. Similarly to niobium pentoxide, aluminum oxide can host TLS defects and impact coherence times.

The first step in evaluating the transmon loss budget is to solve for the electric field distribution at the transmon's fundamental mode and calculate the energy stored in each lossy dielectric/interface region. Subsequently, calculate the participation ratios for each region. The result may resemble the following table as examples:

Interface	**p_i (simulated)**	**$\tan \delta_i$ (typical)**
MS (metal–substrate (oxide)	4×10^5	2×10^{-3}
SV (substrate–vacuum) (native SiO_2)	2×10^{-5}	1×10^{-3}
MV (metal–vacuum) (AlOx)	1×10^{-5}	1×10^{-3}
Bulk Si	3×10^{-3}	1×10^{-6}

Using

$$\frac{1}{Q_{\text{dielectric}}} = \sum_i p_i \cdot \tan \delta_i$$

$$\frac{1}{Q_{\text{dielectric}}} = (4 \times 10^{-5})(2 \times 10^{-3}) + (2 \times 10^{-5})(10^{-3}) + \cdots \approx 1.2 \times 10^{-7}$$

$$Q_{\text{dielectric}} \approx 8.3 \times 10^{6}$$

Now we include conductor and junction loss.

Metal loss (surface resistance of Al): $Q_{\text{metal}} \sim 10^7 - 10^8$

Junction oxide loss (AlOx): $\tan\delta \sim 10^{-2}$. Since the junction volume is very small, typically $Q_{\text{junction}} \sim 10^6$.

$$\frac{1}{Q_{\text{total}}} = \frac{1}{Q_{\text{dielectric}}} + \frac{1}{Q_{\text{metal}}} + \frac{1}{Q_{\text{junction}}}$$

$$\frac{1}{Q_{\text{total}}} \approx \frac{1}{8.3 \times 10^{6}} + \frac{1}{10^{8}} + \frac{1}{10^{6}} \approx 1.33 \times 10^{-6}$$

$$Q_{\text{total}} \approx 7.5 \times 10^{5}$$

To convert Q to coherence time, assume frequency $f = 5\,\text{GHz} \Rightarrow \omega = 2\pi \times 5 \times 10^9$. Hence, $T_1 = \frac{Q_{\text{total}}}{\omega} \approx \frac{7.5\times10^5}{2\pi\cdot5\cdot10^9} \approx 24\,\mu\text{s}$. This would be an example of a lossy transmon, which is why materials studies are so important.

To directly measure the loss tangent, one can use a CPW or lumped-element resonator built mostly from the material of interest. This assumes the material dominates the loss. The resonator must be carefully designed to localize the electric field in the material. Measure Q_i of the resonator at cryogenic temperatures. From this, one can extract the loss tangent using

$$\tan\delta \approx \frac{1}{Q_i}$$

This method can be used for bulk substrates (e.g., Si, sapphire), thin films, and deposited oxides.

Alternatively, if there exists a transmon with known Q_{measured} and simulated participation ratios p_i, one can fit the loss tangents by solving

$$\frac{1}{Q_{\text{measured}}} = \sum_i p_i \cdot \tan\delta_i$$

By testing multiple devices with different geometries and p_i, a linear regression can solve for multiple unknown $\tan\delta_i$ values.

In Nb-based transmon qubits with Al JJs, once losses from Nb surface oxides are mitigated, it becomes unclear which mechanisms dominate the remaining decoherence. Several candidates compete at roughly comparable scales: dielectric loss in the substrate, dissipation in the underlying metal, interface-related TLSs, niobium hydrides [250] formation, and losses intrinsic to the JJ itself. Identifying the next-limiting mechanism is, therefore, essential for pushing coherence times beyond the current state of the art.

A variety of substrate materials are actively used or investigated in this context. These include standard sapphire [139], high-purity HEMEX sapphire, annealed MSE sapphire, and silicon substrates such as intrinsic Si and DSP-Si. Each substrate type presents a different balance of bulk dielectric loss, interface quality, and compatibility with superconducting film deposition. Understanding how these materials behave—especially under different cleaning and annealing workflows—is central to determining which substrate-related losses remain after Nb oxide TLS contributions are reduced.

To disentangle these mechanisms, researchers rely on a broad suite of materials characterization and loss analysis tools. Techniques such as X-ray photoelectron spectroscopy (XPS) and time-of-flight secondary ion mass spectrometry (TOF-SIMS [251]) probe surface chemistry and contamination. THz spectroscopy, electron energy loss spectroscopy (EELS), atom probe tomography (APT), and transmission electron microscopy (TEM) provide nanoscale insight into film structure, interfaces, and defect distributions. Tunneling spectroscopy, magneto-optical imaging, DC transport, and muon spin rotation (μSR) reveal superconducting properties, vortex behavior, and impurity phases. Complementary measurements of dielectric loss in high-Q Nb host cavities offer parts-per-billion sensitivity to bulk and interface dissipation. Together, these tools form a comprehensive framework for isolating and quantifying the microscopic origins of loss in superconducting qubit materials.

5.9 TLS Losses

TLS losses emerge wherever structural disorder, broken bonds, or fluctuating atomic configurations exist. In transmon circuits, the relevant electric fields span metals, oxides, and dielectrics simultaneously, allowing TLS loss in many distinct regions to couple to the qubit. As a result, qubit relaxation and dephasing often reflect the *collective behavior of multiple TLS reservoirs*, each with its own physical origin, density, and dynamical properties [252, 253].

Several material systems are now recognized as particularly important TLS hosts. Native and processed metal oxides, such as Nb_2O_5 or AlO_3, can contain high densities of amorphous bonding configurations that support electric-dipole TLS. Substrate surfaces and near-surface layers, including SiO_2 on silicon or disordered regions of sapphire, introduce additional TLS that participates through strong fringe fields at metal edges. Deposited dielectrics and encapsulation layers, while often essential for protection or wiring, can contribute further loss if amorphous or hydrogen-rich. Even adsorbates such as water, hydrocarbons, or residual process contamination can form fluctuating dipoles that behave as TLS at millikelvin temperatures.

Understanding TLS loss, therefore, requires moving beyond a single-defect picture toward a materials-aware view of the entire electromagnetic environment. Each interface and dielectric volume contributes according to both its intrinsic loss tangent and its participation in the qubit's electric field. The following sections examine the dominant TLS-hosting materials individually—beginning with metal oxides such as Nb_2O_5, then extending to substrate-related dielectrics and deposited films—to clarify how microscopic disorder at these locations ultimately shapes transmon coherence.

Losses in Nb_2O_5 are influenced by the presence of TLS, which are atomic or molecular configurations that exist in two distinct states and contribute to energy loss mechanisms within the amorphous part of the surface oxide. Understanding and mitigating these TLS mechanisms are crucial for enhancing qubit performance. The primary TLS mechanisms in Nb_2O_5 include (Figure 5.2):

1) Tunneling Atoms: Certain atoms within the Nb_2O_5 lattice may occupy positions where they can quantum mechanically tunnel [253] between two adjacent sites. As the atoms oscillate between positions, they absorb energy from the system, leading to dielectric losses.
2) Hydrogen Rotors: Hydrogen atoms or protons present as impurities can become interstitials within the Nb_2O_5 matrix. These interstitials can rotate or reorient between two equilibrium positions, acting as TLSs. Their motion can interact with lattice vibrations and external electromagnetic fields, leading to energy dissipation.
3) Dangling Bonds: Dangling bonds occur when atoms at the surface or within the bulk of Nb_2O_5 have unsatisfied valencies, leading to localized electronic states. These unpaired electrons can transition between two energy states, contributing to dielectric losses. The presence of dangling bonds is often associated with defects or amorphous regions within the material.
4) Collective Motion: Groups of atoms or molecular units within Nb_2O_5 can undergo collective or correlated movements between two configurations. This collective motion can form a TLS where the entire group transitions between states, leading to energy absorption and loss. Such mechanisms are more complex and can involve interactions over larger regions of the material.

Oxygen vacancies [254] in Nb_2O_5 can be an important microscopic source of loss because they create electric dipoles and unpaired electronic spins—two ingredients that couple efficiently to qubits. A TLS arises when a defect can occupy two nearby configurations with similar energy, separated by a small barrier. In Nb_2O_5, an oxygen vacancy can lead to two possible positions for nearby Nb–O bonds, two nearly equivalent charge configurations, or two local lattice distortions. The vacancy, therefore, behaves like a quantum tunneling defect, enabling to switch between configurations under the influence of strain or electric field.

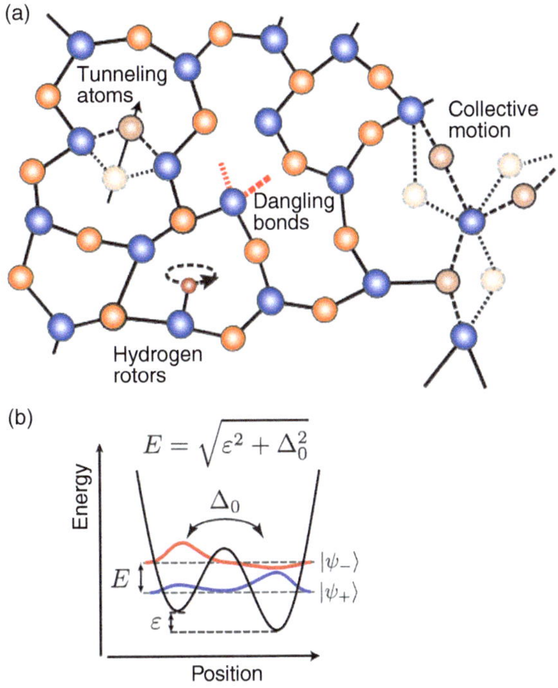

Figure 5.2 (a) TLS formation in disordered materials: structural defects in amorphous oxides and dielectrics give rise to atomic configurations that can tunnel between two nearly degenerate positions, forming TLSs. Various microscopic sources contribute to TLSs, including tunneling atoms (light atoms hopping between metastable lattice sites), hydrogen rotors (localized hydrogen atoms rotating between multiple configurations), dangling bonds (unpaired atomic bonds producing localized electric dipoles), and collective atomic motion involving clusters of atoms undergoing correlated reconfigurations. These defects behave as fluctuating electric dipoles, coupling strongly to the electric fields of superconducting qubit capacitors and resonators, leading to dielectric losses. (b) Energy-level structure of a TLS: the TLS is modeled as a quantum particle in an asymmetric double-well potential with an energy bias; Adapted from Ref. [256] / IOP Publishing.

Because these two configurations typically have different electric-dipole moments, the defect couples strongly to the microwave electric fields present in transmon.

The same vacancy can also trap unpaired electrons. When an oxygen atom is removed, its bonding electrons do not simply disappear; depending on charge compensation, one or two electrons may remain localized on nearby niobium atoms. If an electron is unpaired, it carries spin-1/2, producing a localized magnetic moment that can be measure via muon spin resonance [255]. These magnetic moments can fluctuate in time due to spin flips, or interactions with nearby defects. Such fluctuating spins are believed to be a major contributor to $1/f$ magnetic flux noise, which dominantly affects dephasing (T_2) in superconducting qubits.

TLS can be modeled using a quantum two-state system with the Hamiltonian [256]:

$$H = \frac{1}{2}\begin{bmatrix} \varepsilon & \Delta_0 \\ \Delta_0 & \varepsilon \end{bmatrix}$$

ε is the asymmetry energy (difference between potential wells).
Δ_0 is the tunneling energy.

The eigenvalues of this Hamiltonian give the energy splitting: $E = \pm\frac{1}{2}\sqrt{\varepsilon^2 + \Delta_0^2}$.

This energy gap defines the TLS dynamics, including its interaction with external fields and thermal excitations. The TLS can absorb energy from an external electromagnetic field. The response depends on the transition probability between the states. The probability of occupation of the energy states follows the Boltzmann distribution:

$$P_1 - P_2 = \tan h\left(\frac{E}{2k_BT}\right)$$

where P_1 and P_2 are the occupation probabilities of the two levels and k_BT represents the thermal energy. The energy dissipation due to TLS is proportional to the difference in occupation probabilities and describes the temperature dependence of dielectric losses.

The probability distribution can be derived as follows:

First, the occupation probabilities of the two states are

$$P_1 = \frac{e^{-E_1/k_BT}}{e^{-E_1/k_BT} + e^{-E_2/k_BT}}$$

$$P_2 = \frac{e^{-E_2/k_BT}}{e^{-E_1/k_BT} + e^{-E_2/k_BT}}$$

Factoring out e^{-E_1/k_BT} from the denominator gives

$$P_1 = \frac{1}{1 + e^{-(E_2-E_1)/k_BT}} = \frac{1}{1 + e^{-E/k_BT}}$$

$$P_2 = \frac{e^{-E/k_BT}}{1 + e^{-E/k_BT}}$$

The difference in occupation probabilities is

$$P_1 - P_2 = \frac{1 - e^{-\frac{E}{k_BT}}}{1 + e^{-\frac{E}{k_BT}}}$$

Using the identity: $\frac{1-e^{-x}}{1+e^{-x}} = \tan h\left(\frac{x}{2}\right)$,

$$P_1 - P_2 = \tan h\left(\frac{E}{2k_BT}\right)$$

The net population difference between the two states follows the hyperbolic tangent function. This tan$h$ dependence appears in the response of TLSs interacting with external fields, leading to the well-known dissipation function:

$$Q^{-1} \propto tan\,h\left(\frac{\hbar\omega}{2k_B T}\right)$$

which describes the temperature dependence of losses in TLS systems, where $\hbar\omega$ is the photon energy interacting with the TLS.

5.10 Lifetime Variability

Despite remarkable improvements in average coherence times, transmon qubits remain time-dependent systems [257]. Even when operated under nominally identical conditions—fixed temperature, fixed frequency, and stable control electronics—their coherence properties fluctuate over time. These variations are not experimental artifacts but reflect the presence of microscopic degrees of freedom within the materials themselves. As a result, transmon performance must be understood not only in terms of static loss mechanisms but also through the lens of temporal variability ranging from seconds to months.

On short timescales, spanning seconds to hours, transmons commonly exhibit sudden changes in T_1, intermittent frequency shifts, and fluctuations in dephasing rates. In continuous monitoring experiments, coherence can appear stable for long periods and then abruptly degrade, only to partially recover later. Such behavior is frequently accompanied by noise in the qubit frequency and irregular modulation of Ramsey fringes. These effects point to stochastic microscopic processes rather than macroscopic instability of the circuit.

The origin of this short-term variability lies in the behavior of TLSs embedded in amorphous surface oxides and interfacial layers. These defects, which arise from atoms or atomic groups occupying nearly degenerate configurations, can tunnel between states even at millikelvin temperatures. When a TLS switches configuration, it alters the local dielectric environment seen by the qubit's electric field. This produces sudden changes in dielectric loss, leading directly to fluctuations in T_1, and induces local electric-field shifts that translate into qubit frequency noise. In some cases, an individual TLS may lie close to resonance with the qubit, hybridizing with the circuit mode and creating sharp, time-dependent loss features. As the TLS itself drifts in energy due to strain or interactions with neighboring defects, the strength of this coupling can change abruptly, producing intermittent coherence degradation.

Nonequilibrium quasiparticles (unpaired electrons) provide an additional source of short-term variability. Bursts of quasiparticles generated by cosmic rays [258], environmental radiation, or stray infrared photons can transiently increase relaxation rates and suppress coherence. These events typically appear as sudden drops in T_1, followed by gradual recovery over minutes as quasiparticles recombine or diffuse away.

At intermediate timescales—hours to days—transmon behavior often exhibits slow drift rather than abrupt jumps. Experimentalists frequently observe "good days" and "bad days," during which nominally identical measurements yield systematically different coherence times. These trends are commonly attributed to spectral diffusion within the TLS bath itself. TLSs are not isolated defects; they interact weakly with one another through strain and electric fields. When one defect changes configuration, it slightly shifts the energy landscape of its neighbors. Over time, this collective interaction causes the ensemble of TLS to wander in frequency, leading to slow evolution of the effective loss spectrum sampled by the qubit.

Surface adsorbates can further contribute to variability on these timescales. Residual molecules—such as hydrogen, water, or hydrocarbons—may migrate slowly along metal or substrate surfaces, even at cryogenic temperatures. Although present only at sub-monolayer levels, their motion modifies local dielectric environments and can influence qubit loss. Microwave excitation itself can also reshape the TLS landscape, as strong measurement or control fields saturate certain defects or drive local rearrangements, leading to history-dependent coherence behavior.

On the longest timescales, extending from weeks to months, transmon variability could arise from genuine materials aging. Devices may gradually improve or degrade with time, and changes induced by thermal cycling are often irreversible. These long-term trends reflect the fact that native superconducting oxides are not inert or static materials. In niobium-based systems, for example, the Nb_2O_5 layer can continue to evolve through oxygen migration, suboxide formation, and structural relaxation. Such processes alter the density and distribution of TLS and, therefore, modify dielectric loss over extended periods.

Hydrogen plays a particularly important role in niobium systems. Long known in the accelerator community, hydrogen can diffuse within the near-surface region and interact strongly with oxide layers, forming metastable complexes that introduce additional loss channels. Mechanical strain relaxation following cool-down also contributes, as differential thermal contraction between films, substrates, and oxides slowly redistributes internal stress. Because TLS energies are highly strain-sensitive, even small relaxations can shift defect populations over time.

Qubit lifetime variability is clearly illustrated in Figure 5.3, which plots repeated measurements of T_1 over extended experimental time, revealing large-amplitude fluctuations. In the absence of active control, the measured T_1 (shown in purple)

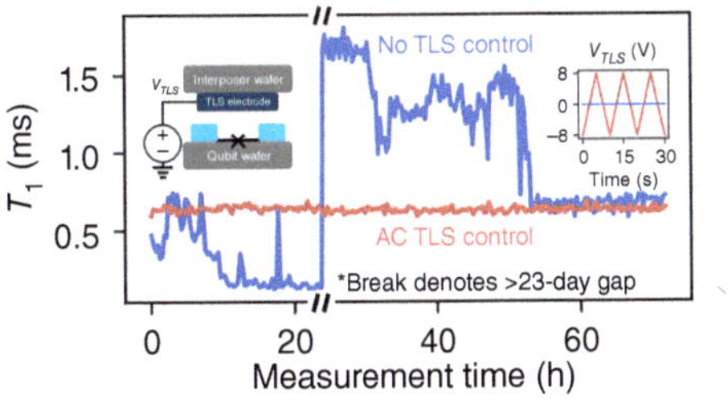

Figure 5.3 Time-resolved measurements of the transmon energy relaxation time T_1 recorded over extended experimental durations. The purple trace shows repeated T_1 measurements performed without TLS control, revealing pronounced temporal fluctuations with variations approaching a factor of two on timescales of hours to days. These fluctuations arise from the stochastic dynamics of microscopic two-level systems (TLSs) in the device's dielectric environment, which intermittently come in and out of resonance with the qubit. The discontinuity marked by the break symbol corresponds to a gap of more than 23 days between measurement sessions, after which similarly large lifetime variations reappear, highlighting the long-lived and slowly evolving nature of the TLS bath. The red trace shows T_1 measured under active AC TLS control, in which an oscillatory electric field is applied to modulate the TLS energies through their dipole moments. This controlled modulation suppresses slow spectral diffusion by dynamically averaging TLS–qubit interactions, resulting in a substantially more stable and reproducible lifetime. The inset illustrates the applied AC TLS bias waveform $V_{\mathrm{TLS}}(t)$, which periodically sweeps TLS frequencies during coherence measurements (Ref. [259] / Dane et al.)

exhibits dramatic excursions—sometimes changing by nearly a factor of two—on timescales of hours to days. These fluctuations persist even when the qubit is nominally idle, demonstrating that they originate from the internal dielectric environment rather than from control electronics or experimental drift.

The figure also highlights a striking feature of TLS-driven variability: the fluctuations are not merely random noise but show long-lived correlations. In the dataset shown, a clear discontinuity separates measurement periods by more than 23 days, yet the same pattern of variability reappears, indicating that the underlying TLS ensemble evolves slowly and retains memory over extremely long timescales. This observation reinforces the idea that TLS dynamics are governed by complex structural rearrangements rather than simple thermal activation.

An important advance presented is the demonstration that active TLS control can dramatically reduce lifetime variability [259]. In this approach, an oscillatory electric field is applied across the qubit capacitor, modulating the TLS energies through their electric-dipole moments. This does not imply that TLSs have disappeared; rather, their fluctuating contribution is being averaged out dynamically. This technique confirms that TLSs are indeed the dominant source of long-term T_1 variability, and it offers a practical tool by suppressing slow TLS fluctuations. What previously required days or weeks of repeated measurements to average over rare TLS configurations can now be achieved on experimental timescales of hours. Improving transmon coherence is not solely a matter of reducing the number of TLS but also of controlling their dynamics.

5.11 Tantalum and Tantalum Capping

Replacing Nb with Ta in 2D transmon qubits have shown higher coherence [260]. The considerable potential of tantalum was demonstrated in a 2D transmon qubit with an exceptional T_1 lifetime of ~500 μs and a qubit dephasing time T_2 of 557 μs [261]. Recent work improved transmon coherence by utilizing tantalum (Ta) as a base layer and high-resistivity silicon (Si) as a substrate 2D transmons with a lifetime (T_1) of up to 1.68 ms [79]. Similarly, recent studies show that coating niobium (Nb) with tantalum (Ta) significantly increases T_1 by factors of two to three with a Si substrate and coatings [262] (Figure 5.4). The median T_1 was above 180 μs, with a maximum of 450 μs. With larger footprint qubit geometries (lower participation ratios for TLS) of Ta-capped Nb on sapphire substrates demonstrated median qubit lifetimes above 300 μs, with several individual measurements in excess of 550 μs. The enhanced coherence observed in the Nb/Ta devices is consistent with reduced dielectric loss arising from the formation of a more stable tantalum oxide surface layer, which suppresses parasitic TLS defects at metal–vacuum and metal–substrate interfaces. These results highlight the

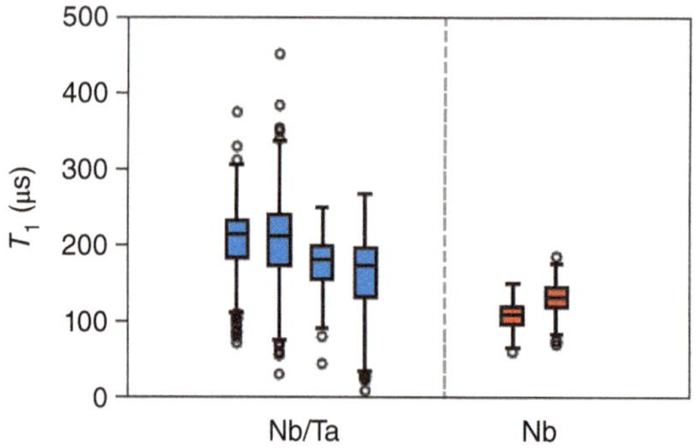

Figure 5.4 Comparison of energy-relaxation times for qubits fabricated with Nb and Nb/Ta capacitive elements. Measured energy-relaxation times T_1 for transmon qubits incorporating either niobium (Nb) or tantalum-coated niobium (Nb/Ta) capacitor structures. Each box plot represents a statistical distribution obtained from repeated cooldowns and multiple nominally identical devices. The central line denotes the median value, the box bounds indicate the interquartile range, and whiskers extend to 1.5× the interquartile range; open circles indicate statistical outliers. Qubits employing Nb/Ta capacitors (left group) exhibit systematically longer T_1 lifetimes compared to devices fabricated with Nb-only capacitors (right group) (Ref. [262] / Springer Nature / CC BY 4.0).

dominant role of capacitor-surface participation in limiting transmon relaxation times and demonstrate that targeted materials engineering of capacitive elements can yield substantial improvements in qubit coherence.

One likely reason for lower TLS losses is that Ta_2O_5 shows fewer suboxides than Nb_2O_5 in the transition region from metal to oxide, as shown by aberration-corrected TEM and EELS [263]. The stoichiometry of suboxides at the niobium surface includes metallic species with corresponding losses. Additionally, Ta_2O_5 exhibits a closer-to-crystalline bonding nature than Nb_2O_5, which has more amorphous regions. The crystalline bonding potentially hinders H atomic diffusion toward the oxide/metal interface.

5.12 Materials for 3D QPU–Niobium Cavities

While several loss mechanisms contribute to decoherence in transmon qubit [247, 254, 263], mounting experimental evidence points to surface TLSs as a dominant limitation. In particular, the amorphous native oxides that form on superconducting metals play a big role by hosting fluctuating atomic configurations that couple strongly to microwave electric fields. Understanding and mitigating these surface-related losses, therefore, require a deeper materials science perspective. In this context, accelerator-based SRF cavities provide a unique platform: their exceptionally high-quality factors, well-controlled surface treatments, and decades of systematic materials studies enable precise isolation of oxide-related dissipation mechanisms. We now turn to a detailed examination of TLS loss in niobium pentoxide (Nb_2O_5), drawing on accelerator cavity experiments that probe how thermal treatments, oxide modification, and surface processing influence microwave losses—offering insights that are directly relevant to improving coherence in superconducting qubits.

In GHz frequency niobium cavities, a decrease in the quality factor at fields lower than $E_{acc} \lesssim 5\,MV/m$ has been a long-standing mystery, called the LFQS. Typically, at 2 K, the Q falls from a few $\times 10^{10}$ to about 10^{10} or less, and at 1.5 K for a higher Q cavity, the Q falls from $>10^{11}$ to about 7–8 x 10^{10}. In groundbreaking work, Q measurements were extended to ultralow temperatures ($T < 20$ mK) using a dilution refrigerator to show that the Q continues to drop from a 7 x 10^{10} at 2 K down to about 3 x 10^{10} at 0.1 K and then the Q stays constant down to 20 mK. Since the contribution to the surface resistance from thermally excited quasiparticles is negligible, below 1 K, the $Q(T)$ curves must be dominated by residual resistance. The temperature dependence fits (Figure 5.5) the $1/\tan h\,[\alpha(\hbar\omega/2kT)$ prediction from the model of TLS losses discussed.

The Q measurements were also extended to ultralow fields to reveal two new features: (i) a saturation of the Q at $E_{acc} \sim 0.1$ MV = m and (ii) stronger Q degradation

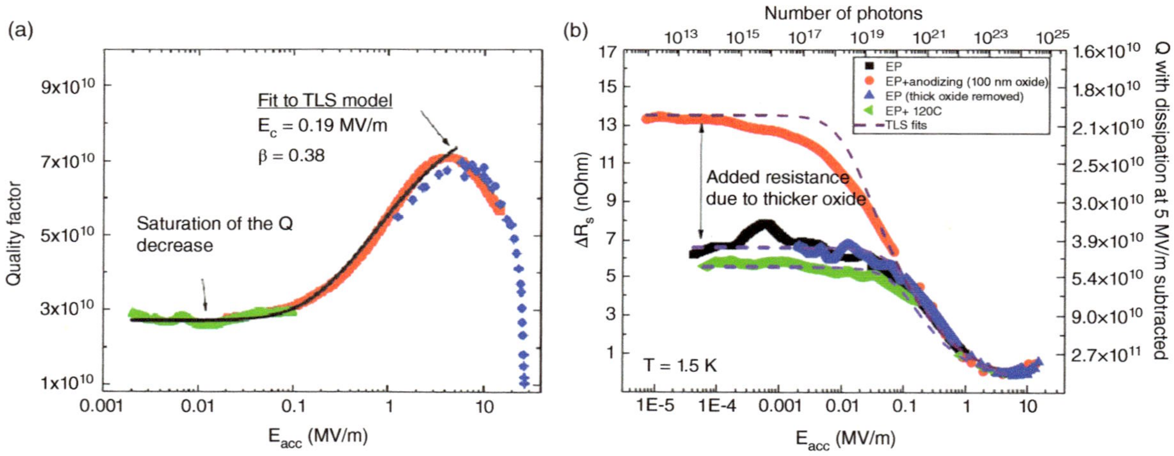

Figure 5.5 (a) The dependence of the intrinsic quality factor Q on the accelerating field E_{acc} over the full accessible field range for an electropolished single-cell niobium SRF cavity. The data were assembled by stitching together two complementary measurement techniques: continuous-wave (CW) measurements at low and moderate fields and single-shot decay measurements at higher fields, allowing accurate coverage from the very low-field regime up to fields exceeding 10 MV/m. At high RF fields, the red curve was processed using a resolution bandwidth (RBW) of 10 kHz in order to mitigate the effects of Lorentz force detuning that would otherwise distort the measured Q. The initial decrease of Q at very low fields is followed by a saturation behavior and then a rise in Q with increasing field, revealing the characteristic "low-field Q-slope." The solid line represents a fit to a two-level system (TLS) loss model, which attributes the low-field Q degradation to dielectric losses from TLS defects in surface oxides and amorphous regions. The good agreement between the model and the data supports the TLS interpretation of the low-field Q behavior in SRF niobium cavities. (b) The plots show the additional microwave surface resistance ΔR_s as a function of accelerating field amplitude E_{acc} for several differently prepared electropolished niobium cavities, all measured at $T = 1.5$–1.6 K. The additional resistance is defined relative to the value measured at $E_{acc} = 5$ MV/m so that ΔR_s highlights the field-dependent loss mechanisms rather than the baseline residual resistance. Different colors correspond to distinct surface conditions, including native oxide, intentionally thickened oxide, oxide removal, and mild baking, allowing direct comparison of how oxide thickness and treatment influence low-field losses. The increase in ΔR_s at very low RF fields indicates enhanced dissipation attributed to two-level systems (TLSs) located primarily in surface oxides and disordered interfacial regions; this effect becomes more pronounced for thicker oxide layers, as indicated. Dashed curves represent fits to the TLS loss model, which quantitatively describe the observed low-field rise and its dependence on surface treatment. The convergence of curves at higher fields shows saturation and eventual reduction of TLS-related losses as the TLS ensemble is driven toward saturation by the stronger RF field.

for thicker (100 nm) niobium pentoxide grown by anodization using a DC voltage of 48 V in ammonia solution. The residual (at low T) surface resistance increased by about 12 nΩ for the thicker film. After removing the thick oxide and regrowing the standard, natural thin oxide layer (3–5 nm), the original higher Q was restored. These features identify TLS losses in the pentoxide as a primary contributor to the LFQS effect (Figure 5.5).

A definitive confirmation for TLS losses in the pentoxide comes from record high-Q measurements obtained after removing the pentoxide by heat treatment (Figure 5.6). About 5-nm-thick Nb_2O_5 pentoxide layer is completely dissolved. Remeasuring the Q without exposure to air (i.e., in the absence of the oxide) increased the Q by large factors. A 1.3-GHz cavity was heat treated at 340 °C in vacuum for several hours, and a 5-GHz cavity was treated at 450 °C. After heat treatment, both cavities showed high-quality factors near 10^{11}. The heat treatment suppressed the residual resistance. This is consistent with the removal of TLS losses hosted by the pentoxide layer. After heat treatment, the 1.3-GHz Nb cavity achieved photon lifetimes of 0.5–2 s corresponding to an improvement of approximately 50–200 times. TOF-SIMS depth profiles (Figure 5.7) obtained on the cavity cutout revealed the changes in the niobium pentoxide after 400 °C in situ heat treatment.

5.13 Qubit Lifetimes and the Scaling Cost of QEC

The connection between improving qubit lifetimes and the need for QEC is subtle but central to the future of superconducting quantum computing. At first glance, it seems that if T_1 becomes very long, perhaps error correction becomes unnecessary. But the real story is nuanced. Besides lifetime, error per gate is also important.

If a single-qubit gate takes 20 nanoseconds and the qubit has $T_1 = 100\,\mu s$, then the relaxation-induced error per gate is roughly 2×10^{-4}. If the lifetime reaches 100 ms, the relaxation contribution to gate error becomes extraordinarily small, on the order of 10^{-7}. Thus, longer lifetimes reduce physical error rates, with powerful consequences.

In conventional 2D planar transmon architectures, lifetimes have improved steadily—from tens of microseconds a decade ago to millisecond-scale devices in leading laboratories (Figure 5.8). The figure shows two important trends: (i) steady improvements in planar 2D transmon coherence driven by materials advances, such as tantalum and surface loss mitigation and (ii) the emergence of ultra-long lifetimes in 3D cavity and bosonic architectures, where reduced dielectric participation and engineered electromagnetic environments enable millisecond-scale coherence.

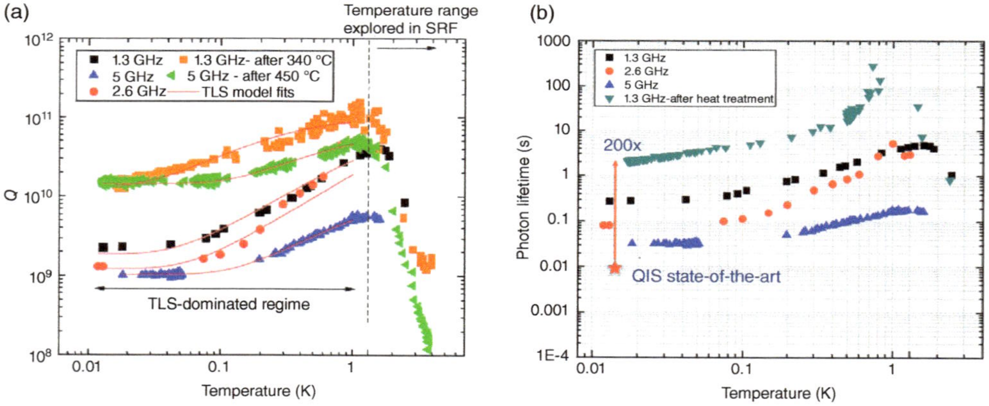

Figure 5.6 (a) Intrinsic cavity quality factor Q of the investigated 1.3, 2.6, and 5 GHz cavities as a function of temperature. Previously, SRF cavities have only been studied at temperatures above about 1.3 K. Below about 1 K, a significant decrease in Q is observed consistent with the TLS dissipation—the red lines show TLS model fits. A dramatic increase in Q associated with the oxide-modifying heat treatment is apparent on the1.3 GHz and 5G Hz cavities. (b) Photon lifetime (in seconds) on a logarithmic y-axis against temperature (in Kelvin) on a logarithmic x-axis. It compares cavities at different operating frequencies (1.3, 2.8, 5 GHz) and the effect of heat treatment. The "QIS state-of-the-art" is indicated at approximately 0.015 K with a photon lifetime of about 0.01 seconds. The data after heat treatment shows a significant, approximately 200×, increase in photon lifetime at low temperatures compared to the initial 1.3 GHz data. [Romanenko SRF 2019 Talk].

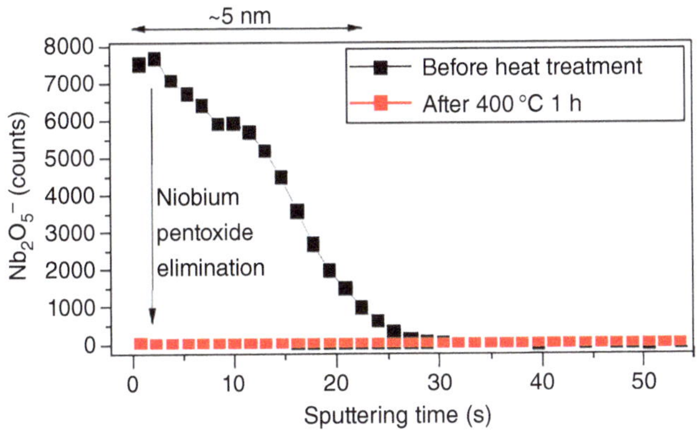

Figure 5.7 TOF-SIMS depth profiles obtained on the cavity cutout revealing the changes in the niobium pentoxide and nitrogen-related signals after 400 °C in situ heat treatment. About 5-nm-thick Nb_2O_5 pentoxide layer is completely dissolved after the 400 °C treatment.

But planar devices are also limited by surface dielectric loss, parasitic TLS, crosstalk, and gate infidelity. In such systems, QEC—typically in the form of surface codes—remains essential. The overhead can be thousands of physical qubits per logical qubit. Better lifetimes reduce the required code distance and thus the number of physical qubits, but they do not eliminate the need for error correction. The code distance is the minimum error chain length that flips the logical qubit.

The situation is particularly interesting in 3D cavity-based or bosonic encodings whose T_1 values already reach the 1–25 ms range (Figure 5.8). Because photon loss is often the dominant error channel, and because bosonic codes can be designed specifically to correct photon loss events, improvements in cavity lifetime directly translate into improved logical protection. If such systems reached $T_1 \sim 100$ ms, the rate of photon decay would become low, and only modest error-correction overhead might be required. In that regime, hardware and encoding begin to work synergistically: the physical system itself does much of the protection.

But even if T_1 reached 100 ms, QEC would not disappear. Energy relaxation is only one source of error. Dephasing (T_2), leakage outside the computational subspace, measurement errors, correlated noise, and—importantly—two-qubit gate errors still limit performance. Large-scale quantum algorithms may require billions of gate operations. Even with physical error rates as low as 10^{-7}, accumulated errors would eventually overwhelm an unprotected computation. Fault tolerance is fundamentally about controlling errors in the asymptotic, large-depth limit.

In bosonic or 3D architectures, logical information is encoded within a single high-quality oscillator, and improved cavity lifetimes reduce the dominant error

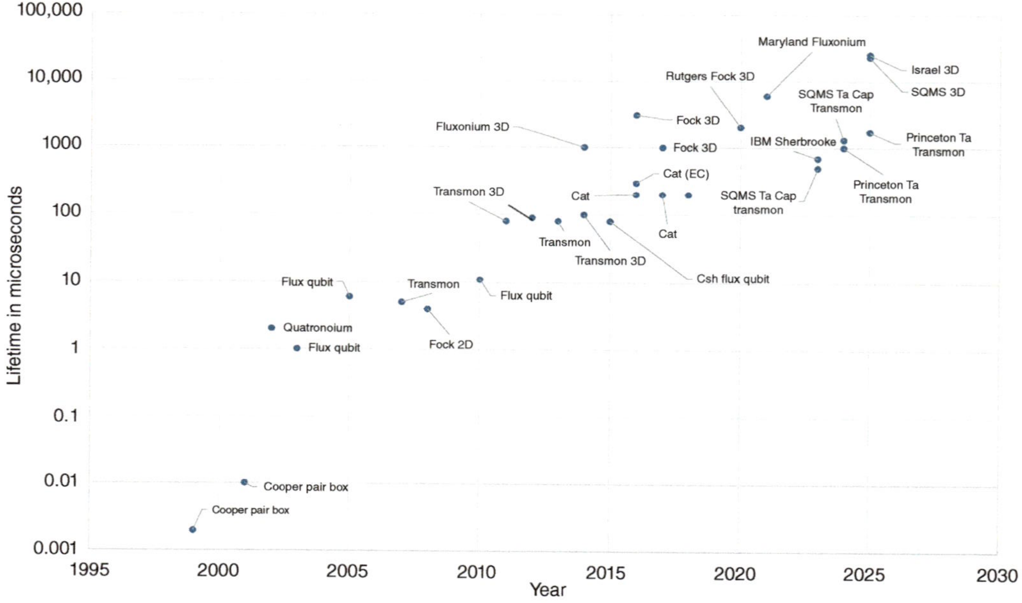

Figure 5.8 Progression of superconducting qubit energy relaxation times (T_1) extended to 2025. Energy relaxation times T_1 (in microseconds) for superconducting qubit platforms as a function of year. The data through ~2020 are adapted from [33], illustrating the historical progression from early Cooper-pair box devices to planar transmons, flux qubits, fluxonium, and three-dimensional (3D) cavity architectures. The figure is extended here with recent reported milestones (2021–2025), highlighting continued improvements in materials engineering and 3D cavity implementations Maryland [264], IBM Sherbrooke [265], SQMS Ta cap[262], Princeton [79, 260, 261], SQMS 3D [202] Rutgers [192], Israel [198].

channel directly. One might move from large 2D surface code patches to much smaller repetition codes or hardware-efficient bosonic encodings. The architectural implications are profound: cryogenic complexity drops, wiring density decreases, and scaling challenges become more manageable.

The bottom line is that increasing T_1 from microseconds to milliseconds—and potentially to 100 ms—does not eliminate the need for QEC. Instead of fighting overwhelming noise with massive redundancy, one begins operating deep below threshold, where QEC becomes lighter, more economical, and more integrated with the hardware. In that regime, the distinction between "physical qubit" and "logical qubit" begins to blur, especially in cavity-based systems. The future of superconducting quantum computing may, therefore, depend on pushing hardware lifetimes into regimes where error correction becomes a refinement rather than a rescue operation.

5.14 Conclusion

The total quality factor of a qubit is determined by the combined contributions of individual loss channels. Participation ratios quantify how much electromagnetic energy is stored in lossy regions, guiding both materials choice and device geometry optimization. For dielectric and substrate losses, oxide layers and interfaces—though microscopically thin—can dominate losses due to high electric field concentrations. Bulk substrate losses depend on the dielectric loss tangent, making careful substrate selection (sapphire vs. silicon) essential. For metal and junction losses, surface resistance of superconducting films (niobium, aluminum) contributes to conductor losses.

Techniques such as XPS, SIMS, TEM, and resonator measurements are critical for identifying loss origins. Fabrication variations—etching, deposition methods, chemical treatments—directly affect coherence performance.

TLSs act as a central loss mechanism. TLSs arise from tunneling atoms, hydrogen rotors, dangling bonds, or collective atomic motion in amorphous oxides such as Nb_2O_5 and Al_2O_3. TLSs interact strongly with qubit fields, producing temperature-dependent dielectric losses that cap coherence times. Tantalum appears to be a superior material. Tantalum films and tantalum-capping layers on niobium have demonstrated record coherence ($T_1 > 500\ \mu s$). Fewer suboxides, more crystalline oxide bonding, and reduced hydrogen diffusion possibly explain tantalum's advantage. JJ barriers (AlOx) introduce variability and host TLS defects that further degrade coherence.

In 3D niobium cavities, the puzzling LFQS is explained by TLS in niobium pentoxide. Heat treatments that dissolve surface oxides dramatically improve Q, confirming TLS as a dominant loss channel.

This chapter makes clear that the quest for high-coherence superconducting qubits is as much a materials science problem as it is a quantum engineering problem. By carefully engineering surfaces, oxides, and substrates—and by adopting alternative materials such as tantalum—the field continues to push coherence toward the thresholds required for fault-tolerant quantum computation.

6

Quantum Computing Applications

6.1 Introduction

The purpose of this chapter is to provide select examples of the emerging applications, highlighting both the theoretical frameworks and the evolving implementations. We explore how quantum algorithms are adapted, developed, and deployed across a range of disciplines including physics, chemistry, materials science, and finance.

This chapter highlights how QC moves beyond theory into real-world applications across multiple domains of science, engineering, and finance. The chapter begins with the central motivation: classical computers struggle with problems that grow exponentially in complexity, while quantum devices can model these systems naturally.

The chapter introduces five core areas of application:

1) **Cryptography** [266]: One of the most striking potential applications of QC lies in the field of cryptography, and it is embodied in Shor's algorithm. Developed by Peter Shor in 1994, this algorithm demonstrated for the first time that a sufficiently powerful quantum computer could factor large integers exponentially faster than any known classical algorithm. The factoring problem sits at the heart of widely used public-key cryptosystems such as RSA. We will discuss this algorithm and its implementation in some detail with simple examples.
2) **High-Energy Physics (HEP)** [267–269]: Quantum simulation of fundamental interactions, with particular emphasis on *QCD*. Lattice QCD, which maps quarks and gluons onto discrete space-time grids, rapidly overwhelms classical supercomputers. Quantum computers can encode color charges and gauge symmetries directly, employing algorithms such as *Trotter–Suzuki decomposition* to simulate nonperturbative regimes.

3) **Chemistry** [270, 271]: One of the earliest and most compelling application areas, where quantum computers natively represent the electronic wavefunctions of molecules. Algorithms like the *Variational Quantum Eigensolver (VQE)* and *Quantum Phase Estimation (QPE)* allow calculation of ground and excited state energies with high accuracy, promising breakthroughs in drug discovery, catalysis, energy materials, and biochemistry.
4) **Materials Science/Condensed Matter Physics:** Quantum devices can tackle many-body quantum systems, such as magnets, superconductors, or correlated electron systems, that defy classical approaches. Using for example the *Ising model*, quantum algorithms simulate spin interactions and capture emergent properties such as phase transitions and entanglement.
5) **Finance and Economics** [272]: Quantum optimization and sampling methods are being tested on portfolio optimization, derivative pricing, and risk management. Tools like *Quantum Amplitude Estimation (QAE)* offer speedups for *Monte Carlo simulations*, while quantum machine learning promises improved pattern recognition in financial data.

6.2 Shor's Algorithm for Factoring Large Numbers

6.2.1 Introductory Remarks

In 1994, Peter Shor showed that a quantum computer could factor large integers in polynomial time by turning factoring into a *period-finding* problem and using the quantum Fourier transform (QFT) to extract that period efficiently. That insight immediately mattered because the security of widely deployed public-key systems (especially RSA) rests on the practical difficulty of factoring a large composite number on classical machines. RSA-2048 uses a number N that is 2048 bits long, roughly 617 decimal digits. It is constructed as the product of two very large prime numbers,

$$N = p \times q$$

where p and q are each 1024-bit primes. So, approximately

$$p \approx q \approx 10^{308}.$$

Factoring RSA-2048 would take millions of years on current classical supercomputers using the best classical algorithm.

The quantum part of Shor's algorithm finds a hidden period using interference. Classical factoring algorithms (e.g., the number field sieve) are *subexponential* but still grow fast enough that RSA key sizes like 2048 bits are safe in practice today; Shor's period-finding step, on a fault-tolerant quantum computer, scales far more gently with key size—so at large N it becomes a qualitative break from classical capabilities.

It's also important to separate toy demonstrations from cryptographically relevant factoring. On real hardware, "full Shor" has been demonstrated only for *very small* integers; 15 is the canonical one; 21 has been demonstrated with photonic implementations [273]. Even factoring 35 with the Shor's algorithm on a QC would probably take 10–20 logical qubits with reliable error correction and with100–1000 gates. This is tantamount to several thousand physical qubits with error rates $\lesssim 10^{-4}$, i.e., a small fault-tolerant quantum computer. Today's QCs are not yet reliable enough. To break RSA-2048 could possibly be done in less than a week with under a million physical qubits [266]. That is far beyond the capabilities of today's machines. Shor invented his algorithm in 1994, and the first experimental quantum factoring of 15 was achieved in 2001, about seven years later.

Error correction on superconducting qubit platforms is steadily improving for increasing code size toward fault tolerance demands [98]. Roadmaps from major superconducting players are now framed around scaling logical qubits to run large numbers of error-corrected gates. IBM, for example, has targeted a fault-tolerant system by 2029 on the order of hundreds of logical qubits with very large circuit capability [274].

Credible estimates for when RSA-2048 might realistically be threatened span a wide range because they depend on error rates, manufacturing scale, and whether architectures will hit their roadmaps. One conservative cybersecurity analysis suggests RSA-2048-breaking quantum capability may be decades away [275]. If superconducting platforms achieve sustained, scalable fault tolerance and can manufacture at the ~million-physical-qubit level, the "danger zone" could plausibly move into the mid-2030s! A new estimate [276] of the resource requirements for Shor's algorithm claims that if you use Low Density Parity Check (LDPC) codes rather than the surface code, it should be possible to break RSA-2048 with fewer than 100,000 physical qubits, which is an order-of-magnitude improvement over the previous estimate [266]. In the LDPC code, each qubit only interacts with a *small number* of other qubits so that the connectivity graph is "sparse." In theory, it requires far fewer qubits for the same protection as the surface code (Section 2.23).

6.2.2 Implementation Concepts

There are two main steps: *modular exponentiation* and *QFT*.

6.2.2.1 Modular Exponentiation

Factoring a large number N is hard. But a quantum computer is good at discovering patterns. Shor's algorithm turns factoring into finding the hidden rhythm in a repeating pattern. Once you know that rhythm, the factors fall out.

Here is the core idea. Imagine you want to find the factors of a number—say 15. Instead of directly splitting 15, Shor uses a clever trick: pick a random number a, and examine how the following sequence behaves.

$$a^1 \bmod N, \quad a^2 \bmod N, \quad a^3 \bmod N, \ldots$$

In case "*modN*" is new to you, here is what a^1 mod$N$ means. Keep only the remainder after dividing by N. Examples: 7 mod 5 = 2 (because 7 divided by 5 leaves remainder 2). 11 mod 4 = 3, 20 mod 6 = 2, and 20 mod 5 = 0. By the way, 4 mod 15 = 4 because 4 divided by 15 goes 0 times, leaving a remainder of 4.

Surprisingly or not, this sequence $a^x \bmod N$ always repeats after some number of steps. This is because there are only finite possible remainders $0, 1, 2, \ldots, N-1$. So as you keep calculating powers of a, eventually one of the remainders must repeat. When the first repetition happens with r, so that

$$a^{x+r} \bmod N = a^x \bmod N,$$

the sequence has period r because $a^{x+r} \bmod N = a^{x+2r} \bmod N = a^{x+3r} \bmod N = \ldots$. Factoring the number N can be done if you know this period.

Now dividing both sides of the above by a^x gives

$$a^r \bmod N = 1$$

So the period r is exactly the number of steps needed for the powers of a modulo N, to come back to 1. When 1 is the remainder, $a^r - 1$ is a multiple of N, say, $m * N$. Furthermore, if r is even, then you can write

$$a^r - 1 = (a^{r/2} - 1)(a^{r/2} + 1).$$

Just as you can factor $x^2 - 1 = (x^{2/2} - 1)(x^{2/2} + 1) = (x - 1)(x + 1)$. Since $N = pq$ (two primes), one of these two numbers must contain the factor p and the other must contain the factor q. More generally, these terms are likely to have a nontrivial common factor with N. So it is necessary to determine the greatest common divisor gcd $(a^{r/2} + 1)$ *and* N, and gcd of $(a^{r/2} - 1)$ *and* N, which can be calculated efficiently using a classical algorithm. Thus, once the period is known, simple relationships let us extract the prime factors of the original number N.

But if r is odd, we throw it away and try again to find a periodic pattern with a different value of a and an even value or r. The chance of success is very good—for a random a, the period is even with probability at least 50%. So usually, a few tries are enough.

There is another case when you also have to try again, i.e., when $a^{r/2} \not\equiv -1 \bmod N$. In this case, the greatest common denominator just returns 1 or N. So we have to start again.

As a concrete and simple example, to factor 15 using Shor's algorithm, start by picking a random number a, such that $1 < a < N$. Let's pick $a = 2$. Now classically compute the greatest common denominator between 2, 15,

$$\gcd(2, 15) = 1$$

That's not a factor, so we continue. We compute powers of 2 and reduce them modulo 15, i.e., $2^x \bmod 15$ (remainder after dividing by 15)

x	2^x	2^x mod(15)
1	2	2
2	4	4
3	8	8
4	16	1 (because 16 mod 15 = 1)
5	32	2 (because 32 mod 15 = 2)
6	64	4
7	128	8
8	256	1

A pattern: 2, 4, 8, 1… emerges for 2^x mod(15). The repeating rhythm has period $r = 4$. Now computing $2^{r/2} - 1$ and $2^{r/2} + 1$ gives

$$2^2 - 1 = 3$$
$$2^2 + 1 = 5$$

These are the factors of 15. Shor discovered how to turn factoring into a pattern-finding problem.

Classically, we computed each $2^x \bmod 15$ one by one. To factor a very large number, you have a billion x values, which means a billion operations. But with a quantum computer, you compute them all at once. The circuit depth grows only logarithmically with x.

To see how you can get failures for some choices of a and how you use the gcd step, it is better to use the example of factoring 21.

The very first thing Shor does (classically) is check $\gcd(a, 21)$. If it's not 1, you've already found a factor and you don't even need the quantum period-finding step. This is the case for the choices $a = 3, 6, 7, 8, 9$.

We look for a periodic pattern for $a = 2, 4, 5, 10$.

For $a = 2$, computing $2^x \bmod 21$ gives $1, 2, 4, 8, 16, 11, 1, \ldots$

So the order is 6, and it qualifies since $2^6 \equiv 1(\text{mod} 21)$.

In this case, the midpoint is $2^{r/2} = 2^3 = 8$, and $8 \bmod 21 = 8 \not\equiv -1$.

So the gcd step works:

$$\gcd(8 - 1, 21) = \gcd(7, 21) = 7, \quad \gcd(8 + 1, 21) = \gcd(9, 21) = 3$$

If we pick a = 4, $4^x \bmod 21$: 1, 4, 16, 1, 4, 16, … So the pattern repeats every =3. Here r is odd, so the choice fails.

If we pick =5, compute 5^x mod 21 to yield

$$1, 5, 4, 20, 16, 17, 1, 5, 4, 20, 16, 17, \ldots$$

So the period is $r = 6$.

Now check the crucial midpoint:

$$5^{r/2} = 5^3 = 125 \equiv 20 \equiv -1(\text{mod}21).$$

Then, $a^{r/2} - 1 \equiv -2,\ \ a^{\frac{r}{2}} + 1 \equiv 0(\text{mod}N)$.

Now $\gcd(-2, 21) = \gcd(2, 21) = 1$ *and* $\gcd(0, 21) = 21$, which are both trivial factors, and the choice fails.

When we pick $a =$ *10, we get the pattern for* 10^x mod 21: 1, 10, 16, 13, 4, 19, 1, 10, 16, 13, 4, 19, … with period $r = 6$.

Check the midpoint: $10^3 = 1000 \equiv 13(\text{mod}21)$.

Once again, $\gcd(13 - 1, 21) = \gcd(12, 21) = 3,\ \gcd(13 + 1, 21) = \gcd(14, 21) = 7$.

And the factors are $21 = 3 \times 7$.

6.2.2.2 Implementation

We now return to factoring 15 and work this out for factoring with a QC. Start with the choice of $a = 2$. We want the quantum computer to find the period of the repeating pattern in 2^x mod 15. Set up two quantum registers. A "register" is just a small group of qubits. The first register will hold different values of x. The second register will hold the values of 2^x mod 15.

For the small demo, we can let x range from 0 to 7, with the hope to find the pattern with just this limited range. The numbers 0–7 can be written in three-bit binary, so we use three qubits in the first register.

$$|000\rangle \to x = 0$$

$$|001\rangle \to x = 1$$

$$|010\rangle \to x = 2$$

$$|011\rangle \to x = 3$$

$$|100\rangle \to x = 4$$

$$|101\rangle \to x = 5$$

$$|110\rangle \to x = 6$$

$$|111\rangle \to x = 7$$

Each qubit holds one bit of the binary expansion of x:

$$x = x_2 x_1 x_0(\text{ binary }) = x_0 \cdot 2^0 + x_1 \cdot 2^1 + x_2 \cdot 2^2.$$

Each x_i is either 0 or 1.

x_0: least significant bit.

x_1: middle bit.
x_2: most significant bit, with the highest power of 2.

Apply Hadamard gates to these three qubits so that the first register is in an equal superposition of all eight numbers:

$$\frac{1}{\sqrt{8}}(|000\rangle + |001\rangle + |010\rangle + |011\rangle + |100\rangle + |101\rangle + |110\rangle + |111\rangle)$$

We also make a second register (another group of qubits) initialized to $|0\rangle$. This second register will eventually hold $2^x \bmod 15$, which needs to be built up. The QC must build this map for every x at once:

$$|x\rangle\,|0\rangle \longrightarrow |x\rangle|2^x \mathrm{mod} 15\rangle.$$

We use the binary bits of x to control modular multiplications. If the bit is $0 \rightarrow$ do nothing. If that bit is $1 \rightarrow$ multiply the second register by a chosen fixed number mod 15. The chosen number depends on 2 raised to the binary expression for x.

$$2^x = 2^{x_0 \cdot 2^0 + x_1 \cdot 2^1 + x_2 \cdot 2^2}$$

$$2^x = 2^{x_0 2^0} \cdot 2^{x_1 2^1} \cdot 2^{x_2 2^2}$$

$$2^x = (2^1)^{x_0} \cdot (2^2)^{x_1} \cdot (2^4)^{x_2}$$

$$2^x = (2)^{x_0}(4)^{x_1}(16)^{x_2}.$$

Therefore, according to the value of the bits,
If $x_0 = 1 \rightarrow$ multiply by $2^1 = 2$ (because $2^1 \bmod 15 = 2$)
If $x_1 = 1 \rightarrow$ multiply by $2^2 = 4$ (because $2^2 \bmod 15 = 4$)
If $x_2 = 1 \rightarrow$ multiply by 2^4 (which means 1, because $2^4 \bmod 15 = 16 \equiv 1$)
If a bit is 0, do nothing.

We proceed with a branch-by-branch evaluation following the above rules for all the eight qubits, starting with the value 1 in each case, and we get the values $2^x \bmod 15$.

Step by Step:

$x = 0, |000\rangle$ do nothing, final value for second register is 1, which agrees with $2^0 \bmod 15 = 1$.

$x = 1, |001\rangle$ for $x_0 = 1 \rightarrow$ multiply by 2^1; $\begin{matrix} x_1 = 0 : \text{nothing} \\ x_2 = 0 : \text{nothing} \end{matrix}$ Final value stays = 2, which agrees with $2^1 \bmod 15 = 2$.

$x = 2, |010\rangle$ $x_0 = 0$: do nothing; $x_1 = 1$ *so* : $\times 4 \rightarrow$ value = 4; $x_2 = 0$: do nothing, so final value stays = 4, which agrees with $2^2 \bmod 15 = 4$.

$x = 3, |011\rangle$ $x_0 = 1$, *so* $\times 2 \rightarrow$ value = 2; $x_1 = 1$: *so 4 x current* value = 4 x2 = 8; $x_2 = 0$: do nothing; final value = 8, which agrees with $2^3 \bmod 15 = 8$.

$x = 4$, $|100\rangle$ $x_0 = 0$: nothing $x_1 = 0$: nothing; $x_2 = 1 : \rightarrow \times 1 \rightarrow$ value stays 1, so final value = 1, matches $2^4 \bmod 15 = 1$.

$x = 5$, $|101\rangle$, $x_0 = 1 \rightarrow \times 2 \rightarrow$ value $= 2$, $x_1 = 0$: nothing; $x_2 = 1 \rightarrow \times$ *current value by* 1, so final value stays 2, which agrees with $2^5 \bmod 15 = 2$.
$x = 6$, $|110\rangle$, $x_0 = 0$: do nothing; $x_1 = 1$ *multiply by* $\times 4 \rightarrow$ value $= 4$;
$x_2 = 1 : \times 1 \rightarrow$ value remains 4, final value $= 4$, which agrees with $2^6 \bmod 15 = 4$.
$x = 7 = |111\rangle$, $x_0 = 1 : \times 2 \rightarrow$ value $= 2$, $x_1 = 1 \rightarrow \times 4 \rightarrow$ value $= 8$,
$x_2 = 1 : \rightarrow \times 1 \rightarrow$ remains 8, final value $= 8$, which agrees with $2^7 \bmod 15 = 8$.

So the procedure gives the period-4 repeating pattern 1, 2, 4, 8, 1, 2, 4, 8 for the final values in the output register. Because the input is a superposition of all $|x\rangle$, these rules happen on all branches in parallel.

After modular exponentiation, the joint state of (x/ second register) is

$$\frac{1}{\sqrt{8}}(|0\rangle|1\rangle + |1\rangle|2\rangle + |2\rangle|4\rangle + |3\rangle|8\rangle + |4\rangle|1\rangle$$
$$+ |5\rangle|2\rangle + |6\rangle|4\rangle + |7\rangle|8\rangle)$$

6.2.2.3 Modular Exponentiation on a Quantum Computer

This is the hardest part of Shor's algorithm.

We discuss how a QC performs modular exponentiation for some concrete cases. First, $a^x \bmod N$ with $a = 2, x = 2, N = 15$; therefore, $2^2 \bmod 15$. The QC is asked to perform a reversible transformation.

The task reduces to starting from the number 1, multiply by 2 twice, and reduce mod 15, using only reversible quantum operations.

Quantum computers usually work in binary, so first the target number: $15 = 1111_2$. We, therefore, need four qubits to store numbers modulo 15.

Now $1 = 0001_2$ stored as $|0001\rangle$. To multiply by 2 in binary is the same as a left shift, so shift all bits left by one position with SWAP operations on a QC. After the shift: $|0001\rangle \rightarrow |0010\rangle$.

Next subtract 1111_2 from the number. Subtraction succeeds if $y \geq 15$; subtraction fails if $y < 15$. If subtraction fails, we undo it. This makes a reversible adder built from CNOT gates and Toffoli gates. These gates subtract 1111_2 from the number.

In short, QC of $a^x \bmod N$ is performed using reversible arithmetic: the value register is shifted to multiply by a, then N is subtracted to test for overflow, and a borrow flag is used to conditionally add N back if the subtraction fails. All of these steps are implemented with reversible logic built from CNOT and Toffoli gates, ensuring that the computation remains coherent and works correctly on quantum superpositions.

For $N = 15$ and $a = 2$, this process works identically for $x = 1$, 2, and 3. The quantum computer computes $2^3 \bmod 15$ by performing three identical reversible cycles: shift left to multiply by 2, subtract 15 to test overflow, and use the borrow flag to conditionally add 15 back. In these cases, the subtraction always fails, so the add-back step restores the value and no modular reduction occurs.

For $x = 4$, a new situation arises. After four shifts, the value reaches 16, which is large enough that subtracting 15 succeeds. This sets the borrow flag to 0, preventing the add-back step and leaving the reduced value $1 = |0001\rangle$. This is the first genuine modular reduction in the reversible arithmetic process. Similarly, for $x = 5$, $2^5 = 32$,and $2^5 \bmod 15 = 2$, and the same logic continues for higher powers.

6.2.2.4 Measurement

Now we measure the second register. If we measure **1**, the first register becomes a superposition of the x's that gave 1: state $\propto |0\rangle + |4\rangle$. It collapses to

$$\frac{1}{\sqrt{2}}(|0\rangle + |4\rangle)$$

If measurement gives 2, the first register collapses to

$$\frac{1}{\sqrt{2}}(|1\rangle + |5\rangle)$$

If measurement gives 4, the first register collapses to

$$\frac{1}{\sqrt{2}}(|2\rangle + |6\rangle)$$

If measurement gives 8, the first register collapses to

$$\frac{1}{\sqrt{2}}(|3\rangle + |7\rangle)$$

In *all* cases, the first register looks like "two values of x, separated by 4." We always get a comb of the form:

$$\frac{1}{\sqrt{2}}(|x_0\rangle + |x_0 + 4\rangle)$$

The above describes the logic of the first stage of the algorithm.

6.2.2.5 QFT Operation

In the next step, we use interference via *QFT* to extract the repeating structure and its period. The QFT is like a magical stethoscope. It amplifies the hidden rhythm (period) and suppresses everything else. A classical computer would need exponential time to find the rhythm. A quantum computer finds it in polynomial time.

QFT takes a pattern of positions and turns it into a pattern of frequencies. If the input is periodic, the output contains only a few sharp peaks.

How is the conversion done from periodic spacing to the telling peaks?

Formally, QFT is a linear unitary transformation on the basis state $|x\rangle$ defined by

$$\text{QFT}\,|x\rangle = \frac{1}{\sqrt{Q}}\sum_{y=0}^{Q-1} e^{2\pi ixy/Q}\,|y\rangle.$$

Here $|y\rangle$ is a binary (or a d-ary) number stored in the qubits representing one of the output values of the QFT. In the QFT part of the algorithm, the input is a superposition of $|x\rangle$ states so that $\text{QFT}[\sum_k |x_k\rangle] = \sum_k \text{QFT}\,|x_k\rangle$. The contribution from each x_k are coherently added to create a final superposition of $|y\rangle$ states. After measurement, you obtain one specific y. This measured value y is used to estimate the fraction

$$\frac{y}{Q}$$

where $Q = 2^n$ is the size of the first register.

As we established above, if we measure the second register and get **1**, the first register becomes a superposition of the x's that gave the result of 1, so it collapses to

$$\frac{1}{\sqrt{2}}(|\,0\rangle + |\,4\rangle)$$

Now we evaluate $\text{QFT}(|\,0\rangle + |\,4\rangle) = \frac{1}{\sqrt{2}}(\text{QFT}\,|\,0\rangle + \text{QFT}\,|\,4\rangle)$ from the general form $\text{QFT}\;|\,x\rangle$.

First, $\text{QFT}\,|\,0\rangle = \frac{1}{\sqrt{8}}(|\,0\rangle + |\,1\rangle + |\,2\rangle + |\,3\rangle + |\,4\rangle + |\,5\rangle + |\,6\rangle + |\,7\rangle)$.

All y values appear equally.

Next $\text{QFT}\,|\,4\rangle = \frac{1}{\sqrt{8}}\sum_{y=0}^{7} e^{2\pi i(4y/8)}\,|\,y\rangle = \frac{1}{\sqrt{8}}\sum_{y=0}^{7}(-1)^y\,|\,y\rangle$.

So $|\,0\rangle, |\,2\rangle, |\,4\rangle, |\,6\rangle$ get $+\,1$, and $|\,1\rangle, |\,3\rangle, |\,5\rangle, |\,7\rangle$ get $-\,\mathbf{1}$ as signs of coefficients.

When we add the two results, something special happens: For even y, the two terms add → constructive interference. For odd y, they cancel → destructive interference. This gives

$$\text{QFT}(|\,0\rangle + |\,4\rangle) = \sqrt{\frac{2}{8}}(|\,0\rangle + |\,2\rangle + |\,4\rangle + |\,6\rangle)$$

Four basis states survive (0, 2, 4, 6). The others vanish. This is the QFT output state. $\text{QFT}(|x_0\rangle + |x_0 + r\rangle)$ causes destructive interference for some y and constructive interference for others. From the ratios y/Q, we will recover $r = 4$.

Generally, when a function repeats with a fixed period (e.g., 4), its QFT does not produce a single sharp outcome, but rather a structured pattern of peaks. When the QFT acts on this evenly spaced superposition, the amplitudes from all these basis states interfere with one another. Constructive interference occurs only at those Fourier frequencies whose phases align across every term in the sum. Because the inputs are separated by four, the phases line up precisely when the Fourier index corresponds to multiples of $Q/4$, where $Q = 8$ is the size of the register. At those points, all probability amplitudes add coherently. Everywhere else, the phases disagree and largely cancel. The peaks appear at indices corresponding to fractions $0/4$, $Q/4$, $2Q/4$, and $3Q/4$, so we recover the period of 4.

$$\frac{y}{Q} = \left\{\frac{0}{4}, \frac{1}{4}, \frac{2}{4}, \frac{3}{4}\right\}$$

It does not matter whether the second register measurement was

$$1 \rightarrow \text{pair } \{0, 4\}$$
$$2 \rightarrow \text{pair } \{1, 5\}$$
$$4 \rightarrow \text{pair } \{2, 6\}$$
$$8 \rightarrow \text{pair } \{3, 7\}$$

All have spacing = 4. All produce y ∈ {0,2,4,6} after QFT. All imply $r = 4$.

6.2.2.6 Quantum Computer Implementation

Next we discuss how a quantum computer implements the QFT using quantum gates (Figure 6.1). The QFT breaks down into three types of gates and carries out the operations on q_2, q_1, q_0 in order, since these bits are ordered as most significant bit (MSB), middle but and least significant bit (LSB). The gates are

Hadamard gates (H): Turn a basis state into an equal superposition, introducing the first layer of phases.

Controlled-phase rotation gates (R_6): In general, these apply a small phase depending on other qubits' values.

$$R_k = \begin{pmatrix} 1 & 0 \\ 0 & e^{2\pi i/2^k} \end{pmatrix}$$

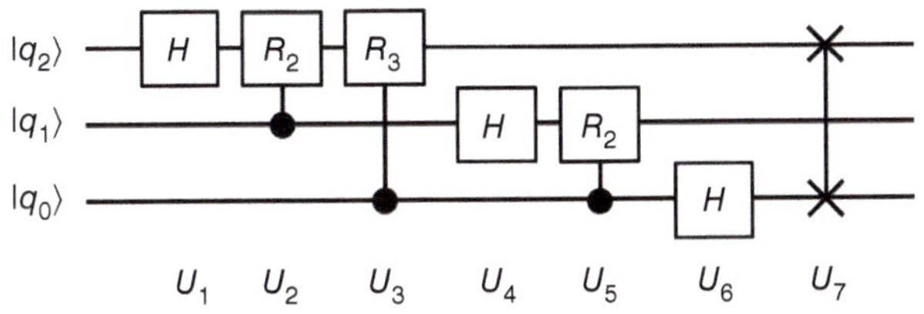

Figure 6.1 Circuit implementation of the quantum Fourier transform acting on a three-qubit register (q_2, q_1, q_0). The QFT maps computational basis states into superpositions whose relative phases encode discrete Fourier frequencies. The transformation begins with a Hadamard gate applied to the most significant qubit q_2, followed by controlled phase rotations R_2 and R_3, which apply conditional phases of $e^{i\pi/2}$ and $e^{i\pi/4}$, respectively, depending on the states of the lower-order qubits. This sequence entangles the qubits through progressively smaller phase increments, encoding binary-weighted phase information. The procedure is then repeated for qubit q_1, which undergoes a Hadamard gate followed by a controlled R_2 rotation conditioned on q_0. Finally, a Hadamard gate is applied to the least significant qubit q_0. The controlled rotation gates implement the phase correlations required to construct the discrete Fourier transform in the computational basis, with rotation angles decreasing exponentially with qubit significance. The final pair of SWAP operations reverses the qubit order, converting the natural output of the QFT—produced in bit-reversed order—into the standard binary ordering.

The label k (1, 2, 3) denotes how much phase rotation is applied, not to be confused with qubit label. The angles $2\pi i/2^k$ are chosen by QFT theory to produce sharp frequency peaks corresponding to the periodicity. This matrix acts differently on $\mid 0\rangle$ and $\mid 1\rangle$.

Action on $|0\rangle$:

$$R_k \mid 0\rangle = \begin{pmatrix} 1 & 0 \\ 0 & e^{2\pi i/2^k} \end{pmatrix} \begin{pmatrix} 1 \\ 0 \end{pmatrix} = \begin{pmatrix} 1 \\ 0 \end{pmatrix} = \mid 0\rangle$$

So no phase, no change.

Action on $|1\rangle$:

$$R_k \mid 1\rangle = \begin{pmatrix} 1 & 0 \\ 0 & e^{2\pi i/2^k} \end{pmatrix} \begin{pmatrix} 0 \\ 1 \end{pmatrix} = \begin{pmatrix} 1 \cdot 0 + 0 \cdot 1 \\ 0 \cdot 0 + e^{2\pi i/2^k} \cdot 1 \end{pmatrix} = \begin{pmatrix} 0 \\ e^{2\pi i/2^k} \end{pmatrix} = e^{2\pi i/2^k} \mid 1\rangle.$$

$$R_k \mid 1\rangle = e^{2\pi i/2^k} \mid 1\rangle.$$

Specifically, the gates are given in Table 6.1.

Finally, Swap gates at the end: QFT reverses qubit order, so the output must swap qubit 1 $\leftrightarrow$ qubit n. This swapping step can often be omitted as it is related to the labeling of the qubits and can be done in a classical postprocessing.

The QFT is applied qubit by qubit from MSB to LSB, q_2, $q_1 q_0$ q_0 as shown in Figure 6.1 with the following operations:

U_1 Hadamard on q_2; supplies the (1/2)(π-phase) contribution
U_2 Controlled-R_2 (control: q_1, target: q_2); adds the 1/4 phase contribution (if $q_1 = 1$)
U_3 Controlled-R_3 (control: q_0, target: q_2); adds the 1/8 phase contribution (if $q_0 = 1$)
U_4 Hadamard on q_1
U_5 Controlled-R_2 (control: q_0, target: q_1)
No R_3 here because there is no qubit below q_0.
U_6 Hadamard on q_0 (no controlled rotations, because there are no less significant qubits)

The above procedure is designed so that Hadamard gives the largest phase chunk for that output bit, while the controlled R_k gates add *smaller and smaller* binary fractions, as each qubit only "listens" to qubits below it.

Table 6.1 Phase rotation gates.

Gate	Phase applied to $\|1\rangle$
R_1	$e^{2\pi i/2} = -1$
R_2	$e^{2\pi i/4} = i$
R_3	$e^{2\pi i/8} = e^{i\pi/4}$

Next we carry out these gate operations after the measurement that gave the result 1, so the first register collapses to

$\frac{1}{\sqrt{2}}(|0\rangle + |4\rangle)$ which in binary is $\frac{1}{\sqrt{2}}(|000\rangle + |100\rangle)$

We expect the result:

$$\text{QFT}(|0\rangle + |4\rangle) = \sqrt{\frac{2}{8}}(|0\rangle + |2\rangle + |4\rangle + |6\rangle)$$

Apply H on q_2 (on the most significant bit first)

$$\frac{|000\rangle \xrightarrow{H_{q2}} (|000\rangle + |100\rangle)}{\sqrt{2}}$$

$$\frac{|100\rangle \xrightarrow{H_{q2}} (|000\rangle - |100\rangle)}{\sqrt{2}}$$

Add them and the $|100\rangle$ cancels: $|\psi_1\rangle = |000\rangle$.

Next perform the controlled R_2 with $q_1 \rightarrow q_2$.

This gate only does something if control $q_1 = 1$. But in $|000\rangle$, $q_1 = 0$, so nothing happens:

$$|\psi_2\rangle = |000\rangle$$

Similarly for controlled – R_3 with $q_0 \rightarrow q_2$, nothing happens:

$$|\psi_3\rangle = |000\rangle$$

Moving on to the next significant digit, apply H on q_1 (middle bit)

$$|000\rangle \xrightarrow{H_{q1}} \frac{1}{\sqrt{2}}(|000\rangle + |010\rangle)$$

Followed by controlled-R_2 with $q_0 \rightarrow q_1$. This only acts if $q_0 = 1$. But both $|000\rangle$ and $|010\rangle$ have $q_0 = 0$. So no change:

The result for middle bit is $\frac{1}{\sqrt{2}}(|000\rangle + |010\rangle)$.

Next, apply H on q_0. Apply H to the last bit of each term:

The result becomes $\frac{1}{2}(|000\rangle + |001\rangle + |010\rangle + |011\rangle)$,

$$|000\rangle \xrightarrow{H_{q0}} \frac{|000\rangle + |001\rangle}{\sqrt{2}}$$

$$|010\rangle \xrightarrow{H_{q0}} \xrightarrow{\frac{|010\rangle + |011\rangle}{\sqrt{2}}}$$

giving

$$\frac{1}{\sqrt{2}}\left(\frac{|000\rangle + |001\rangle}{\sqrt{2}} + \frac{|010\rangle + |011\rangle}{\sqrt{2}}\right)$$
$$= \frac{1}{2}(|000\rangle + |001\rangle + |010\rangle + |011\rangle)$$

U_7 Finally SWAP $q_2 \leftrightarrow q_0$(to fix the bit order), $\frac{1}{2}(|000\rangle + |010\rangle + |100\rangle + |110\rangle)$

which means the output of the QFT is peaked at 0, 2, 4, 6 (see the list of equations repeated below), which indicates the period $r = 4$.

$$|000\rangle \rightarrow x = 0$$
$$|001\rangle \rightarrow x = 1$$
$$|010\rangle \rightarrow x = 2$$
$$|011\rangle \rightarrow x = 3$$
$$|100\rangle \rightarrow x = 4$$
$$|101\rangle \rightarrow x = 5$$
$$|110\rangle \rightarrow x = 6$$
$$|111\rangle \rightarrow x = 7$$

At this point, measuring gives 0, 2, 4, 6 with equal probability.

6.2.3 Another Measurement Case

Let us say the measurement of the second register reveals

$$|\psi_0\rangle = \frac{1}{\sqrt{2}}(|001\rangle + |101\rangle) = \frac{1}{\sqrt{2}}(|1\rangle + |5\rangle)$$

1) Apply H on q_2

$|001\rangle$ has $q_2 = 0$:

$$|001\rangle \xrightarrow{H_{q2}} \frac{|001\rangle + |101\rangle}{\sqrt{2}}$$ $|101\rangle$ has $q_2 = 1$:

$$|101\rangle \xrightarrow{H_{q2}} \frac{|001\rangle - |101\rangle}{\sqrt{2}}$$

Now combine:

$$|\psi_1\rangle = \frac{1}{2}[(|001\rangle + |101\rangle) + (|001\rangle - |101\rangle)] = |001\rangle$$

Interference cancels the $|101\rangle$ term.

2) and 3) Nothing happens because

$$q_2 = 0 \text{ and } q_1 = 0$$

$$\psi_3\rangle = |001\rangle$$

4) Apply H on q_1:

$$|001\rangle \xrightarrow{H_{q1}} \frac{|001\rangle + |011\rangle}{\sqrt{2}}$$

$$\psi_4\rangle = \frac{|001\rangle + |011\rangle}{\sqrt{2}}$$

5) Controlled-R_2 from $q_0 \rightarrow q_1$

In $|001\rangle$: $q_1 = 0 \rightarrow$ no phase

In $|011\rangle$: $q_1 = 1 \rightarrow$ multiply by $e^{i\pi/2} = i$

$$|\psi_5\rangle = \frac{|001\rangle + i|011\rangle}{\sqrt{2}}$$

6) Apply H on q_0. Here $q_0 = 1$ in both terms, and

$$|\psi_6\rangle = \frac{1}{\sqrt{2}}\left(\frac{|000\rangle - |001\rangle}{\sqrt{2}} + i\frac{|010\rangle - |011\rangle}{\sqrt{2}}\right)$$

$$|\psi_6\rangle = \frac{1}{2}(|000\rangle - |001\rangle + i|010\rangle - i|011\rangle)$$

This is the pre-SWAP QFT output.

7) SWAP $q_2 \leftrightarrow q_0$ (for the "bit-order fix"). Swap first and last bits:

$$|\psi_7\rangle = \frac{1}{2}(|000\rangle - |100\rangle + i|010\rangle - i|110\rangle)$$

In numerical order,

$$\frac{1}{2}(|000\rangle + i|010\rangle - |100\rangle - i|110\rangle) \equiv \frac{1}{2}(|0\rangle + i|2\rangle - ||4\rangle - i|6\rangle)$$

Even though some amplitudes have phases $(-1, i, -i)$, the measurement probabilities are insensitive to phases are exactly equal to 1/4 for 0, 2, 4, and 6, which is exactly what you expect for period $r = 4$ with $Q = 8$.

6.3 High-Energy Physics

Quantum computers offer a powerful new approach to tackling some of the most challenging problems in physics and fundamental science. Because they are based on quantum systems, they are uniquely suited to simulate other quantum systems more efficiently than classical computers. This opens up transformative possibilities in high-energy phenomena. Among the most compelling applications is the simulation of QCD through a method known as lattice QCD.

QCD is the fundamental theory describing the behavior of quarks and gluons, the building blocks of protons, neutrons, and other hadrons. It governs the strong nuclear force, one of the four fundamental forces of nature, responsible for binding quarks together into nucleons and holding together atomic nuclei. Unlike QED, where the interaction strength allows for perturbative techniques to work well, QCD becomes strongly coupled at low energies, making perturbative methods ineffective. As a result, understanding QCD in the nonperturbative regime—where some of the most interesting phenomena occur, such as quark confinement and hadron mass generation—requires a different computational strategy.

In lattice QCD, space-time is discretized into a finite grid or lattice. Quarks occupy the lattice sites, while gluons, which mediate the strong force between quarks, are placed on the links connecting the sites. This framework transforms QCD from a continuous field theory into a large but finite numerical problem, allowing simulations for the quantum dynamics. However, the price of discretization is steep: the number of variables involved increases exponentially with the size of the lattice. Even small lattices in 4D space-time quickly exceed the capabilities of the most powerful classical supercomputers. The sheer size of the Hilbert space, along with the need to represent color charge and gauge invariance, makes direct simulation an enormous computational challenge. Several theoretical studies [277, 278] have highlighted the advantages of directly encoding HEP problems into qudits, requiring fewer operations and potentially better noise tolerance [279].

Quantum computers offer a fundamentally different path forward. Because qubits can represent quantum information directly, they can encode the internal degrees of freedom at each lattice site, including color charge, spin, and field amplitudes. The evolution of the system over time is governed by a Hamiltonian H, and the quantum computer simulates its dynamics by applying the unitary operator $U(t) = e^{-iHt}$. Since the Hamiltonian is typically composed of multiple noncommuting terms, a technique known as *Trotter–Suzuki decomposition* [280, 281] is employed. This method breaks the total evolution into a sequence of smaller steps, each of which corresponds to a simpler piece of the Hamiltonian. Quantum gates are used to implement these steps on the hardware, enabling the simulation to proceed incrementally.

Once the system has evolved to the desired point in time, measurement operations extract key observables. These include correlation functions, which are crucial for determining the masses of hadrons, and structure functions, which reveal how quarks are spatially distributed within protons and neutrons. They can also be used to investigate phase transitions, such as the transition from ordinary hadronic matter to a quark–gluon plasma, which occurs under extreme temperatures and densities similar to those found in the early universe or inside neutron stars.

Looking ahead, quantum computer simulations of the strong force would allow physicists to address questions such as why protons and neutrons have the masses they do. It could also shed light on the behavior of nuclear matter at ultrahigh densities, helping us understand the interior of neutron stars and possibly revealing the exotic phases of matter governed by strong interactions.

This ambitious endeavor is actively pursued by a number of leading institutions around the world. Research efforts are underway at CERN, Fermilab, and several U.S. national laboratories, as well as by private QC companies such as IBM and Google. The roadmap paper [282] provides the status of high-energy physics QCs and gives examples of theoretical and experimental target benchmark applications, which can be addressed in the near future.

6.4 Chemistry

One of the most promising applications of QC lies in the simulation of molecular and electronic structures. At the core of quantum chemistry [283] is the challenge of solving the Schrödinger equation for systems of interacting electrons and nuclei, a task that becomes exponentially difficult as the number of particles increases. The exponential growth of the quantum state space forces classical computers to rely on approximations, which often fail when dealing with strongly correlated electrons, transition states, or large, complex molecules. Classical computational chemistry methods like Hartree–Fock theory, and density functional theory (DFT) become unreliable when treating excited states, or bond-breaking processes, transition metal complexes, reaction intermediates, and photochemical pathways.

Quantum computers directly represent quantum states using qubits, naturally handling superposition and quantum interference—essential features of chemical systems. Some of the specific algorithms designed for chemical simulation include VQE—a leading method for approximating ground-state energies on today's noisy quantum devices [284]. VQE works by preparing a parameterized quantum state using a quantum circuit, measuring the expected energy, and then using a classical optimizer to adjust the parameters to minimize the energy (Figure 6.2). This hybrid quantum-classical loop allows VQE to find low-energy states efficiently, although it can require many repetitions to succeed [285].

Another powerful algorithm is QPE [1, 281], which provides a way to extract eigenvalues of a quantum system with high precision and leads to the energy levels of a molecule. However, QPE demands significantly more qubits, long coherent operations, and error-corrected quantum processors, making it less accessible with current technology compared to VQE. As quantum hardware advances, QPE is expected to become widespread for high-precision quantum chemistry simulations.

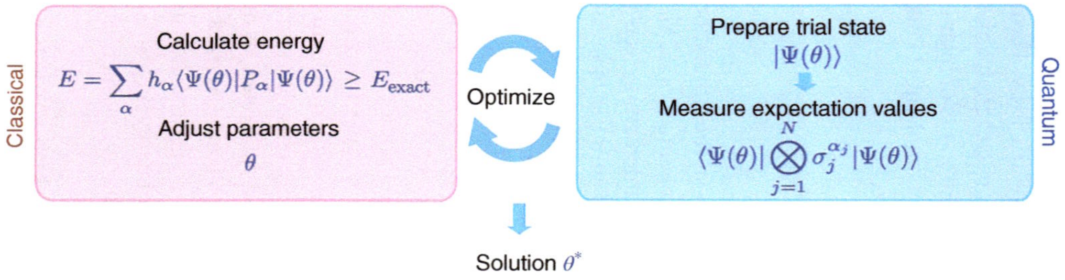

Figure 6.2 Schematic of the VQE algorithm illustrating the closed optimization loop between a classical optimizer and a quantum processor. On the quantum side (blue), a parameterized trial state $|\Psi(\theta)\rangle$ is prepared using a variational quantum circuit. Expectation values of the molecular or problem Hamiltonian are obtained by decomposing the Hamiltonian into a weighted sum of Pauli operators and measuring each term individually to evaluate $\langle \Psi(\theta) | P_j | \Psi(\theta) \rangle$. These measurements are performed repeatedly to estimate expectation values from finite sampling statistics. On the classical side (pink), the measured expectation values are combined to compute the total energy estimate, which serves as the objective function for a classical optimization routine. Based on this energy estimate, the optimizer updates the circuit parameters $\boldsymbol{\theta}$ and sends new values back to the quantum processor (Figure 6.3). This iterative loop continues until convergence, yielding an optimal parameter set $\boldsymbol{\theta}^*$ that minimizes the energy and approximates the ground-state energy E_0 (Ref. [284] / IOP Publishing Ltd / CC BY 3.0).

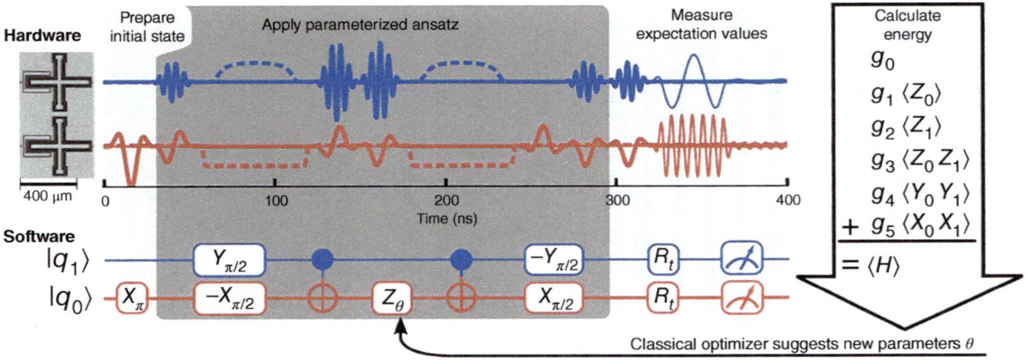

Figure 6.3 [32] Hybrid quantum–classical implementation of the quantum variational eigensolver (QVE/VQE) for the H_2 ground-state problem. Schematic illustration to estimate the ground-state energy of the hydrogen molecule using superconducting transmon qubits. The top panel (hardware layer) shows the physical realization on two Xmon transmon qubits, including microwave pulse sequences used to implement single-qubit rotations (thick colored waveforms), DC flux pulses used to enact two-qubit entangling operations (dashed segments), and microwave spectroscopy tones employed for qubit readout (thin oscillatory signals). The time-domain pulse structure highlights how abstract quantum gates are translated into experimentally applied control waveforms (Ref. [289] / American Physical Society / CC BY 3.0).

In the long term, QC may allow simulation of complex biological molecules, the optimization of reaction pathways at the quantum level, and the design of materials with properties tailored from first principles. Accurate calculation of molecular ground- and excited-state energies could lead to precise predictions of reaction thermodynamics, enabling chemists to design better drugs, catalysts, and materials. Understanding reaction mechanisms at the quantum level would allow for the modeling of enzyme catalysis, photochemical transformations, and key industrial reactions like nitrogen fixation with high fidelity. Moreover, quantum simulation [286] could play a crucial role in the discovery of new materials for energy storage, solar cells, and quantum technologies. Solving problems like the precise mechanism of photosynthesis [287], the behavior of high-temperature superconductors, or the creation of new catalysts for green energy could become feasible.

Some of the landmark implementation examples include IBM simulations of H_2, LiH, and BeH_2 using VQE on real quantum hardware [288]. Another example is Google that used their Sycamore processor to simulate isomerization of diazene.

We discuss one example that shows how superconducting transmon qubits, even in small numbers, have impacted chemistry by solving the H_2 molecular ground-state problem. For H_2, this means solving the electronic Schrödinger equation for two electrons bound to two nuclei. The methodology—Hamiltonian mapping, ansatz preparation, VQE optimization—is a template for quantum chemistry on superconducting platforms.

IBM and Google pioneered (2017) this work using 2D superconducting transmon qubits. At the time, their devices had coherence times of tens of microseconds. The H_2 Hamiltonian was expressed in second-quantized form with nuclei fixed and moving electrons.

For two electrons in H_2, the Hamiltonian basically says: each electron has kinetic energy $\left(-\frac{1}{2}\nabla^2\right)$. Each electron feels attraction to the two protons $(-1/|\mathbf{r}_i - \mathbf{R}_A|)$. The two electrons repel each other $(+1/|\mathbf{r}_1 - \mathbf{r}_2|)$. And finally, the two protons repel each other.

$$\hat{H}_{\text{el}} = -\frac{1}{2}\sum_{i=1}^{2}\nabla_i^2 - \sum_{i=1}^{2}\sum_{A=1}^{2}\frac{Z_A}{|\mathbf{r}_i - \mathbf{R}_A|} + \frac{1}{|\mathbf{r}_1 - \mathbf{r}_2|}$$

Electrons belong to a set of allowed orbitals (mathematical “rooms” they can sit in). For H_2 in the simplest case, there are four spin-orbitals: bonding-up, bonding-down, antibonding-up, and antibonding-down. Each orbital can be either empty or occupied by one electron. Now we switch from “continuous wavefunctions” to occupation numbers. Instead of writing equations that describe electrons moving around, one can track which electron is in which orbital. If orbital p is empty, you add an electron by using a creation operator $a_p^\dagger$. If orbital p is occupied, the annihilation operator a_p is used to remove the electron. The Hamiltonian becomes a set of rules about how electrons can hop and interact:

$$H = \sum_{p,q} h_{pq} a_p^\dagger a_q + \frac{1}{2} \sum_{p,q,r,s} \langle pq \mid rs \rangle a_p^\dagger a_q^\dagger a_s a_r$$

The first term says: electrons can hop between orbitals (with strength (h_{pq})). The second term says: electrons interact—if you put one in orbital p and another in q, they affect each other through Coulomb repulsion. This process converts differential equations in space to algebra with raising/lowering operators. Then it becomes possible for quantum computers to work with "is orbital p occupied or not?" rather than to directly solve wave equations in 3D space. Once the Hamiltonian is converted to creation/annihilation form, it can be mapped onto qubits, because each orbital = one qubit (0 = empty, 1 = occupied).

In short, fermionic operators are then mapped onto qubits using the Jordan–Wigner transformation (a standard method to encode electron creation/annihilation into spin operators). The fermionic creation and annihilation operators are $a_i^\dagger$ and a_i. The first creates an electron in orbital I. The second annihilates (removes) an electron from orbital i. These operators obey fermionic anticommutation rules:

$$\left\{a_i, a_j^\dagger\right\} = \delta_{ij}, \{a_i, a_j\} = 0$$

meaning that two electrons cannot occupy the same state (Pauli exclusion principle) where anticommutation operation of two operators is defined as $\{a, b\} = ab + ba$.

As discussed in Chapter 3, transmons are two-level spin-½ systems, described by Pauli operators. But electrons are fermions, not spins, so their algebra is different. There is a way to map fermions → qubits, such that the correct physics is preserved. The Jordan–Wigner transformation is a systematic method (from 1928!) to encode fermionic operators into Pauli spin operators acting on qubits. After applying the transformation, all fermionic operators become products of Pauli matrices that can be implemented directly on transmon qubits with microwave gates. The Hamiltonian of H_2 in a minimal basis maps into a sum of Pauli strings like:

$$H = \alpha_0 I + \alpha_1 Z_0 + \alpha_2 Z_1 + \alpha_3 Z_0 Z_1 + \alpha_4 (X_0 X_1 + Y_0 Y_1)$$

Z, X, Y are Pauli matrices acting on qubit 0 or 1. Z_0, Z_1 measure occupancy (are orbitals filled?). X_0X_1, Y_0Y_1 are entangling terms (representing electron correlations). Each term is now just a Pauli operator acting on qubits, measurable on a transmon device. The coefficients α_i depend on the bond length between the hydrogen nuclei from quantum chemistry integrals for the orbitals. Thus, the scary differential equations of electrons around nuclei become compressed into Pauli gates acting on a handful of qubits.

Each hydrogen atom has one atomic orbital (the 1s). Two hydrogens together combine into two spatial orbitals: the bonding orbital with lower energy (σ_g) and antibonding orbital with higher energy (σ_u). Every orbital can hold two electrons, one with spin up and one with spin down. In total, there are four spin-orbitals.

A two-qubit transmon system was used by IBM in 2017. Each spin-orbital is mapped to one qubit. So in principle, H_2 needs four qubits to describe all possibilities. Because the molecule has two identical atoms, some excitations are redundant or forbidden. H_2 has two electrons. Instead of tracking four orbitals with four qubits, the relevant configurations are both electrons in the bonding orbital or one electron excited into the antibonding orbital. Those two possibilities can be encoded as a TLS. So we can describe the essential physics with two qubits instead of four. Qubit 0 represents bonding orbital occupancy. Qubit 1 represents antibonding orbital occupancy. This reduction makes the simulation feasible on a small transmon-based device while still capturing the essential chemistry of H_2. Each transmon qubit is a superconducting circuit that behaves like a TLS. It is manipulated with microwave pulses tuned to its transition frequency to implement single-qubit rotations: $R_x(\theta)$, $R_y(\theta)$, $R_z(\theta)$. A rotation means apply a microwave pulse for a controlled time so that the qubit rotates by angle θ on the Bloch sphere.

Real electrons are correlated: when one is in orbital A, the other has a probability of being in orbital B. Entangling the two qubits captures electron correlation, i.e., superposition of bonding-only and mixed bonding/antibonding states. In practice, this is usually done with a CNOT or CZ gate between those two active qubits. Recall that CNOT is built from transmon couplers that let one qubit's state flip another.

Prepare the trial state $|\psi(\theta)\rangle$, also called the ansatz, since we don't know the exact ground state of H_2.

$$|\psi(\theta)\rangle = U(\theta)|00\rangle$$

Tune θ so that the wavefunction approaches the ground state of the molecular Hamiltonian. The electrons don't behave identically in the two orbitals—their probability amplitudes can be different. By allowing independent rotation angles, the ansatz can flexibly adjust how much each orbital is occupied.

On hardware, it reduces to start with both qubits in the ground state $|00\rangle$ (both orbitals empty). Apply $R_y(\theta_1)$ to qubit 0 and $R_y(\theta_2)$ to qubit 1. The occupation of orbital 0 depends on θ_1. The occupation of orbital 1 depends on θ_2. Apply a CNOT (qubit 0 controls qubit 1). Entanglement builds correlated states so that when one orbital is occupied, the other is not. The entangled state represents the bonding/antibonding electron pairs in H_2. This creates a flexible trial state $|\psi(\theta)\rangle$.

The next step is to estimate the energy expectation value:

$$E(\theta) = \langle\psi(\theta) \mid H \mid \psi(\theta)\rangle$$

Here H is the molecular Hamiltonian as described above. Instead of measuring the full H at once, you measure the expectation value of each operator. The quantum computer natively measures in the computational basis ($|0\rangle$,$|1\rangle$), which corresponds to the eigenbasis of the Pauli-Z operator. So operators involving Z can be directly measured. To measure X, apply a Hadamard (H) gate before measurement, $HXH = Z$. Measuring in the Z-basis *after* H is equivalent to measuring in the X-basis. Similarly, to measure X_0X_1 or Y_0Y_1, we need to rotate the measurement basis first. For multiqubit terms, apply H to both qubits, then measure in Z-basis. In each case, the measurement results (bitstrings) are converted into ±1 values and averaged to estimate the expectation value over multiple shots.

The next step is to minimize $E(\theta)$ over circuit parameters θ, via a classical optimizer, to approach the ground-state energy E_0. For two qubits with independent rotation angles, instead of a single parameter θ, there is a parameter vector: $\theta = (\theta_0, \theta_1)$ for the rotation angles for qubit 0 and qubit 1. The energy expectation spans a 2D surface. The optimizer's job is to find the lowest valley on that surface. The optimizer determines whether the energy goes up or down if the angles are changed a little. One can also use derivates to make better guesses on the next choices. Based on which direction of θ change lowers the energy, the optimizer updates the whole vector, until the energy settles into the minimum.

The VQE approach exploits the strengths of both platforms: the quantum device efficiently represents entangled many-body wavefunctions, while the classical computer performs the global optimization, making the method well suited for near-term NISQ hardware.

For small molecules like H_2, classical computational chemistry can solve the equations exactly. But as molecules get larger, the Hilbert space grows exponentially, and classical methods require enormous resources. QC provides a natural way to encode wavefunctions directly into qubits.

The bottom panel (software and circuit layer) depicts the corresponding quantum circuit representation. The protocol begins with preparation of the Hartree–Fock reference state, followed by application of a parameterized unitary coupled-cluster (UCC) ansatz that generates a correlated electronic trial state $|\Psi(\theta)\rangle$. Expectation values of the molecular Hamiltonian are obtained through efficient partial tomography using basis-rotation operations R_t, allowing measurement of individual Pauli terms such as Z_0, Z_1, Z_0Z_1, X_0X_1, and Y_0Y_1.

The measured expectation values are combined to compute the total electronic energy according to the quantum eigensolver equation (QEE). This energy estimate is then passed to a classical optimizer, which updates the variational parameters $\boldsymbol{\theta}$ and feeds new values back to the quantum processor. The closed feedback loop is repeated until convergence, yielding an approximation to the molecular ground-state energy. Figure 6.3 emphasizes the layered structure of VQE: hardware-level pulse control, software-level circuit abstraction, and classical optimization operating together within a hybrid quantum–classical framework.

6.5 Materials Science [286]

Condensed matter physics explores systems composed of an enormous number of interacting particles—including solids, liquids, magnets, superconductors, and many other phases of matter. What makes this field remarkable is its focus on emergent properties: large-scale behaviors like magnetism, superconductivity, and crystallization that arise from the collective quantum interactions of individual particles, typically electrons and nuclei. At the heart of condensed matter, problems are quantum many-body systems, where the number of possible quantum states grows exponentially with the number of particles involved. Even a modest-sized system can have more quantum states than there are atoms in the universe, making direct simulation using classical computers extraordinarily difficult.

Because they obey quantum mechanical principles, quantum computers can naturally encode the full wavefunction of a many-body system using qubits. In this way, quantum computers have the potential to solve problems that are fundamentally intractable for classical supercomputers, offering a window into behaviors and phases of matter previously beyond reach.

One of the central challenges in condensed matter physics is finding the lowest-energy state, or ground state, of a system. The ground state determines key properties such as whether a material behaves as a metal, an insulator, a magnet, or a superconductor. Classical optimization algorithms often struggle to find ground states because they can become trapped in local minima, energy configurations that are stable but not truly the lowest possible. Quantum computers can employ specialized algorithms such as the VQE to find ground states more efficiently. VQE works by preparing trial quantum states, measuring their energies, and optimizing parameters to minimize the energy.

The Ising model [290] provides an instructive example. The Ising model captures how spins—tiny magnetic moments associated with particles—interact with one another. In its classical form, each spin can point either up (+1) or down (–1), and the total energy of the system, or Hamiltonian, depends on the alignment of neighboring spins. Adding a magnetic field modifies the Hamiltonian to include terms that favor spins aligning in a particular direction.

In the quantum Ising model, spins are represented by qubits, which exist in a superposition of up and down simultaneously. Interactions induce spin flips, represented mathematically by operators like the Pauli σ_3 matrix. The energy landscape captures phenomena like quantum phase transitions and entanglement that are invisible to classical treatments.

Quantum computers simulate the Ising model by mapping each spin to a qubit, constructing circuits that apply quantum gates corresponding to the spin–spin interactions and the influence of the magnetic field. Through the application of time evolution operators, often broken into small steps via techniques like Trotter–Suzuki decomposition, the quantum computer evolves the system's state. Measurement of the final qubit states allows researchers to extract physical observables, such as magnetization, spin correlations, and signatures of phase transitions.

Quantum computers enable the direct exploration of quantum critical phenomena, entanglement growth, and real-time quantum dynamics. The ability to model the quantum behavior of many-body systems without uncontrolled approximations will profoundly enhance our understanding of condensed matter physics, leading to insights into materials with exotic properties, the behavior of strongly correlated electrons, and potentially the discovery of new phases of matter with transformative technological applications.

6.6 Finance and Economics

In finance and economics, some of the most challenging problems involve high-dimensional optimization, stochastic modeling, and complex simulations. Stochastic modeling is a mathematical approach to model processes that are uncertain or random over time, which makes it ideally suited for describing the inherent unpredictability of financial markets. These areas are notoriously difficult for classical computers to handle, especially as the size and complexity of the datasets grow. By leveraging quantum optimization and quantum-enhanced sampling techniques, researchers and financial institutions are beginning to explore how quantum computers can address critical problems such as portfolio optimization, derivative pricing, Monte Carlo simulations, and credit risk assessment. There is also growing interest in the potential for quantum-enhanced trading models, which could adapt more dynamically to market conditions by processing information in ways that classical algorithms cannot easily replicate.

Finance naturally involves massive volumes of data, coupled with inherent uncertainty and intricate, nonlinear relationships between variables. These characteristics align well with the strengths of quantum algorithms. Unlike classical algorithms, which often struggle with combinatorial explosion, quantum algorithms are capable of processing enormous combinatorial spaces more efficiently.

One of the most significant applications under investigation is portfolio optimization. In the classical world, portfolio optimization often boils down to solving a form of quadratic optimization under a variety of constraints, including expected returns, risk tolerance, diversification requirements, and transaction costs. Many formulations of this problem can be mapped to a quadratic unconstrained binary optimization (QUBO) problem, which is particularly amenable to quantum approaches, especially as the number of assets grows large and the market conditions introduce nonlinear complexities.

Another major area where QC shows promise is in the acceleration of Monte Carlo simulations, which are widely used in finance for tasks such as calculating value at risk (VaR), pricing complex derivatives, and assessing portfolio sensitivity to various market scenarios. Classical Monte Carlo methods require an enormous number of random samples to achieve accurate results, especially for rare-event probabilities or complex payout structures. Quantum algorithms like QAE offer a quadratic speedup over classical Monte Carlo methods, potentially allowing financial analysts to achieve the same level of precision with far fewer samples. This improvement could dramatically reduce the computational cost and time required for risk management and pricing calculations.

Quantum machine learning is another emerging tool with applications in finance. By leveraging quantum enhancements in pattern recognition and data classification, quantum machine learning algorithms could potentially identify hidden patterns in financial data, predict market trends more effectively, and construct more adaptive trading strategies. Practical implementation remains in early stages.

While many of these applications remain at the proof-of-concept stage due to current hardware limitations, the trajectory is clear. As quantum processors become more powerful and error rates continue to drop, QC is poised to profoundly reshape the landscape of financial analysis, optimization, and risk management. In a future where financial systems are even more complex and interconnected than today, quantum computers may provide the only feasible path to solving problems at the scales and speeds that global markets demand. Some of the key players are JPMorgan, Goldman Sachs, HSBC, BBVA, and Multiverse Computing.

6.7 Conclusions

QC is emerging as a transformative tool across several major scientific and technological domains. While present-day quantum hardware is not yet capable of running Shor's algorithm at cryptographically relevant scales, its theoretical existence threatens to reshape the security landscape. Shor's algorithm motivated the development of post-quantum cryptography and provided a clear benchmark for what "useful" quantum advantage would mean in practice. In this sense, Shor's algorithm is more than a mathematical curiosity; it defines both a challenge and a goal for the future of QC and secure communication.

In HEP, it offers a fundamentally new way to address problems that have resisted classical computation. Quantum computers provide a potential pathway to simulate QCD in its fully nonperturbative regime, opening the door to first-principles studies of quark confinement and the generation of hadron masses. Techniques such as mapping QCD onto lattice formulations and using Trotter–Suzuki decomposition to step through Hamiltonian dynamics lie at the core of these efforts. Success in this direction could yield deep insights into the structure of neutron stars, the behavior of quark–gluon plasma, and the properties of exotic phases of matter.

In chemistry, QC directly attacks the exponential computational cost of solving the Schrödinger equation for many-electron molecules. Algorithms such as the VQE already provide hybrid quantum–classical approaches that are compatible with today's noisy devices, while QPE promises chemically accurate energies on future error-corrected hardware. The potential applications are broad and compelling, including drug design, understanding enzyme catalysis, photochemical reactions, nitrogen fixation, and the development of catalysts for green energy. Early landmark demonstrations by IBM on molecules, such as H_2, LiH, and BeH_2, and Google's simulation of diazene isomerization illustrate the promise of these approaches.

In materials science and condensed-matter physics, quantum algorithms enable the simulation of strongly correlated many-body systems whose ground states and excitation spectra are difficult or impossible to capture using classical approximations. Models such as the quantum Ising model allow direct exploration of spin correlations, magnetization behavior, and quantum-phase transitions. These capabilities may guide the design of new materials, including high-temperature superconductors, exotic magnetic phases, and novel quantum devices.

Beyond the physical sciences, QC is also poised to influence finance and economics. Quantum optimization techniques allow portfolio allocation problems to be framed as QUBO tasks. Monte Carlo simulations—ubiquitous in pricing and risk—can benefit from quadratic speedups through QAE. Quantum machine learning methods offer the potential to uncover subtle structure in financial data, informing trading strategies, risk assessment, and adaptive market models. Several major financial institutions, including JPMorgan, Goldman Sachs, and Multiverse Computing, are already actively exploring these directions.

Taken together, these developments underscore that QC is not confined to a single application area. Instead, it represents a broadly enabling technology with the potential to reshape how we tackle some of the most challenging problems in science, engineering, and finance.

This chapter makes clear that applications of QC span both fundamental science and practical industries. From probing the deep structure of matter and simulating molecular chemistry to designing next-generation materials and optimizing global finance, quantum algorithms promise to address problems unreachable by classical computing. While hardware limitations persist, the trajectory is unmistakable: QC is set to become a cornerstone of discovery and innovation across many disciplines.

References

1 Nielsen, M. and Chuang, I. (2000). *Quantum Computation and Quantum Information*. Cambridge: Cambridge University Press.

2 Feynman, R.P. (1982). Simulating physics with computers. *Int. J. Theor. Phys.* 21: 467.

3 Braginsky, V.B. and Khalili, F.Y. (1996). Quantum nondemolition measurements: the route from toys to tools. *Rev. Mod. Phys.* 68: 1–11.

4 Roffe, J. (2019). Quantum error correction: an introductory guide. *Contemp. Phys.* 60 (3): 226–245.

5 Devitt, S.J., Munro, W.J., and Nemoto, K. (2013). Quantum error correction for beginners. *Rep. Prog. Phys.* 76: 076001.

6 Mermin, N.D. (1990). Simple unified form for the major no-hidden-variables theorems. *Phys. Rev. Lett.* 65: 3373.

7 Shor, P.W. (1994). Algorithms for quantum computation: discrete logarithms and factoring in Proc. In: *35th Annu. Symp. Found. Comput. Sci*, 124–134. Los Alamitos, CA: IEEE Computer Society Press. https://doi.org/10.1109/SFCS.1994.365700.

8 Boixo, S., Isakov, S.V., Smelyanskiy, V.N. et al. (2016). Characterizing quantum supremacy in near-term devices (arXiv:1608.00263v1).

9 Nakahara, M. and Ohmi, T. (2008). *Quantum Computing: From Linear Algebra to Physical Realizations*. Boca Raton, FL: CRC Press.

10 Kwon, S., Tomonaga, A., Bhai, G.L. et al. (2021). Gate-based superconducting quantum computing. *J. Appl. Phys.* 129: 041102.

11 McMahon, D. (2008). *Quantum Computing Explained*. Hoboken, NJ: Wiley-Interscience.

12 Kockum, A.F., Soro, A., Vikstål, P., et al. (2023). Lecture notes on quantum computing. (ArXiv). https://arxiv.org/abs/2311.08445.

13 Berberich, J. and Fink, D. (2023). Quantum computing through the lens of control: a tutorial introduction (ArXiv). https://doi.org/10.1109/MCS.2024.3466448.

14 Abhijith, J., Adedoyin, A., Ambrosiano, J. et al. (2018). Quantum algorithm implementations for beginners. *ACM Trans. Quantum Comput.* https://doi.org/10.1145/3517340.

15 Rieffel, E. and Polak, W. (2011). *Quantum Computing: A Gentle Introduction*. Cambridge, MA and London: MIT Press.

16 Huang, H.-L., Wu, D., Fan, D., and Zhu, X. (2020). Superconducting quantum computing: a review. *Sci. China Inf. Sci.* 63: 180501.

17 Chow, J.M. (2010). Quantum information processing with superconducting qubits. PhD Dissertation. Yale University.

18 Ezratty, O. (2025). *Understanding Quantum Technologies*, 8e (Version 8.1).

19 DiCarlo, L., Chow, J.M., Gambetta, J.M. et al. (2009). Demonstration of two-qubit algorithms with a superconducting quantum processor. *Nature* 460: 240–244.

20 Grover, L.K. (1998). Quantum computers can search rapidly by using almost any transformation. *Phys. Rev. Lett.* 80: 4329. arXiv:quant-ph/9712011.

21 Arute, F., Arya, K., Babbush, R. et al. (2019). Quantum supremacy using a programmable superconducting processor. *Nature* 574: 505–510.

22 IBM (2022). 'IBM Unveils 400 Qubit-Plus Quantum Processor and Next-Generation IBM Quantum System Two'.

23 Preskill, J. (2018). Quantum computing in the NISQ era and beyond. *Quantum* 2: 79.

24 Watrous, J. (2009). *Quantum Computational Complexity, Encyclopedia of Complexity and System Science*. Berlin: Springer.

25 Aaronson, S. (2008). The limits of quantum computers. *Sci. Am.* 298: 62–69.

26 Aharonov, D., Van Dam, W., Kempe, J. et al. (2004). Adiabatic quantum computation is equivalent to standard quantum computation. *Proceedings of 45th Annual IEEE Symp. on Foundations of Computer Science* 42–51.

27 Bravyi, S., Dial, O., Gambetta, J.M. et al. (2022). The future of quantum computing with superconducting qubits. *J. Appl. Phys.* https://doi.org/10.1063/5.0082975.

28 Devoret, M.H. and Schoelkopf, R.J. (2013). Superconducting circuits for quantum information: an outlook. *Science* 339 (6124): 1169–1174.

29 You, J.Q. and Nori, F. (2005). Superconducting circuits and quantum information. *Phys. Today* 58 (11): 42–47.

30 Schoelkopf, R.J. and Girvin, S.M. (2008). Wiring up quantum systems. *Nature* 451: 664.

31 Clarke, J. and Wilhelm, F.K. (2008). Superconducting quantum bits. *Nature* 453: 1031–1042.

32 Wendin, G. (2017). Quantum information processing with superconducting circuits: a review. *Rep. Prog. Phys.* 80: 106001.

33 Kjaergaard, M., Schwartz, M.E., Braumüller, J. et al. (2020). Superconducting qubits: current state of play. *Annu. Rev. Condens. Matter Phys.* 11 (1): 369–395.

34 Girvin, S.M. (2014). Circuit QED: superconducting qubits coupled to microwave photons. In: *Proc. of the 2011 Les Houches Summer School on Quantum Machines* (ed. M.H. Devoret et al.). Oxford: Oxford University Press.

35 Josephson, B.D. (1962). Possible new effects in superconductive tunnelling. *Phys. Lett.* 1: 251.

36 Josephson, B.D. (1974). The discovery of tunnelling supercurrents. *Rev. Mod. Phys.* 46: 251.

37 Anthony, J. (1989). *Leggett, Quantum Mechanics at the Macroscopic Level*, 187. Berlin: World Scientific.

38 Yang, C.-P., Chu, S.-I., and Han, S. (2003). Possible realization of entanglement, logical gates, and quantum-information transfer with superconducting-quantum-interference-device qubits in cavity QED. *Phys. Rev. A* 67: 042311.

39 Yang, S.C.-P. and Han, S. (2004). Quantum information transfer and entanglement with squid qubits in cavity QED: a dark-state scheme with tolerance for nonuniform device parameter. *Phys. Rev. Lett.* 92: 117902.

40 You, J.Q. and Nori, F. (2003). Quantum information processing with superconducting qubits in a microwave field. *Phys. Rev. B* 68: 064509.

41 You, J.Q. and Nori, F. (2003). Quantum information processing with superconducting qubits in a microwave field. *Phys. Rev. B* 68: 064509.

42 Kis, Z. and Paspalakis, E. (2004). Arbitrary rotation and entanglement of flux SQUID qubit. *Phys. Rev. B* 69: 024510.

43 Blais, A., Huang, R.-S., Wallraff, A. et al. (2004). Cavity quantum electrodynamics for superconducting electrical circuits: an architecture for quantum computation. *Phys. Rev. A* 69: 062320.

44 Devoret, M.H., Wallraff, A., and Martinis, J.M. (2014). Superconducting Qubits: A Short Review (arXiv:cond-mat/0411174).

45 Wallraff, A., Schuster, D.I., Blais, A. et al. (2004). Strong coupling of a single photon to a superconducting qubit using circuit quantum electrodynamics. *Nature* 431: 162–167.

46 Siddiqi, I. (2021). Engineering high-coherence superconducting qubits. *Nat. Rev. Mater.* 6 (10): 875–891.

47 Leibfried, D., Blatt, R., Monroe, C. et al. (2003). Quantum dynamics of single trapped ions. *Rev Mod Phys* 75: 281–324.

48 Blatt, R. and Roos, C.F. (2012). Quantum simulations with trapped ions. *Nat. Phys.* 8: 277–284.

49 Cirac, J.I. and Zoller, P. (1995). Quantum computations with cold trapped ions. *Phys. Rev. Lett.* 74: 4091–4094.

50 Haffner, H., Roos, C., and Blatt, R. (2008). Quantum computing with trapped ions. *Phys. Rep.* 469: 155–203.

51 Bruzewicz, C.D., Chiaverini, J., McConnell, R., and Sage, J.M. (2019). Trapped-ion quantum computing: progress and challenges. *Appl. Phys. Rev.* 6 (2): 021314.

52 Knill, E., Laflamme, R., and Milburn, G.J. (2001). A scheme for efficient quantum computation with linear optics. *Nature* 409: 46.

53 Pittman, T.B., Jacobs, B.C., and Franson, J.D. (2001). Probabilistic quantum logic operations using polarizing beam splitters. *Phys. Rev. A* 64: 062311.

54 Franson, J.D., Donegan, M.M., Fitch, M.J. et al. (2002). High-fidelity quantum logic operations using linear optical elements. *Phys. Rev. Lett.* 89: 137901.

55 Alsing, P., Battle, P., Bienfang, J. C. et al. (2022). Accelerating Progress Towards Practical Quantum Advantage: The Quantum Technology Demonstration Project Roadmap. (ArXiv). https://arxiv.org/abs/2210.14757.

56 Imamoglu, A., Awschalom, D.D., Burkard, G. et al. (1999). Quantum information processing using quantum dot spins and cavity QED. *Phys. Rev. Lett.* 83: 4204–4207.

57 Petta, R., Johnson, A.C., Taylor, J.M. et al. (2005). Coherent manipulation of coupled electron spins in semiconductor quantum dots. *Science* 309: 2180–2184.

58 Hanson, R., Kouwenhoven, L.P., Petta, J.R. et al. (2007). Spins in few-electron quantum dots. *Rev. Mod. Phys.* 79: 1217–1265.

59 Lewenstein, M., Sanpera, A., Ahufinger, V. et al. (2007). Ultracold atomic gases in optical lattices: Mimicking condensed matter physics and beyond. *Adv. Phys.* 56: 243–379.

60 Jaksch, D. and Zoller, P. (2005). The cold atom hubbard toolbox. *Ann. Phys.* 315: 52–79.

61 Gross, C. and Bloch, I. (2017). Quantum simulations with ultracold atoms in optical lattices. *Science* 357: 995–1001.

62 Bloch, I., Dalibard, J., and Zwerger, W. (2008). Many-body physics with ultracold gases. *Rev. Mod. Phys.* 80: 885–964.

63 Saffman, M., Walker, T.G., and Mølmer, K. (2010). Quantum information with Rydberg atoms. *Rev. Mod. Phys.* 82 (3): 2313–2363.

64 Weiss, D.S. and Saffman, M. (2017). Quantum computing with neutral atoms. *Phys. Today* 70 (7): 44–50.

65 Gyger, F., Ammenwerth, M., Tao, R. et al. (2024). Continuous Operation of Large-Scale Atom Arrays in Optical Lattices (ArXiv). https://doi.org/10.1103/PhysRevResearch.6.033104.

66 Hanson, R., Gywat, O., and Awschalom, D.D. (2006). Room-temperature manipulation and decoherence of a single spin in diamond. *Phys. Rev. B* 74: 161203.

67 Dutt, M.V.G., Childress, L., Jiang, L. et al. (2007). Quantum register based on individual electronic and nuclear spin qubits in diamond. *Science* 316: 1312–1316.

68 Pezzagna, S. and Meijer, J. (2021). Quantum computer based on color centers in diamond. *Appl. Phys. Rev.* 8 (1): 011308.

69 Barenco, A., Bennett, C.H., Cleve, R. et al. (1995). Elementary gates for quantum computation. *Phys. Rev. A* 52: 3457–3467.

70 Shor, P.W. (1997). Polynomial-time algorithms for prime factorization and discrete logarithms on a quantum computer. *SIAM J. Comput.* 26: 1484–1509.

71 Farhi, E., Goldstone, J., Gutmann, S., and Sipser, M. (2000) Quantum Computation by Adiabatic Evolution (arXiv:quant-ph/0001106).

72 Farhi, E., Goldstone, J., Gutmann, S. et al. (2001). A quantum adiabatic evolution algorithm applied to random instances of an NP-complete problem. *Science* 292: 472–475. *Rep. Prog. Phys.* 80 (2017) 106001.

73 Albash, T. and Lidar, D.A. (2018). Adiabatic quantum computation. *Rev. Mod. Phys.* 90: 015002.

74 Raussendorf, R. and Briegel, H.J. (2001). A one-way quantum computer. *Phys. Rev. Lett.* 86: 5188.

75 Raussendorf, R., Browne, D.E., and Briegel, H.J. (2003). Measurement-based quantum computation on cluster states. *Phys. Rev. A* 68: 022312.

76 Farhi, E., Goldstone, J., and Gutmann, S. (2014). A Quantum Approximate Optimization Algorithm (ArXiv). https://arxiv.org/abs/1411.4028.

77 Otterbach, J.S., Manenti, R., and Alidoust, N. et al. (2017). *Unsupervised Machine Learning on a Hybrid Quantum Computer*. arXiv:1712.05771 [quant-ph].

78 Oliver, W.D. and Welander, P.B. (2013). Materials in superconducting quantum bits. *MRS Bull.* 38: 816.

79 Bland, M. P., Bahrami, F., Martinez, J. G., et al. (2025). 2D transmons with lifetimes and coherence times exceeding 1 millisecond (ArXiv). https://arxiv.org/abs/2503.14798.

80 Maldonado, T.J., Flick, J., Krastanov, S., and Galda, A. (2022). Error rate reduction of single-qubit gates via noise-aware decomposition into native gates. *Sci. Rep.* 12 (1): 6379. https://doi.org/10.1038/s41598-022-10339-0.

81 Paudel, H.P., Syamlal, M., Crawford, S.E. et al. (2022). Quantum computing and simulations for energy applications: review and perspective. *ACS Eng. Au* 2: 151–196.

82 Deutsch, D. (1985). Quantum theory, the Church–Turing principle and the universal quantum computer. *Proc. R. Soc. A* 400: 97.

83 Zwitser, L. (2017). *Quantum Computing, Bachelor's thesis, Leiden Institute of Advanced Computer Science (LIACS)*. The Netherlands: Universiteit Leiden, Leiden.

84 Gottesman, D. (1998). The Heisenberg Representation of Quantum Computers (ArXiv). https://arxiv.org/abs/quant-ph/9807006.

85 Majer, J., Chow, J.M., Gambetta, J.M. et al. (2007). Coupling superconducting qubits via a cavity bus. *Nature* 449 (7161): 443–447. https://doi.org/10.1038/nature06184.

86 Wei, K.X., Lauer, I., Pritchett, E. et al. Native two-qubit gates in fixed-coupling, fixed-frequency transmons beyond cross-resonance interaction. *PRX Quantum 5* 020338. https://doi.org/10.1103/PRXQuantum.5.020338.

87 Wei, K.X., Lauer, I., Pritchett, E. et al. (2024). Native two-qubit gates in fixed-coupling, fixed-frequency transmons beyond cross-resonance interaction. *PRX Quant* 5: 020338.

88 Rigetti, C. and Devoret, M.H. (2010). Fully microwave-tunable universal gates in superconducting qubits with linear couplings and fixed transition frequencies. *Phys. Rev. B* 81: 134507.

89 Chow, J.M., Córcoles, A.D., Gambetta, J.M. et al. (2011). Simple all-microwave entangling gate for fixed-frequency superconducting qubits. *Phys. Rev. Lett.* 107: 080502. https://doi.org/10.1103/PhysRevLett.107.080502.

90 Magesan, E. and Gambetta, J.M. (2020). Effective Hamiltonian models of the cross-resonance gate, *Phys. Rev. A* 101, 052308. https://doi.org/10.1103/PhysRevA.101.052308.

91 Paik, H., Mezzacapo, A., Sandberg, M. et al. (2016). Experimental demonstration of a resonator-induced phase gate in a multiqubit circuit-QED system. *Phys. Rev. Lett* 117: 250502. https://doi.org/10.1103/PhysRevLett.117.250502.

92 Paik, H., Mezzacapo, A., Sandberg, M. et al. (2016). Experimental demonstration of a resonator-induced phase gate in a multiqubit circuit-QED system. *Phys. Rev. Lett.* 117: 250502. https://doi.org/10.1103/PhysRevLett.117.250502.

93 Einstein, A., Podolsky, B., and Rosen, N. (1935). Can quantum-mechanical description of physical reality be considered complete? *Phys. Rev.* 47: 777.

94 Lihm, J.-M., Noh, K., and Fischer, U.R. (2018). Implementation-independent sufficient condition of the Knill-Laflamme type for the autonomous protection of logical qudits by strong engineered dissipation. *Phys. Rev. A* 98: 012317. https://doi.org/10.1103/PhysRevA.98.012317.

95 Lihm, J.-M., Noh, K., and Fischer, U.R. (2018). Implementation-independent sufficient condition of the Knill–Laflamme type for the autonomous protection of logical qudits by strong engineered dissipation. *Phys. Rev. A* 98: 012317. https://doi.org/10.1103/PhysRevA.98.012317.

96 Reed, M.D., DiCarlo, L., Nigg, S. et al. (2012). Realization of three-qubit quantum error correction with superconducting circuits. *Nature* 482: 382–385.

97 Mariantoni, M., Martinis, J.M., and Cleland, A.N. (2012). Surface codes: towards practical large-scale quantum computation. *Phys. Rev. A* 86: 03232. https://doi.org/10.1103/PhysRevA.86.032324.